Nature's End

Nature's End

History and the Environment

Edited by

Sverker Sörlin

Professor of Environmental History, Royal Institute of Technology, Stockholm, Sweden

and

Paul Warde

Reader in Early Modern History, University of East Anglia, UK

palgrave
macmillan

First published in hardback 2009, and in paperback 2011 by
PALGRAVE MACMILLAN

Palgrave Macmillan in the UK is an imprint of Macmillan Publishers Limited,
registered in England, company number 785998, of Houndmills, Basingstoke,
Hampshire RG21 6XS.

Palgrave Macmillan in the US is a division of St Martin's Press LLC,
175 Fifth Avenue, New York, NY 10010.

Palgrave Macmillan is the global academic imprint of the above companies
and has companies and representatives throughout the world.

Palgrave® and Macmillan® are registered trademarks in the United States,
the United Kingdom, Europe and other countries.

ISBN 978–0–230–20346–4 hardback
ISBN 978–0–230–20347–1 paperback

This book is printed on paper suitable for recycling and made from fully
managed and sustained forest sources. Logging, pulping and manufacturing
processes are expected to conform to the environmental regulations of the
country of origin.

A catalogue record for this book is available from the British Library.

A catalog record for this book is available from the Library of Congress.

10   9   8   7   6   5   4   3   2   1
20   19   18   17   16   15   14   13   12   11

Printed and bound in Great Britain by
CPI Antony Rowe, Chippenham and Eastbourne

Contents

Preface

This book is the result of an ambition to formulate an environmental history that cares for the environment, cares for the salience of all knowledge including the sciences and cares for a history that is ultimately about the human condition.

With this book we try to say two things. The first is that environmental history is expanding the realm of history to comprise nature and the environment and to weave together as far as possible human experience with the workings of nature into a fuller understanding of the past and how it affects human decisions and destinations. In this regard environmental history is changing historiography and has the potential to change history as well by providing narratives that are novel and different, with more explanatory factors, and more vivid images: a new understanding. The second thing we try to say is that this can only be done if environmental history retains and relentlessly refines its status as history, belonging in the humanist tradition of studying complex phenomena with respect for humans as, ultimately, persons with intentions and morals, and belonging in the realm of the polity.

It is this dual ambition that has guided us as editors. We wanted to see if it was possible to embrace the widening agenda, and the broadening scales, of environmental history and at the same time allow it to address the issues as 'history', in frames and with concepts that make sense to the historical community in its largest extent and that will attract both scholars and students. A further ambition has been to make the end result accessible in a form that should be readable not just to historians, but to anyone in the social sciences or the sciences with an interest in how we could re-articulate the ways and means that we use to understand what we, as humanity, are up to in the never ending coexistence with nature.

We set out during 2004 and 2005 when we were both affiliated with the Centre for History and Economics at Cambridge University and organized a series of seminars and workshops to which we invited a range of eminent speakers and commentators. We also presented a paper that summed up our own position, subsequently published in the journal *Environmental History* (2007:1) as 'The Problem of the Problem of Environmental History: A Re-reading of the Field'. Finally, we invited some of the best people we could think of to help us and asked them to present papers at a workshop in Cambridge in January 2006 which was co-hosted by the Centre for Research in the Arts and the Social Sciences, CRASSH, and by the Department of Geography. After the workshop we continued working with several of the contributors who put in tremendous efforts into revising the papers to serve the purpose and design of this volume. As we went along we invited more contributors to get a reasonable coverage of themes and world regions.

Without trying to be exhaustive we would like to thank several people who have contributed, with papers, comments, or as referees, to help shape the project and this volume to what it finally became. Apart from the authors of this volume, who deserve our special gratitude, we are thinking of: Peter Alagona, Alan Baker, Dorothee Brantz, Peter Burke, Christopher Bayly, Michael T. Bravo, Mark Cioc, Stefania Gallini, Wilko Graf von Hardenberg, Gerry Kearns, Mike Lewis, Emma Rothschild, Chris Smout, and Gareth Stedman Jones. For economic support we would like to extend our heartful thanks to CRASSH and the Centre for History and Economics, where Inga Markan tirelessly served us with all kinds of logistical and emotional support.

Cambridge, March 2009
Sverker Sörlin
Paul Warde

Notes on Contributors

William M. Adams is the Moran Professor of Conservation and Development at the University of Cambridge. He works on relations between society and nature, particularly on rural development and conservation, and with a particular focus on Africa. Books include *Against Extinction: The Story of Conservation* (Earthscan, 2004), and *Green Development: Environment and Sustainability in a Developing World* (3rd edition, Routledge, 2008).

Peter Burke is Professor Emeritus of Cultural History at the University of Cambridge. He has been a leading figure in the development of cultural history, especially of the Renaissance and Early Modern Europe. Research books include *A Social History of Knowledge* (Polity, 2000) and *Languages and Communities in Early Modern Europe* (CUP, 2004).

Tim Cooper lectures in history at the University of Exeter (Tremough Campus). His interests lie in the fields of urban, environmental and political history, and especially the combined histories of waste and environmental politics in the nineteenth and twentieth centuries.

Vinita Damodaran is Senior Lecturer in South Asian History at the University of Sussex. She has written extensively on the political, social and environmental history of South Asia. Together with Richard Grove and S. Sangwan. She has also edited *Nature and the Orient, the environmental history of south and southeast Asia* (OUP, 1998).

Althea Davies is a research fellow in the School of Biological and Environmental Sciences at the University of Stirling. Her research deals with environmental archaeology and palaeoecology, focusing on long-term ecological change and land-use history in upland environments. She has published widely in edited volumes and journals such as *The Holocene* and *Landscape History*.

Robert A. Dodgshon was Gregynog Professor of Human Geography at Aberystwyth University until the end of 2007. Now Emeritus Professor, he has written very extensively on the spatial aspects of social change, and on mountain communities, especially the Scottish Highlands and islands. Publications include *Society in Time and Space: A Geographical Perspective on Change* (CUP,

1998), and *The Age of the Clans. The Highlands from Somerled to the Clearances* (Historic Scotland, 2002).

Mark Elvin is one of the leading historians of China, with a particular interest in environmental history, demography, and the history of science. His many publications include *The Retreat of the Elephants: An Environmental History of China* (Yale, 2004), and *The Pattern of the Chinese Past* (Methuen, 1973).

Georgina Endfield is Reader in Environmental History at the University of Nottingham. Her research interests include: climate and environmental history; historical conceptualisation of climate change; social responses; and adaptation to climate change and extreme weather events. She has published widely on the climate and economic history of Mexico in journals such as *Climatic Change* and *The Journal of Historical Geography*.

Matthew Evenden is an Associate Professor at the Department of Geography, University of British Columbia. His interests focus especially on the environmental history of river development and fisheries issues in western Canada. Publications include *Fish versus Power: An Environmental History of the Fraser River* (CUP, 2004).

Richard Grove is a leading practitioner of environmental history working at the Australian National University and The Centre for World Environmental History at the University of Sussex. He has published very widely in the field, with particular attention to the global history of ideas about the environment and relationships between the history of science, colonialism, and environmental change. Works include *Green Imperialism: Colonial Expansion, Tropical Island Edens and the Origins of Environmentalism, 1600–1860* (CUP, 1995).

Alistair Hamilton is an ecologist and Programme Leader for Environmental courses at the Scottish Agricultural College. His research interests are in the applied management of upland and woodland habitats, as well as in exploring interdisciplinary studies in furthering our current and historical understanding.

Nick Hanley is Professor of Environmental Economics at the University of Stirling. He has published widely in the field of environmental economics, environmental valuation and agricultural economics. Books include (with A. Owen) *The Economics of Climate Change* (Routledge, 2004).

Kirsten Hastrup is Professor of Anthropology at the University of Copenhagen. Her range of interests spans environmental history, the anthropology of theatre, Arctic societies, and historical and epistemological theory. Publications include *Nature and Policy in Iceland 1400–1800* (Clarendon, 1990), and *Action* (Museum Tusculanum Press, 2004).

Holger Nehring is Lecturer for Contemporary European History at the University of Sheffield. He has published widely on the comparative history of social movements in post-World War II Western Europe and is currently working on a monograph on the transnational history of the British and West German protests against nuclear weapons in the late 1950s and early 1960s. He has edited, together with Florian Schui, *Global Debates about Taxation* (Palgrave, 2007).

Libby Robin is Senior Fellow at The Fenner School of Environment and Society. She is a leading writer on Australian environmental history with interests spanning Australian and comparative global environmental history, the history of the environmental sciences, and the history of ornithology. Books include *How a continent created a nation* (UNSW Press, 2007), and (with Tom Griffiths), *Ecology and Empire: Environmental History of Settler Societies* (University of Washington Press, 1997).

Sverker Sörlin is Professor of Environmental History at the Royal Institute for Technology, Stockholm. His main research interests lie in the intersection of the history of ideas, the history of scientific practice, and environmental history. Publications include (with Anders Öckerman) *Jorden en ö: En global miljöhistoria* [The world is an island: a global environmental history] (Natur och Kultur, 1998), and (with Michael Bravo) *Narrating the Arctic: A Cultural History of Nordic Scientific Practices* (Science History Publications, 2002).

Paul Warde is Reader in Early Modern History at the University of East Anglia. His research focuses on the economic and environmental history of Europe, with a particular focus on wood resources and energy. Books include *Energy Consumption in England & Wales 1560–2000* (CNR, 2007), and *Ecology, Economy and State Formation in Early Modern Germany* (CUP, 2006).

Fiona Watson is a historian and was the Director of the AHRC Centre for Environmental History at the University of Stirling. She has published numerous works on Scottish medieval and environmental history, including *Scotland: A History* (Tempus, 2001), and (with Chris Smout and Alan Macdonald) *The Native Woodlands of Scotland: An Environmental History, 1500–1900* (Edinburgh UP, 2005).

Graeme Wynn is Professor of Geography at the University of British Columbia and is a leading and widely published environmental historian and historical geographer of Canada. His wide research interests focus on the development of new world societies and the environmental impacts of European expansion around the world. Works include *Canada and Arctic North America: An Environmental History* (ABC-CLIO, 2007), and *Timber Colony: A Historical Geography of Early Nineteenth-Century New Brunswick* (Toronto UP, 1980).

Making the Environment Historical – An Introduction

Sverker Sörlin and Paul Warde

'History is a nightmare, from which I am trying to awake.' Stephen Dedalus' words in James Joyce's novel *Ulysses* are not just the kind of words that sound familiar to a student in school who just cannot get the point of learning about all the familiar details of history: wars (mostly wars), battles (surprisingly important), royalty, leading politicians and their deeds, lineages of the welfare states, Raj and Mandarins, liberation struggles here and there, reforms and revolutions. Joyce was part of a modernist revolution of thought and form, of literary and artistic expression in early 20th century Europe, and in the mind of modernists, futurists and nihilists history was indeed, if not a nightmare, at least a burden, an unnecessary set of fetters that constrained human ingenuity and human deed. Friedrich Nietzsche, in *On the Advantages and Disadvantages of History for Life* (1874), wanted to free history from its backward looking obsession and nostalgia. History was part of making a new world; only feebleminded *Ressentiment Menschen* would be interested in the fine tuning of ephemeral bygones. History should be part of action, and therefore the worst enemies of future-embracing history were the historians.

It is not surprising that Nietzsche's nihilistic credo has not had much success as historiography *per se*; his text has instead remained as a stimulating chapter in the history of the philosophy of history. But if we take a broader view his dissent, shared with Joyce and much of modernism, was nonetheless important. It heralded, long before it was realized by most historians, a historiography that departed from the ornamental national narratives that had become the mainstay of 19th century histories and that continued, often in popularized form, in the 20th. In that sense history *was* a nightmare, insofar that it seemed to tie eternally peoples and communities to the atrocities of war, to nations pitted against each other in eternal conflict, and to the heroism of *my* nation, not those of the others.

Long before it was generally discovered and discussed among professional historians, the modernists realized that there were more histories out there that

1

needed to be told and that had to do with realities, entire continents of human experience, that were not present in the moralist and metaphysical historiographies that dominated the 19th century. Even more importantly, they were right in assuming that history fundamentally reflected the preoccupations, interests, and anxieties of the society in which it was written. History was a nightmare because the world as it was known was a nightmare, and history did not provide a way out of it, rather the opposite: history built the minds of men that, just because of their minds, produced precisely these nightmares: above all the First World War, which Joyce lived through, as so many others, never to become the same again.

This may seem an odd digression to introduce a book on environmental history. It would be much more commonplace, as has been done in most collections and textbooks in the field so far, to start out with the long list of environmental concerns and challenges that face humanity and argue, as we certainly also do, the case for addressing these concerns with a deeper understanding of how they are products of history, of human action, and why we therefore have to learn about them as historical realities, and how they can be managed, even solved, by just the same sort of humans as those who created them. In that sense the editors of this book are Nietzscheans: history is not just about understanding, it is also about action and about moral predicaments and determinations to guide action.

We do digress because we wish to underpin a bolder statement that we believe follows from our kinship with those rebel modernist thinkers. History today, no less than in the previous century, does ultimately reflect the preoccupations, interests, and anxieties of our times. And our times are marked by the increasing role of the environment, as the Earth of wonders and resources, as the threatened Planet, as a Nature full of surprises who can 'hit back' on ignorant humans, or as the material Context of our everyday lives wherever we are. Environmental history is a meaningful enterprise for just that reason: it seeks to provide the history that can tell us how we arrived here and what we need to know to handle our global environmental predicament.

The overriding image of the Earth in the last several decades – you may say since the environmental 'awakening' heralded by Rachel Carson's iconic bestseller *Silent Spring* (1962), although certainly not caused by it – has been one of gloom and pessimism. The title of this book, *Nature's End*, echoes that, and harks back to books with similar, more or less apocalyptic titles over the past few decades. The sorrow and despair and sense of loss that is communicated in titles such as *The End of Nature* (Bill McKibben in 1990) and *The Ends of the Earth* (Donald Worster in 1988) must of course be respected. But we do not believe that nature has reached its end, nor that this will ever happen. Rather we understand nature's end as the beginning of the environment, or even the age of environment. Nature needs no humans, but there is an environment only where humans live and where humans have entered into a self-conscious relationship with their surroundings. That is why the past of nature precedes the history of the environment – and why it will persist beyond the epoch of humans (if we can imagine such a future).

To talk of nature's end is therefore to talk of the purpose and uses of nature in the human context. Human societies are always dependent on nature, use its resources, and emit their refuse into it. Indeed, we humans ourselves are *of* nature. All that naturalness, from the physical laws of the universe to the inner workings of our own brains and hearts, is studied and can be understood scientifically. The environment must not be mixed up with nature in this respect. The environment is, in some instance, a human product, an alloy of nature, and the impacts of human labour which emerges as a historical category. It is a category of reality which has a history, and it is a history that it is possible, even necessary, to separate from nature's past. In this sense, there is and should be an end to nature. What we are saying comes close to what Roland Barthes prophetically wrote more than half a century ago, that

> Progressive humanism...must always remember to reverse the terms... constantly to scour nature, its 'laws' and its 'limits' in order to discover History there, and at last to establish Nature itself as historical.[1]

Writing in the 1950s, before the 'environmental' revolution, he does not use that concept. We would argue that when nature is established as historical, it has become environment. From the human(istic) point of view that is, in all senses of the word, nature's end. Geographically there may be no boundary, as the distinction is conceptual; we cannot look for the zone or site where nature ends and environment begins. The boundary itself is historical and it moves with time.

The protection of nature, a central feature of modern environmental politics, wants to preserve certain features of nature in a given area. We may, with some qualifications, accept that there is 'nature' there. At the same time, by defining it as a matter of politics, and by making it part of deliberate zoning of territories, and as one of the welfare instruments of society, it is also part of our environment. As such it is part of history. A nature reserve is also an institution; it has buildings, cabins, perhaps even a research station. It has roads, it may border on a major nature exploitation zone, with major infrastructure such as dams, power plants, electric transportation facilities, logging roads and camps, or oil rigs. They stand at nature's end, and they are formidable parts of the environment.

---

[1] Barthes (1957), 101.

When we talk of sustainability as one of the most crucial environmental concepts in our time, codified by the United Nations Commission on Environment and Development in 1987, headed by Gro Harlem Brundtland, it is clear that we mean the environment. Nature cannot be unsustainable – can it? It is when we in societies transform it and create *environment* that we create the possibility of unsustainability. This quest for survival and sustainability is historical, the (eternal) sustainability of (pure) nature is not.

The deep roots of a vast field

History has already shifted towards thinking about the environment. It was Fernand Braudel who famously likened history with a house. Historians had always written about the finest people and the salons in the upper floors. In the 20th century they increasingly started to work themselves downward, to issues of economy, science, social welfare, food, health, and population. All of these topics and many more have become sub-fields of history. Around 1970 the environment arrived as another topic, creating environmental history as a sub-discipline, spurred on by the newly aroused political interest in environmental issues and ecology in the 1960s.

Environmental aspects of history had of course been treated before that, by economic and social historians – we may think of Karl Wittfogel's work on 'hydraulic cultures' in Asia, *Oriental Despotism* (1957), of works by the *Annales* school, for example Lucien Febvre's prophetically insightful *L'évolution et la terre humaine* (1922). Even Karl Marx's *Capital* (1880–87), despite its inevitable Euro- and anthropocentric leanings, is also an attempt to locate human societies as inextricably interwoven with nature and the economy as a giant transformation of external resources into social value. One could stretch the argument somewhat and claim that several of the major historians of the 19th century – Wilhelm von Riehl or Heinrich von Treitschke in Germany, Jules Michelet in France, Erik Gustaf Geijer in Sweden – did in fact include the land as a significant factor in shaping peoples and nations. But one should also be cautious to realize that nature's role in determining the *Volkscharakter* is not the same as identifying it as a dynamic historical category and a legitimate object of study as such. Nonetheless, if it is only a matter of finding the land as a historical factor the list of forerunners is clearly long indeed and should also include 18th century economic philosophers like Adam Smith, and the enormously rich discourse of early modern agricultural propagandists.

Environmental history *avant le mot* that comes closer to contemporary practice was written in the 20th century by specialists in other disciplines like anthropology, geography, archaeology, or ecology. If we wish to identify precursors of the modern, post-1970, environmental history we should look to scholars such as Carl Sauer, the Berkeley geographer, for his synthetic descriptions in the 1940s through 1960s of the American landscapes under the influence of native Indians and European settlers, or his arch antagonist James Malin, the decidedly pro-settler Kansas ecologist.[2] Another strand of environmental writing can be found in the climate histories of Hubert Lamb in the United Kingdom, starting in the 1940s, or in the regionalist tradition, going back to 19th century German (Ritter) or 20th century French (de la Blache) geographers, and which was carried forward by the Scottish sociologist Patrick Geddes and his American disciple Lewis Mumford. Both of the latter connected the rural focus of regionalism with a strong interest in the city and its region as a metabolic entity, or 'conurbation' as Geddes called it. Regionalist environmentalism became particularly strong in northern Europe, in Scandinavia, and in the United Kingdom with geographers like W.G. Hoskins or H.C. Darby, but certainly had its followers in other parts of Europe and Russia as well.[3]

An intellectual history of Western ideas of the environment was compiled over a period of several decades after the Second World War by another Berkeley scholar, Clarence Glacken and published in 1967 as *Traces on the Rhodian Shore*, a seminal work which was labelled historical geography, since it was not history of science nor even intellectual history *strictu sensi*, and because Glacken himself taught in a geography department. And, most importantly, there was no 'environmental history', under that name, there to claim the work or for Glacken to identify with.

Defining the environmental historically

This short genealogy is well recognized among environmental historians, who understandably have found it important to establish a longer and deeper lineage of 'the environmental' as a factor in the historical understanding of societies than the conventional wisdom that points to the 1960s and the growth of environmental movements and politics. Both the shorter popular version and the slightly longer version are relevant. It is a fact that as an institutionalized specialty environmental history did not see the light of day until the early 1970s and it is essential that it was preceded by a political discourse of the environment, under that word. It is equally crucial to see that pioneering work

---

[2] Kingsland (2005).

[3] Excellent introductions to these aspects of early environmental history, often with leanings toward the spatial, are Buttimer (1982), Hall (1988), Livingstone (1992). Nonetheless, the intellectual history of environmental history in Europe is, surprisingly, not yet covered very well and important names are likely to occur as the history of geography, anthropology, ethnology in different countries is revisited in the search for early attempts at environmental history.

was done in a range of other disciplines, and it was a process that happened in many countries.[4]

However, there is a peculiar gap between this relatively recent and limited discipline-based historiography and the omnipresence of nature and natural resources as fundamentals of human societies that goes back to the very roots of human existence. We would like to suggest that there is a useful middle ground in understanding the importance of the environmental and that it can be understood historically as the emergence of self-conscious discourses of the environmental.

Several of the chapters in this volume sustain such a broader definition of environmental history. Richard Grove and Vinita Damodaran demonstrate that there is a colonial discourse, with important connections to British academia and to colonial intellectual centres, notably in India, that were preoccupied with understanding man–nature relationships in tropical settings. In the early decades of the 20th century this growing interest in climate, vegetation, and successionist ecology became increasingly professionalized among geographers, archaeologists, anthropologists, and of course, among scientists, not least ecologists, who were either based or did field work in colonial areas.

It was, as Grove and Damodaran point out, a 'de facto' environmental history that emerged from the colonial context, although the term itself was not used. It had clear applications, for example in wildlife preservation and protection of nature reserves, which became core missions of some of these early 'environmental historians'. Similarly, the environmental dimension appears as an element of regional histories such as Fernand Braudel's of the Mediterranean that started to appear from 1949. With this approach environmental history has a colonial and regional genealogy rather than a national one. Both chronologically and geographically it is therefore less an American product of the 'environmental sixties' than usually perceived, although Grove and Damodaran observe early examples of de facto environmental history there too. Similarly Bill Adams finds a discourse of wildlife and nature that is transformed into issues of nature reserves and national parks. Or rather, he finds a range of discourses that occur in different parts of the world, the United States, Africa, Europe, Asia, again with significant colonial bearings and with global interconnections: it was concerns of big game hunters from Britain that pushed the early creation of nature reserves in Africa and parts of Asia.

Both topics have been visited before. Richard Grove wrote importantly on nature protection as part of 17th and 18th colonial science discourse in *Green Imperialism* (1995) and Richard Drayton followed on his heels in *Nature's*

---

[4] See for example Sörlin and Öckerman (1998/2002) for Sweden. In more general terms the same observation is well argued in Hughes (2006), Chapter 1, with numerous references, most of them to the English-speaking world.

Government (2000). In this and the history of nature parks the role of knowledge as a structuring element in environmental understanding is exemplified: colonial science for sure, but as several authors in this volume demonstrate, using local and indigenous knowledge as well. These 'environmentalizations' of ever growing parts of the world are thus at the same time part of a globalization of the environment, a process that has been going on for centuries.

Still, it is surprising how little these insights seems to have affected our understanding of the environmental as a category in time and space. We may reason on several time scales here. One very long one may depart from the observation that human labour stands in interaction with 'raw nature'. Marx importantly noted that it is through labour that man enters history. But the wisdom of the papers by Grove and Damodaran, and Adams respectively is how they manage to demonstrate that element of structured knowledge in colonial scientific discourses that turned the environment from a philosophical category – nature that is transformed by human action into the realm of the human – into a historical category as part of the polity, and often at the service of the state. In that sense our common understanding of environmental history as a late 20th century invention with a few forerunners is not just too short on the historiographical time scale, it has not sufficiently integrated our new insights into how the environment has been shaped as an object of political thought and action.

Thus, the history of the environmental is therefore not a linear chronology, but rather a constantly growing set of historicizing projects, emerging in different fields of social and political discourse. Sverker Sörlin's chapter on the historiography of climate change is another case in point. Although there has always been a climate, and climate speculation was burgeoning already in Antiquity, the notion of historical climate with a distinct influence on human societies is comparatively modern, though still much older than the global warming phenomenon that has become scientifically acknowledged in the late 20th century. Twentieth century climate concern, that has reached gale force in the 21st, has in itself been a driver of the historicizing of the environmental. That climate has been part of history since at least the 19th century discovery of ice ages is obvious also because it is constantly part of controversy. Since the 1930s there has been a protracted argument as to whether climate change is forced by human societies. And well before that there was a discussion, with roots in older societies around the world, about the possible influence on climate on human psyche and culture. Climate determinism, with or without warming, is a political idea. Taking precautions against climate change, caring for places that are likely to be hit, mitigating risk through green taxes, are all measures that will be deliberated and fought out in the public realm.

The contributions in the first section of the book may be read as a sympathetic revision of the more hard-nosed constructivism of the 1980s and 1990s, that followed the theoretical inroads of Foucault, Luhmann, and other historians and social scientists, stating that the environment, as well as nature, was produced socially and in the realm of ideas and only thus emerged in *discourses* of science, politics, and economics. But these contributions may even more be read as a revision of the earlier positivist notion of the environment as the outside 'other'. Although this notion became established in historiography only since the 19th century and perhaps most precisely in classic works of agricultural history from the middle of the 20th, Paul Warde demonstrates how this artefactual externality called 'environment' actually emerged in agricultural discourse since the 16th century. In a paradoxical way there is an evolution of the 'environment' as something on which the farm and farmer are of course dependent, while at the same time the ecological principles and the principles of soil chemistry and recycling, sustainability and circulation, are applied to the farm itself. With time, and with growing insights in what has been called the environmental sciences[5] (meteorology, ecology, geography, geology, etc.), the environment was increasingly taken into consideration, but then chiefly as an intervening force, ' "exogenous", that is, thunderbolt-like blasts of natural power from skies that farmers hoped would be forever blue'. Remarkably, in this line of agricultural history, the environment became as external to the individual household as the other unpredictable force that guided farmers, the market, another unquantifiable noun.

Historicizing the environment as the external other is, also paradoxically, quite similar to the discursive invention of the environment that postmodernists have advocated. Warde, however, suggests that there is a more productive middle ground: farmers are always interacting with whatever is on their farms (soils, cattle) and whatever is geographically farther afield (air, waters, forests, commons, markets). Agriculturalists are, writes Warde, 'environing' as they go along with their business. The environment does not start where the farmland ends, it is constantly being produced by the combined economic and cognitive practices that farmers and their families and farmhands undertake. Agriculture thus is a serious contributor to the rise of the environmental, but in ways that some strands of previous historiography have not fully appreciated. Even worse, agricultural histories, both those written by agriculturalists and those written later by professional historians, have restricted the realm of human action to the farm and, albeit involuntarily, pushed the environmental into a 'scientific' surrounding where history has had little mandate to speak. By thinking in terms of environing as a historical process we are invited to define much more of the 'natural environment' into the societal. The rise of the environmental is therefore also the parallel revising of the societal to mean a field of

---

[5] Bowler (1992).

activities and spaces that are also imprinted with the environmental, and vice versa.

The mutual imprint of the societal and environmental, and how the environment emerged in recent political discourse in West Germany, the nation where the Green movement has perhaps achieved greatest political recognition (if one takes the limited criteria of votes and ministerial seats), is undertaken in the contribution of Holger Nehring. His study demonstrates how political discourses engaging the environment belonged not only to an environmentalist social movement, or developments within the natural sciences, but were linked to the structure of political power and especially the role of planning within government. The 'environmental moment' was thus simultaneously a crisis of planning, which at times demonstrably failed to account for environmental risk, but also provided renewed impetus to a planning with a mission to spread environmentally conscious discourse throughout the activities of the state that themselves could facilitate that spread. Although Nehring's chapter ends before the heyday of 'sustainability', what is the latter concept if not a challenge and invitation to planning? We see the challenge of 'the environment' reaching roughly its contemporary form in political discourse: 'Environment thus served as the site at which the complexity of risk societies was negotiated.'

Perhaps the most rewarding insight drawn from these papers is that the rise of the environmental has a much more complex chronology, even if we restrict ourselves to the western context. It is neither a recent and merely political feature, as suggested by the '1962' (Carson) chronology, nor a discursive invention of '1864', the year of George Perkins Marsh's sweeping yet highly influential geographical tract *Man and Nature*, where man was elevated to the dubious role of principal agent of physical change. In thinking about the environmental as a much wider category, as the product of several environing processes that societies have undertaken over centuries, we suggest that the history of the environmental can be redefined in such a way as to both make it less narrow and specialist and at the same time reconnect it to wider strands of history. The implications of this is not just a longer, and we would argue much more interesting, chronology of the environmental as a historical category, but also a perhaps useful hint at how the environment can be studied historically.

Environmental history expands towards the sciences

We have seen that most of the development of the understanding of 'the environment', both as a political category and as the object of historical analysis, has taken place with relatively little input from historians. Indeed, those who studied history have until recently had very little to learn about 'the environment' from their professors. Syllabi have not allowed much space for it, required reading did not favour it. Texts considered path breaking or canonical in the self-understanding of the historical discipline, like E.H. Carr's *What Is History?* (1961), or R.G. Collingwood's *The Idea of History* (1946) do not carry a single reference to nature or the environment. Their universe is that of human life and human mind, not the world that they formed or that formed them.

This has begun to change. Today there are already a number of textbooks and collections that cater to the increasing demand for guides to the emerging field, and reviews of current knowledge. The first of these appeared in the United States, where 'environmental history' as an academic sub-discipline first took shape: among the first collections to appear was Donald Worster's *The Ends of the Earth* (1988). It included both older texts – the earliest from 1950 by Swedish economic historian Gustaf Utterström on the demise of the Norse settlement on Greenland – and recent contributions. The geographical coverage was United States–Europe–Russia, in that order of significance, a priority which reflected the limitations of the actual research rather than any particular analytical perspective. Africanists, Indianists, Australasianists, or world historians were not as yet subsumed under the environmental banner. What is perhaps more significant is that few 'ordinary' historians contributed. Later collections have added in particular to the global coverage, and regional collections have followed for e.g. India, Australia, Canada, and the urban environment.[6] In languages other than English there are now an impressive number of environmental histories, be they national, global, or regional.

The composition of authors and topics in this growing corpus of literature is of course diverse. However, it remains remarkably dominated by non-specialists in history, a fact perhaps even more prominent among the popular 'environmental histories', such as the work of Jared Diamond, where writers with training in the natural sciences have played a leading role. Even to this day, environmental history – as can be judged from the most seminal collections in the field – is frequently written by geographers, archaeologists, anthropologists, and even ecologists. To take one recent example we may look at *Sustainability or Collapse: An Integrated History and Future of People on Earth* (2006), edited by Robert Costanza, L.J. Graumlich, and William Steffen, and not presented as an environmental history *per se*. This seems to us a tradition of 'environmental history', unusually for a field of historical study, incorporating a significant volume of work from scientists. In the case of *Sustainability or Collapse* this is because the initiative and design of the volume, and the Dahlem workshop in Berlin that it was drawn from, came from IHOPE (The Integrated History and Future Of the People on Earth), a network of, principally, earth system scientists and archaeologists seeking to establish the long-term impact of humans on ecologies and other earth systems on all spatial and chronological scales. It seeks to

---

[6] Guha and Gadgil (1992); Griffiths and Robin (1997).

integrate environmental history into the broader understanding of how inter-active social–ecological systems have developed over long periods of time, and contribute to a growing theory of how such change occurs. Indeed, there is a gathering impetus behind such theory that first and foremost comes from scientists in the so-called Global Change community.[7]

However, this work can raise a spectre in the imaginations of historians. The spectre has a double aspect. Firstly, if the study of the 'environment' (as defined by whoever is undertaking the study) is based largely on interpreting the data on long-term change provided by the natural sciences, do historians have a useful role at all? Secondly, if these 'histories' primarily study human response to environmental conditions and change, are we at risk of falling into 'environmental determinism', which has its own venerable but largely discredited (and at times disgraceful) past? Yet historians have always been aware that the constraints that individuals and societies are faced with confront them not with absolute necessities of action, but with the necessity of choice. As a discipline history has rightly been suspicious of approaches that seem to deprive human beings of agency in shaping their world. It is important to assert that environmental history does not retreat from this position, indeed it can demonstrate how human action can have a significant role in processes commonly thought to be 'autonomous', or belonging to the realm of pure 'science'. Darwin, after all, devoted the first chapter of the *Origin of the Species* to the domestication of animals, and these evolutionary pressures are exerted today at an ever-faster pace.[8]

Nevertheless, particular environments present people with particular sets of challenges and risks. A case in point is the history of human life in the mountains, areas all too frequently treated as marginalia in traditional histories of metropolitan centres, empire- and nation-building. The chapter of Robert Dodgshon provides case studies of Scotland and Switzerland that demonstrate both the particular risks with which mountain communities have wrestled down the centuries, but equally the profound role that humans have played in shaping these supposedly wild and untrammelled places.

Environmental history is thus well placed to undertake a role as a bridge builder between the humanities and other disciplines, and its role in com-municating with the natural sciences stands out as particularly crucial. Few other disciplines can take that role, bringing together facts from a wide range of specialties, providing social relevance in a time of environmental cri-sis, and developing a compelling narrative that is the hallmark of historical scholarship.

---

[7] Folke (2006); Berkes and Folke (1998); Robin (2007).
[8] Russell (2003).

An intersticing role

There has been a concern among environmental historians that their work has not reached the audience that it would merit. In the eyes of their fellow historians, those studying the environment can be tarred by association with the supposedly deterministic world of science. This was observed by William Cronon in a widely cited article as early as 1993 and it has been a recurring element of discussion since.[9] As environmental historians move into dialogue with scientists and anthropologists, and as broad brush historical syntheses are presented as our 'new history of the environment', it seems as if this inbuilt tension of environmental history is growing all the more acute. We do not wish to argue that there is anything like an either-or in this situation. On the contrary, we argue that environmental history may well serve its mission best if it manages to span the worlds held apart by methodological and discursive differences. Still, this mission can only be met if environmental history has a contribution that is distinctly its own and one that cannot be readily be provided by anybody else.

How, then, should historians and those with training in the natural sciences bring their expertise together? This problem has been addressed head on in cross-disciplinary studies developed at the University of Stirling and discussed in the chapter of Hamilton, Watson, Davies, and Hanley. Their approach, importantly, is 'cross-disciplinary'. It relies on the pooled expertise of the practitioners of very different approaches to data, in this case from history, palynology (the study of pollen), and ecology. It is not 'interdisciplinary' in the sense of individuals combining insights from a number of fields that they themselves have sought to master, or at the very least engage with.

History is a narrative enterprise. It can collect facts everywhere and weave them together into meaningful stories. It can connect what seems apart. It can use the findings of scientists and locate them into the broader analytical framework of societies and even have something to say about how societies handled crises and challenges in the past. Equally, historians are, or at least should be, sensitive to the dangers of fitting data to an already-existing narrative, of 'cherry-picking' information that appears to fit the story that one expects to be able to tell. This is a reflective standpoint that the sciences generally do not take and which is why history is an inescapable element of a synthetic understanding of the past. When scientists start writing complex narratives of cause and effect in human societies they become historians and will have to enter that game on the same principles as anybody else.

The model of work in Stirling, perhaps paradoxically, has sought initially to keep the disciplines apart. It has been an important point of principle that the disciplinary approach to particular data is not 'polluted' by expectations built

---

[9] Cronon (1993).

up in dialogue with colleagues from different backgrounds. In this model, the disciplines work best together when they are consciously kept at a distance. This is a challenge to understandings of interdisciplinarity that have been developed in the humanities in recent years. It is also a challenge to the traditional approach of historians, where there is no 'bad' data that should not be taken into account in developing an interpretation. Should environmental historians thus develop a different standpoint to their sources? Rather than simply playing a translatory role, or providing narrative direction, should historians also be adapting their own methods to those developed in the laboratory or on fieldwork?

The translation also works in the other direction, however. The sciences themselves, the institutions that sustain them and the manner in which funding has been directed have developed historically within particular contexts and been shaped by particular narratives. Not least of these is the sense of 'crisis' and 'decline'. Libby Robin's chapter on the history of conservation biology in Australia shows us how competing narratives of environmental change and theories of sustainability emerged in a particular national and international context, shaped by the prevalent scientific discourses, competition for space between those with different claims to it, and the units of analysis that as often as not were spaces demarcated by political, rather than 'scientific' decisions. She demonstrates how that history of defining 'the environmental', expounded in the chapters of Grove and Damodaran, Adams, and Sörlin, did not end after the first half of the 20th century. We remain, albeit more reflectively, part of an ongoing process.

History and humanity

A further virtue of the study of environmental history is that its topics easily transcend political and administrative boundaries. Although states have sometimes demonstrated a certain taste for defining their 'natural' boundaries according to coastlines, rivers, or mountain ranges, many (though by no means all) of these features, and the degree to which humans exploit or enjoy them, remain thoroughly impervious to the virtual world of bureaucracy. Environmental history also works on different time scales – days (the cataclysmic event), decades, centuries, millennia – and on different spatial scales from the local over regional and national scales to the continental and the global, including ecological and other physical zones in between. This is partly what makes it uniquely equipped to deal with different kinds of scientific data, which are increasingly collected in massive international databases.

It also may be a way of prioritizing humans and societies without being primarily entrapped within the 'wall-sized culturescapes of the nation', in Clifford

Geertz's expression.[10] The polity-based writing of history, from village and town to state and crown, can be complemented by a history that is 'humanity-based', taking its directions and agenda from the issues that link the fate of individuals and societies to the fate of the earth's physical and life-sustaining systems. Indeed, this can allow historians to study phenomena in their proper context, according to the relationships and decisions that actually affect them, rather than within the constraints of particular boundaries that have little direct relevance to the matter at hand.

Of course, historians of economies, gender, sexuality, ideas, and so on have been doing this for a long period of time, even if many of these histories in practice remain bound to particular political units. Environmental history brings new themes to this mix. One is 'waste', addressed here in the work of Tim Cooper, an issue that stretches from each room of our houses (most of which in the developed world are now provided with a wastepaper basket), to the politics of streets, municipalities, landfill, and sewage systems, and on to the international politics of dumping, resource transfer, and pollution. Much history has been written about the necessity of producing value, or desirable goods, and the manner in which that has shaped our world. What if we also examine the other side of that coin: how have systems dedicated to producing waste affected the ordering of our world during the last two centuries? In doing so, and highlighting the periodic crises that have punctuated discourses about waste, Cooper also calls into question boundaries that are too easily replicated in historical writing and study: between the urban and the rural, or even between what is 'valuable' and what is not.

Nevertheless, we are not calling into question the essential role that institutions, governments, and states have played in shaping how humans act, in influencing environmental processes, and in how we have come to think about such issues. Interrogating this double-edged sword, of how state boundaries have shaped engagement with the environment and our understanding of the environmental, is the task taken up in the chapter provided by Graeme Wynn and Matthew Evenden. The particular object of their attention is Canada, a neglected (at least outside of its own borders!) point of comparison to its powerful and more populous southern neighbour. Their historiographical survey demonstrates how the land and waters which demarcate Canada have provided considerable explanatory power in narratives of national identity and development. Equally, they show how the landmass south of the 49th parallel (from a landscape point of view, surely one of the most arbitrary boundaries in world history) has provided both an irresistible draw and an essential point of

---

[10] Geertz (2000).

differentiation in Canadian writing, and indeed environmental management. Environmental history may be able to transcend boundaries, but it ignores them at its peril.

We are familiar of course with the fact that nations, or advocates of them, have frequently rested claims to legitimacy on their supposed 'naturalness', or alternatively, on their ability to shape a particular, desirable 'natural world': an argument made forcefully in David Blackbourn's recent environmental history of modern Germany, *The Conquest of Nature* (2006). Indeed, Blackbourn's own study was in part a reaction to the new environmental historians of the American West, notably Donald Worster. Worster argued that far from America being built as a 'frontier society' (according to the famous thesis of Frederick Jackson Turner, forged through the experience of the pioneer), the creation of the American West was in fact an enterprise strongly shaped by centralized and technocratic institutions that created an infrastructure, especially in the management of water resources, for modern settlement. Thus we discover that the mobilizing forces of traditional historiography – the state, the nation, the corporation – sit squarely within, and can be reinvigorated by, environmental approaches. We can equally discover that mobilizing particular perceptions of the environment and how they have shaped it have been more important to these institutions than historians have generally realized.

Blackbourn's work seeks to highlight that the intersection of institution, identity, and environment, along with the relatively heavy hand of the state in shaping environments, is not just a phenomena of ancient societies (where the 'hydraulic' civilizations of the Far and Near East have long been an object for study), or of colonial expansion into supposedly pristine regions of the globe. Even at the heart of the Old World, where the landscape has unquestionably been 'cultural' for many centuries, the state has played a central role in major transformation associated with landscape planning, drainage, river straightening, dam construction, and infrastructure development.

At the other end of the old world, Mark Elvin makes a similar and provocative argument in this book about China. In the Chinese example, however, environmental transformation was not necessarily intentional, but saw the development of a certain path dependency determined by the taxation regime, and the 'lock-in' of irrigation and flood prevention measures that once created, could not easily be altered. Indeed, in this account the environment might be seen so much as a function of policy that Elvin can question the utility of an 'environmental history' at all. It may be that environmental history can just wither away once its insights are unquestionably absorbed by the historical mainstream. Yet it seems to us that (like the very idea of nature itself) it denotes a 'problem' that will stubbornly refuse to disappear for some time yet.

Elvin asserts that a major characteristic of Chinese agricultural and environmental development, involving long-term intensification and powerful (though far from all-powerful) centralized institutions, was not simply the consequence of a given environment and the eternal verities of a rice-based agriculture. His environmental history thus gives pre-eminence to a certain institutional and infrastructural development, akin to the account of world environmental history provided recently by Joachim Radkau, in his *Nature and Power* (2008, first published in German in 2002).[11]

Yet environmental history, or for that matter any other kind of history or study of human action, cannot convincingly be reduced to a study of institutions or institutional constraints. As with any other narrow approach to study, this risks circularity of argument. We must always remember that it is a certain irreducibility of nature and the choices with which it presents humanity that demands a 'politics of nature', an understanding of the 'environmental', that institutions are tasked to resolve. Indeed, it is Elvin's deep knowledge of changes in the Chinese environment that allows him to pinpoint the particular impact of institutions.

These issues are brought out explicitly in Georgina Endfield's study of early modern Mexico, where she employs comparative history to examine complex interaction of different landscapes and vegetative regimes, climatic impacts that vary over time, the developing colonial government, and local economies. The rhythms of plant growth, drought, and animal behaviour presented humans with the necessity of certain choices and dilemmas, and tantalizingly, she suggests that the both pre-Columbian and the Spanish colonial regimes may at times have come up with similar answers. Here environmental history can look across the chasm that opens up with the colonization and expropriation of the American peoples in traditional historiography. If different societies were perhaps limited in their ability to respond to particular environmental challenges, Endfield also describes how the more 'dynamic' aspects of historical change, such as social unrest, the development of welfare responsibilities, or legal systems responded directly but differentially to environmental challenge according to the choices made.

We argue thus that environmental history remains a history of humanity; a history (to quote Elvin) that is 'nowhere' in particular because its subject matter is 'everywhere'. Humanity is a huge word, almost too large to utter. It is centuries old, but in its modern usage to think and act on behalf of humanity has been part of political discourse only in the 20th century. It is a word closely akin with the institution of the United Nations and the Declaration of Human Rights from 1948. To make history speak to the issue of the fate of humanity

---

[11] Radkau (2008).

cannot just be done as a slight revision of the 'world history' that followed from the project of writing an evolutionary history of world cultures, a pet project of the early UNESCO under Joseph Needham and Julian Huxley.[12] To speak of 'humanity' does not mean that we think that all humans have, or have had, collective destinies and experiences.

The evolutionism is gone, what has replaced it is perhaps a humble feeling that if history could have something to say in the transition to an ever more *interdependent* world (and this includes ecological relationships too), it must break out of the domains where it served mobilizing purposes – the state, the bureaucracy, the nation, and even the academy. Environmental history provides one such avenue for such thinking. It is, as Kirsten Hastrup argues, a perspective that has as its perhaps chief virtue the insight that destiny is not the only word to evoke when the environment is talked about, nor is disaster. Destination may serve us better, to denote the intention that is always present in 'things human'. It is not always easy to reach one's destination. Disaster may intervene, destiny may have it otherwise. But things human are never limited to being the objects of nature's forces.

Hastrup's story centres on Iceland, an island in the north Atlantic that was the home of Norse settlers, who despite their hardiness came into deep distress following the Black Death, severe climatic cooling, and in the face of the constant irregularities of natural conditions. It would be tempting to write a natural history of Iceland with humans as part of a rich tapestry of geophysical and ecological features. But Hastrup is an anthropologist, and wishes to uncover the environmental history, which is human. She looks for destinations rather than destinies, which is also why she is not surprised that despite generations of unwise behaviour the gradually emerging Icelandic nation also managed to learn from its mistakes, to turn to the sea rather than rely stubbornly on agriculture, and finally to create what was, until very recently, one of the richest societies on earth, quite a feat for the wretched fragment of a people that clung to their bare lands only a few generations ago.

Hastrup's way of presenting her argument includes linguistic analysis of the concepts Icelanders used, the texts with which they shaped their land and gave it meaning as much as they used plough and axe. Quoting Tim Ingold, Hastrup maintains that humans are a doing and faring kind, they are 'waymakers' in an environment which they turn into a 'taskscape' as much as they live in a landscape of their own making. When the environment confronts them, or hits them mercilessly, it is not just a force from the outside that is acting. They have already, as a society and as acting individuals, shaped the circumstances under which that outer pressure is supposed to work on them.

---

[12] Robin and Steffen (2007).

The insights of the environmental history that she holds up for us are not visible if we just look at the traces out there, in the field. It takes things human to discover them:

> We cannot ask contemporary Icelanders about their reasons for responding to circumstance as they did, gradually cutting themselves off from potential sources of supplementary income. But by paying attention also to other sources, such as literature, autobiographies, learned treatises on society, and poetry, we find significant evidence of the Icelandic self-perception being a contributing factor in the decline.

Hastrup strikes a chord that we as editors have tried to keep alive all through our project with this book and its preparations in a series of seminars and workshops, and in intense exchange with our contributors. At its very core it has to do with the recognition that with all its openness to perspectives and insights from the sciences, environmental history has also made us acutely aware that the environment that we care about is always, at some instance, about what Hastrup, echoing Jürgen Habermas, calls the 'life world'. It could be a small world, of a household, family, or village. Combined it could be the humanity of billions affecting life systems covering the entire planet.

'Environmental history is not an externally induced destiny to which people are victims; people are destined also to make decisions that propel environmental changes in certain directions,' writes Hastrup. That is also why we have responsibility, each in every life world we inhabit. Sustainability is a decision, not a destiny. History is not what the past forces us to do in the future. If Stephen Dedalus was still around he might even wake up from his nightmare.

References

Barthes, R. (1972 [1957]) *Mythologies*, London.

Berkes, F. and Folke, C. (eds) (1998) *Linking Social and Ecological Systems: Management Practices and Social Mechanisms for Building Resilience*, Cambridge.

Blackbourn, D. (2006) *The Conquest of Nature: Water, Landscape, and the Making of Modern Germany*, London.

Bowler, P. (1992) *History of the Environmental Sciences*, London.

Buttimer, A. (1982) 'Musing on helicon: root metaphors and geography', *Geografiska Annaler*, 64B, 89–96.

Carr, E.H. (1961) *What Is History?*, London.

Carson, R. (1962) *Silent Spring*, Boston, MA.

Collingwood, R.G. (1946) *The Idea of History*, Oxford.

Cronon, W. (1993) 'The uses of environmental history', *Environmental History Review*, 17, 1–22.

Costanza, R., Graumlich, L.J. and Steffen, W., (2006) *Sustainability or Collapse: An Integrated History and Future of People on Earth*, Cambridge, MA.

Drayton, R. (2000) *Nature's Government; Science, Imperial Britain and the 'Improvement' of the World*, New Haven.

Febvre, L. (1922) *L'évolution et la terre humaine: introduction géographique à l'histoire*, Paris.

Folke, C. (2006) 'Resilience: The emergence of a perspective for social–ecological systems analysis', *Global Environmental Change*, 253–67.

Geertz, C. (2000) *Available Light: Anthropological Reflections on Philosophical Topics*, Princeton, NJ.

Glacken, C. (1967) *Traces on the Rhodian Shore; Nature and Culture in Western Thought, from Ancient Times to the End of the Eighteenth Century*, Berkeley.

Griffiths, T. and Robin, L. (eds) (1997) *Ecology and Empire: Environmental History of Settler Societies*, Keele.

Grove, R. (1995) *Green Imperialism. Colonial Expansion, Tropical Island Edens and the Origins of Environmentalism, 1600–1860*, Cambridge.

Guha, R. and Gadgil, M. (1992) *This Fissured Earth; towards an Ecological History of India*, Oxford, India.

Hall, P. (1988) *Cities of Tomorrow: An Intellectual History of Urban Planning and Design in the Twentieth Century*, Oxford.

Hughes, J.D. (2006) *What Is Environmental History?*, Cambridge.

Joyce, J. (1932) *Ulysses*, Hamburg.

Kingsland, S.E. (2005) *The Evolution of American Ecology, 1890–2000*, Baltimore.

Livingstone, D. (1992) *The Geographic Tradition: Episodes in the History of a Contested Enterprise*, Oxford.

Marsh, G.P. (1864 [2003]) *Man and Nature*, republished Seattle, WA and London.

Marx, K. (1887) *Capital. A Critical Analysis of Capitalist Production*, London.

McKibben, B. (1990) *The End of Nature*, Harmondsworth.

Nietzsche, F. (1980 [1874]) *On the Advantages and Disadvantages of History for Life*, trans P. Preuss, Indianapolis.

Radkau, J. (2008 [2002]) *Nature and Power*, Cambridge.

Robin, L. (2007) *How a Continent Created a Nation*, Sydney.

Robin, L. and Steffen, W. (2007) 'History for the Anthropocene', *History Compass*, 5, 1694–1719.

Russell, E. (2003) 'Evolutionary history: Prospectus for a new field', *Environmental History*, 8, 204–28.

Sörlin, S. and Öckerman, A. (1998/2002) *Jorden en ö. En global miljöhistoria*. Stockholm.

Wittfogel, K. (1957) *Oriental Despotism. A Comparative Study of Total Power*, New Haven.

Worster, D. (ed.) (1988) *The Ends of the Earth. Perspectives on Modern Environmental History*, Cambridge.

Part I
The Rise of the Environmental

1
Imperialism, Intellectual Networks, and Environmental Change; Unearthing the Origins and Evolution of Global Environmental History

Richard Grove and Vinita Damodaran

On 28 August 2005 Hurricane Katrina burst the levees of New Orleans.[1] The storm and resulting flood brought about one of the three most serious natural disasters in American history.[2] In comparison to other past natural disasters perhaps only the Lisbon earthquake of 1755 and the Asian tsunami of 26 December 2004 will have made such a decisive psychical impression on popular memory and attitudes. But the 2005 flood had two almost equally sizeable predecessors; the Mississippi floods of 1927 and 1937; events that certainly loomed large in American history and in landmark works in the written history of the global environment. In particular these floods prompted Gordon East to write *The geography behind history*, published the year after the 1937 flood.[3] In it East warned that:

> If only by its more dramatic interventions, a relentless nature makes us painfully aware of the uneasy terms on which human groups occupy and utilise the earth. The common boast that man has become master of his world has a hollow ring to it when we recall the recurrent floods and famines which afflict the peasants of northern China, the devastating floods of the Mississippi in 1937, the more recent destruction by ice of View Falls bridge across the Niagara river, the assertion that in Central Africa 'the desert is on the move', the widespread soil erosion in parts of Africa and the Middle West of the United States and finally the continual threat of drought

---

[1] *Sydney Morning Herald*, 28 August 2005. This chapter is based on a larger study.
[2] Mathur and da Cunha (2001); Barry (1997); Barry (1979); Afinson (2003). Apart from the 2005 event the Galveston hurricane flood of 1900 and the San Francisco earthquake of 1906 claimed comparable numbers of victims, that is, at least 6000 apiece.
[3] East (1938), 11.

which hangs over the great grain lands of the world – alike in the United States, Canada and South Russia. These and similar happenings or forebodings serve to emphasise the fact that, even for peoples which have reached high levels of material culture, the physical environment remains a veritable Pandora's Box, ever ready to burst out and scatter its noxious contents.

Born, like East's book, out a fear of extreme natural events, environmental history is in fact now much more extensive than that.[4] As we use the term today it is the historically documented and inter-disciplinary part of the story of life and death not only of human individuals but of societies and of species, both others and our own, in terms of their relationships with the physical and biological world about them.[5] Its intellectual origins as a self-conscious domain of enquiry can probably be traced to the encounter of the seventeenth- and eighteenth-century Western Europeans, especially naturalists, medical officers and administrators with the startlingly unfamiliar environments of the tropics and with the damage done to these environments in the course of resource extraction by European empires. From the mid-nineteenth century until the mid-twentieth the discipline developed primarily in the form of 'historical geography'. This period culminated in the publication of a famous volume entitled *Man's role in changing the face of the earth* published in 1956 under the editorship of W.L. Thomas.[6] This landmark volume brought together papers from a meeting convened at the University of Chicago. From about 1956, the year of the Suez Crisis, a process began in which historical geography was eclipsed and partly displaced by a more consciously global environmental history.[7] This was also a period in which European decolonisation began in earnest and in which writings about the possibility of global nuclear catastrophe and pesticide pollution (particularly by Rachel Carson in the USA and Kenneth Mellanby in the United Kingdom) both helped to stimulate the early green shoots of a worldwide and populist environmental movement which finally came to full bloom in the early 1970s.[8] While it would be invidious to point to one

---

[4] For useful bibliographic surveys of environmental history at various stages in its history, see Nash (1972); White (1985); Hughes (2001a); White (2001); McNeill (2003).

[5] Until the early 1970s 'environmental history' was a term conventionally used by quaternary geologists and by archaeologists and rarely dealt with human interactions; for example, see Hamilton (1982). But such books are of great utility in the new discipline. Incidentally in this essay we assume that environmental history and ecological history are for practical purposes now identical scholarly pursuits, albeit with slightly separate intellectual origins.

[6] Thomas (1956).

[7] We use the terms 'global environmental history' and 'world environmental history' interchangeably here as we can see no useful distinction between them.

[8] Carson (1951); Carson (1962); Marco (1987); Mellanby (1967).

single event as changing public consciousness in the post-war period the wrecking of the oil supertanker Torrey Canyon on 18 March 1967 in the Scilly Isles, and the subsequent spilling of 31 million gallons of oil, would certainly be a good candidate.[9] In the wake of the New Orleans events of 2005 it seems quite likely that this year may be a breakpoint in state perceptions of global warming, at least in the United States, but no doubt, too in making global environmental history an acceptable paradigm amongst the mainstream historians. As we shall see, elements of such a publicly responsive history have been around for a long time.

Until about 1970 'environmental history' was a term used chiefly by Quaternary geologists and archaeologists. As such it rarely dealt with historic human interactions with the environment. The breakthrough came in 1969 when an American, Henry Bernstein, taught a course called 'Environmental history' at Strawberry Hill College in the University of London.[10] This course, the first ever taught in the subject, was inspired largely by Bernstein's own fieldwork experiences in India while preparing a dissertation on the history of steam navigation and fuel-wood consumption on Ganges.[11] In arrogating to themselves a term previously used by geologists, historians drew a veil over a period in which most of them had tended to view as exotic the environmental concerns of historical geographers and archaeologists, whom they had undoubtedly treated as second-class academic citizens. After 1969, however, historians began to take historical geographers rather more seriously than heretofore and to start to borrow heavily for them, before actually invading their academic space.[12] Nevertheless, the term 'environmental history' is somewhat misleading in understanding the largely borrowed historiography of the discipline, if such it is. Global environmental history is not new. In many aspects its development predates the term 'environmental history' and, in a very real sense, was already being written by the mid-nineteenth century. It is worth stating this since a deep misconception has emerged that historians and others only started to integrate narratives of human history into ecological contexts in the early 1990s.[13]

---

[9] Petrow (1968).

[10] Bernstein, Personal Interview, University Centre, Cambridge, 10 August 2003.

[11] Bernstein (1956); Bernstein (1960).

[12] Powell (1996).

[13] For example, see Foltz (2003). Foltz cites Clive Ponting's *A green history of the world; the environment and the collapse of great civilizations*, as the pioneer. In fact this journalistic treatment was based exclusively on existing texts by environmental and agrarian historians. Adopting largely outdated archaeological stereotypes, Ponting attributed sudden pre-industrial societal collapses to self-generated ecological crises of over-consumption. In fact rapid collapses have historically been more frequently connected with extreme climate events.

However, the circumstances of decolonisation, a growing angst about the global proliferation of nuclear weapons (and their global destructive capability) and a linked and growing series of perceptions about global environmental crisis all served to stimulate the development of an environmental history which was, from the start, global in its concerns. This is not surprising. Concerns that typified the environmental movement, such as population growth, habitat destruction, climatic change and atmospheric alteration, were intrinsically planetary in scope so that the global dimension, if not explicit, was always high on the typically activist agenda of most environmental historians. Decolonisation and the defeat of France and the United States in the Vietnam War seem to have acted as extra stimulants to a tendency to look back and reflect on the experience of empire, not least in an environmental sense.[14] Ultimately this led to some very fertile intellectual explorations since any reflection on global environmental change led quite quickly to the notion that the environmental impact and speed of environmental change in the imperial context had been some of the most significant in written history as well as often being very well documented. Even the early preoccupation with the 'Western frontier' in American environmental history can be seen as being integral to the expansion of the American land-based empire in the West, even if it was often not conceived as being imperialist as such.[15] If the Vietnam defeat was seen as the first blow to a post-1945 and post-Hiroshima hegemony it could also be seen as a major stimulant to thinking about the enormous environmental impacts of that hegemony. Nevertheless it is surprising how little the environmental impacts of imperialism have touched the consciousness of such mainstream imperial historians as Linda Colley, Niall Ferguson or Christopher Bayly. Indeed, it was possible for Colley to deliver a major lecture at the Australian National University in August 2005 entitled 'Difficulties with empire' without mentioning the words 'environment' or 'ecology' at all.

Global environmental thinking in the 1930s

The rise of a *de facto* environmental history can be traced to the 1930s, where we can observe an innovative convergence of analytical and descriptive writings by geographers, anthropologists, archaeologists and ecologists, many of them taking a global, increasingly anxious and prescriptive view of human–environment interactions. The contributions made by archaeologists and anthropologists started to become a useful one. As well as the ambitious surveys of Gordon Childe, Daryll Forde's major collection of writings published in 1934, and entitled *Habitat, economy and society*, now started to add a highly

---

[14] Engel (2004).
[15] Nobles (1997).

structured and global approach to environmental essays on history and human societies, drawing globally on case studies from Malaya, the Kalahari, Siberia, the Arctic, West Africa, the Pacific, many of them the work of a new breed of anthropologists and geographers in colonial employ.[16] Through Forde's work ran a strong streak of Boasian cultural relativism, an approach that placed far more stress on the cultural relationships of people with their environments than on their evolutionary stages of development. Moreover Forde's studies concentrated on economic interactions with the environment rather than on matters of ritual. This too was an important new direction. Although based in North America, at UC Berkeley, Forde's work ranged very widely geographically. His first book, a history of ancient shipping routes was followed by a treatise on Yucatan Indians. Later he specialised in explaining the relations between environment and the evolution of African societies.

It is important to note that colonial environmental change agendas were decisive in defining the form of global environmental geography, history and anthropology in the 1930s. But they were also affecting historical understandings of the environment at the metropole. Henry Darby, for example, started writing on Rhodesia but then developed a specialist interest in Domesday colonisation, medieval landscape histories and Fenland drainage. But he also worked during the Second World War in Naval Intelligence, collating strategically important geographical information worldwide. This was paralleled in his systematic examination of Domesday land utilisation and was based methodologically on research being carried out by colonial scholars, some of whom we might today classify as environmental historians, and many of whom were now concerned with both climate and land-use history.[17] Perhaps the most interesting of these was L. Dudley Stamp, who was employed as an oil geologist in Burma before becoming the first Professor of Geography at Rangoon University.[18] Like Jan Smuts, Stamp was an adherent of the theories of Arthur Tansley, the British ecologist, who published his first paper in the newly founded *Journal of Ecology* in 1923.[19] In 1930 Stamp published *The vegetation of Burma; from an ecological standpoint*.[20] On the basis of this work he then started to take a more detailed interest in the evolution and climatic history of the Burma landscape. In so doing, Stamp followed in the footsteps of the climatic historian J.C. Mackenzie who had published an extraordinary pioneering paper in 1913

---

[16] Forde (1934); Forde (1927); Forde (1931).
[17] Darby (1952).
[18] Dudley Stamp was admitted to Kings College London at the age of 15 and gained a first-class degree in geology. He was commissioned into the Royal Engineers and served in France during the First World War, where he gained some insight into the potential of air photography. He gained a doctorate in 1921 and then took the first Honours examination in Geography of the University of London, gaining a first-class degree.
[19] Stamp and Lord (1923).
[20] Stamp (1930).

on 'Climate in Burmese history', relating Pagan societal collapse to prolonged episodes of drought.[21]

Between 1923 and 1944 Stamp published a myriad of papers on the history of relations between Burmese hill tribes and land-use, forests, shifting cultivation and climate.[22] However, his main emphasis consisted in charting the shifting relations between people and land-use through time. To this specialism Stamp added a very new technique impossible before 1914; the use of the air photography in surveying land-use change, a technique he reported in the *Journal of Ecology*.[23] This First World War-developed technology now found a peacetime use. Underlying Stamp's surveys in both Burma and Britain (where he developed the official Land Utilisation Survey between 1930 and 1940) continued to be a Tansleyite preoccupation with evolutionary changes in the succession of vegetation types over time.[24] But Stamp's use of air photography in France and Burma clearly stimulated his extensive approach to land-use change. This approach epitomised the repatriation of colonial innovations and academic methods to the metropolitan centre. For many years Stamp's approach also influenced the philosophy and methods of post-war colonial natural resource survey in other colonies. For example, in the Gold Coast Robert Wills carried out a decade-long study on agriculture and its evolution in the Gold Coast (Ghana), adopting the methods of both Stamp and Darby.[25] The Gold Coast Soil Survey then became effectively a global model as Gold Coast staff, mainly geographers were dispersed at decolonisation.[26] This was an important development since the globalisation of colonial soil and land-use survey methods long after decolonisation encouraged the efficient use of land.

---

[21] Mackenzie (1913).

[22] Stamp (1924a); Stamp (1930a); Stamp (1940); Stamp (1942); Stamp (1961).

[23] Stamp (1924b).

[24] The first Land Utilisation Survey was carried out by amateurs, including schoolchildren! The second LUS was carried out after the Second World War and was still largely unofficial.

[25] Wills was a Cambridge graduate. He worked under C.H. Charter, a chemist who became a soil scientist and formed and directed the Gold Coast Survey. He contracted elephantiasis, so that Wills had to write the volume, which is based on regional soil survey reports.

[26] For example, H. Brammer and P.M. Ahn, both geographers, were other members of the G.C. Survey. The President's Gold Medal in 1978 and The B.R. Sen Award for 1981 were presented to Brammer for his work in promoting the optimum use of land, specifically for training and advising on soil survey and land-use planning in Bangladesh bringing with him the experience gained as a member of the Soil and Land Use Survey Organisation in Ghana, which had gained international recognition as a pioneer in tropical soil surveys and land-use recommendations. Brammer established the Bangladesh National Soil Survey; completing reconnaissance soil survey of Bangladesh at high resolution, in particular, allowing the successful promotion of high-yielding wheat and rice varieties at grassroots level.

Successionist ecology, by contrast, not only contributed to the emergence of writings in *de facto* environmental history by colonial researchers but it also led to a growing sense of environmental crisis and conscience among those confronted by systematically gathered evidence of environmental change. It was no accident that ecologists such as Arthur Tansley, Frank Fraser Darling, Alfred Steers and other specialists who had served in the colonies were instrumental in pioneering National parks and protected areas legislation in immediate post-war Britain.[27] This is not to say that historians have been entirely absent from the process; Keith Hancock's Australian writings and G.M. Trevelyan's strident 1920s environmentalism are cases in point.[28] Nevertheless, specialists who had been employed overseas in tropical colonies tended to be familiar with the evidence of rapid environmental change worldwide, and had a sense of the global context and speed of the process.

The rising utility of anthropologists, geographers and ecologists for colonial governments in the immediate pre- and post-Second World War period made some of these specialists relatively assertive and prescriptive in their opinions about the connections between environments and historical change. A particularly didactic and environmentalist exemplar of this was Gordon East (1902–1998), who (as noted at the beginning of the paper) published *The geography behind history* in 1938. East, a lecturer at Birkbeck College London, introduced his book with a quote from Heylyn's *Cosmographica* of 1649; 'if joined together [history and geography] crown our reading with delight and profit; if parted they threaten both with a certain shipwreck'. East called the first chapter of his book 'Geography as an historical document'. The claim of 'geography to be heard in the councils of history', he felt, 'rests on the firm basis that it alone studies comprehensively and scientifically by its own methods and technique the setting of human activity and further that the particular characteristics of this setting serve not only to localise but also to influence part at least of the action'. Once more quoting Heylyn, East noted that just as 'geography without history seemeth a carcasses without motion, so history without geography wandreth as a vagrant without a certain habitation'. But

---

[27] Arthur Tansley (1871–1955) was profoundly affected by his experiences in Ceylon, Malaya and Egypt in 1900 and 1901. These undoubtedly encouraged him in developing a globally applicable system of ecological explanation. Immediately after his return to the United Kingdom he founded the *New Phytologist*. Frank Darling worked extensively in East Africa on ungulate conservation while Alfred Steers worked in the West Indies on coral reef geomorphology and conservation. Moreover all these men were affected by their experiences of tropical island colonies, island biogeography becoming a life-long obsession with Darling, who at one time retreated for some years to the uninhabited Scottish island of North Rona. Alfred Steers colonial academic contributions include Steers (1940a); Steers (1940b); Steers, Chapman, Colman and Lofthouse (1940c).

[28] Trevelyan (1929).

East needed the confirmation of a last word upon the subject and a widely respected academic authority to clinch his arguments for a new kind of history. He found it in the work of J.L. Myres who had suggested that 'it has been nature rather than Man hitherto, in almost every scene, that has determined where the action shall lie. Only at a comparatively late phase of action does man in some measure shift the scenery for himself.'

By the time East's book appeared, war clouds were gathering and man was about to do more than simply shift the scenery or act as a geographic agent. Within seven years a nuclear device would be exploded at Alamagordo in the desert lands of New Mexico, rapidly followed by its use at Hiroshima. East's narratives in *The geography behind history* certainly reflected the great anxieties to which these kinds of events gave rise; indeed, in his co-opting of the authority of an historian of ancient history, East seems to have sounded out the deepest apprehensions of the ability of man not just to destroy nature but to destroy himself and society as well. As we have seen the historical analyses of contemporary man–nature interactions had been closely affected by responses to the cataclysm of the First World War. Such reactions were, if anything, even more pronounced after the Second World War. In his book East used the frightening lessons of early civilisational collapse and crisis, just as G.P. Marsh had done, to cajole his colleagues into looking at history in a new way. For as the old imperial powers stumbled somewhat brazenly to gather up their reluctant colonies after 1945, a veritable avalanche of works appeared inventorying the natural world of the colonies and categorising and theorising man–environment relations, especially in the tropical world.[29]

Post-war environmental history

In the mid-1950s the dictums of the forgotten environmentalists of a century earlier[30] were once more exhumed and brought to the light of day by the generation of 1950s scholars who pioneered post-war environmental history.

---

[29] Incidentally one may use the term 'man and nature' quite advisedly since, despite the pioneering efforts of such pioneer women historical geographers as Ellen Churchill Semple and Jean Mitchell, it was still a man's world in colonial environmental writing; or at least it was until the advent of two brilliant inter-disciplinary economic anthropologists; Audrey Richards in Central Africa and Polly Hill in West Africa. See Semple (1911); Semple (1931).

[30] Here we are referring to environmentalists such as in French colonial Mauritius (the Isle de France) Pierre Poivre and Philibert Commerson who framed pioneering forest conservation legislation designed specifically to prevent rainfall decline n the 1760s. In India William Roxburgh, Edward Balfour, Alexander Gibson and Hugh Cleghorn (all Scottish medical scientists) wrote alarmist narratives relating deforestation to the danger of climate change. Their distinctively modern environmentalist views owed a great deal to the precocious commentaries of Alexander von Humboldt in his *Personal Narrative* and in the *Cosmos*. See Grove (1995).

Prominent among them was David Lowenthal, a polymathic geographer also interested in the problems of small West Indian islands, and early texts on the rise and fall of empires.[31] In his *Versatile Vermonter*, a 1958 biography of George Perkins Marsh, Lowenthal highlighted the way in which Marsh utilised accounts of environmental degradation in the Mediterranean world, the decline of Roman and other empires in the region, and the degradation of colonial island colonies, to colour dour warnings about the contemporary unsustainable use of resources in the United States, particularly on its colonial Western frontier.[32] It was a global knowledge of the history of environmental change and an acquaintance with the colonial (especially French and British) literature on the imperial ecological impact, Lowenthal observed, that fired Marsh's environmentalist discourses. The theme of the Western colonial frontier was continued by Samuel Hays in his 1959 biographical treatment of Theodore Roosevelt's 'Progressive' conservationism.[33] Roosevelt, his environmentalism apparently masquerading as a 'gospel of efficiency', had soon created a veritable empire of public lands, National Parks and National Forests on a scale never equalled before or since in any non-communist state, with the notable exception of British India's forest reserve system, which was in practice Roosevelt's major conservation model.[34]

While Lowenthal and Hays reflected on the origins of conservationism in the context of their own increasingly environmentally conscious post-war milieu, another pioneer of environmental history, Clarence Glacken, was, throughout the decade 1954 to 1965 struggling to produce his magisterial but confusingly named masterpiece, *Traces on the Rhodian Shore; nature and culture in western thought, from ancient times to the end of the eighteenth century*.[35] Glacken's first book, *The Great Loochoo*, based on his doctoral dissertation, was an historical geography of Okinawa.[36] For Glacken this work had served to emphasise the historical significance of insular environmental constraints, so that the intellectual jump from consideration of island to world probably seemed a logical one, as it had to Marsh in his writings a century earlier. It was no coincidence that Glacken was an academic on the Berkeley campus in the mid-1960s or that his

---

[31] Montesquieu (1734[1965]).

[32] Lowenthal (1958).

[33] Hays (1957).

[34] Geologist William McGee had in fact been the ghost-writer of many of Roosevelt's conservationist dictums and legislation; and to an extent Hays did not realise, was a substantial author of Roosevelt's progressive conservationism. The original US Forest Department, however, was based on Congress accepting a global reading of destructive environmental change, published by Franklin Benjamin Hough, a enthusiastic follower of John Croumbie Brown, the Cape Colonial Botanist; see Hough (1878–79). See also Lacey (1979).

[35] Glacken (1967).

[36] Glacken (1955).

work was published by the University Press of California, a state that nurtured the evolution of the Sierra Club, its activist breakaway group Friends of the Earth, hippies and the strongest elements of the anti-Vietnam war movement. Like them *Traces* represented an attitudinal step-change and an entirely new kind of internationalist retrospective on human perceptions and stewardship of the planet and its inhabitants.

As such Glacken's work would never have been completed had it not been for his intellectual protection from the critique of sceptical colleagues at Berkeley by Carl Sauer, the doyen of American cultural geography and human ecology. Sauer, with his interests in the globalising history of plant domestication and cultural landscape evolution, quickly saw the vital importance of Glacken's work in making the history of ideas about the environment a project on a global scale and on a canvass that, moving beyond European thought, encompassed imperial thinking about nature right through from the Akkadian to the European maritime empires. In particular he explored the emergence of ideas associated with man as a dominant environmental agent. Glacken's Classical training as well as his army career in East Asia equipped him for the enormous task he set himself. Like Lowenthal, Glacken sought to exhume the work of previously largely ignored thinkers by drawing attention, for example, to the prodigious writings of John Croumbie Brown, Colonial Botanist of the Cape Colony and the influence on the flurry of French post-Revolutionary writers on deforestation, flooding and 'torrents', the backdrop to much nineteenth-century French state conservationism.[37]

Although at first prevented by its over-erudite title from receiving an appropriate popular reception, Glackens *Traces* almost immediately spun off a group of studies inspired and taking themes directly from different parts of his periodisation of global environmental ideas and environmentalism. For the Classical period the work of Russell Meiggs on *Trees and timber in the ancient Mediterranean world* was notable, while Donald Hughes pursued further the ecological history, proto-environmentalism and the alleged ecologically caused demise of Classical Greece and Rome. A Classicist like Glacken and Meiggs, Hughes was a scholar who had taught for years at the University of Athens, a life experience comparable to G.P. Marsh's diplomatic career in Turkey, Greece and Italy, and Huntington's teaching career in Turkey. Perhaps predictably, Hughes has gone on to produce the first useful book-length essay on *The environmental history of the world*.[38] Having read Glacken, Keith Thomas, an English early modernist historian, interrupted his more mainstream interests and researched *Man and the natural world* (1984), a work which usefully filled in some yawning

---

[37] Grove (1989); McPhee (1999).
[38] Meiggs (1982); Hughes (2001).

gaps left by Glacken in his treatment of English environmental ideas.[39] But, like Glacken, Thomas also felt compelled to halt his essay at 1800 leaving a lacuna in environmental history for the whole of the nineteenth century which has not been properly treated by any environmental historian either for Britain, America or the globe. Instead disciples of Glacken such as John McNeill have opted, at least for now, to jump to the territory of twentieth-century environmental history.[40] Their grounds for doing so are that the real step-change or turning point in modern environmental history occurred in the twentieth rather than the nineteenth century, a view with which many may disagree, choosing instead to locate the most significant turning point in nineteenth-century industrialisation.[41]

The Watershed years

The years 1955 and 1956 emerge as a critical watershed years in the evolution of post-war environmental history, in particular with the holding of W.L. Thomas's landmark meeting at Chicago on 'Man's role in changing the face of the earth', the beginnings of Indian ecological history, the appearance of Glacken's first book *The Great Loochoo* and the publication of W.G. Hoskins' *The making of the English landscape*, the latter being perhaps the first major environmental history to be written by an historian.[42] The Chicago meeting, which opened on 16 June 1955, was sponsored by the Wenner-Gren Foundation for Anthropological Research and had been planned for three years, and was encouraged particularly by Carl Sauer and Lewis Mumford. The Foundation's interest in calling the 1955 Symposium was 'to keep abreast of all the means at man's disposal to affect deliberately or unconsciously the course of his own evolution; in this case what man has done and is doing to change his physical-biological environment on the earth'.[43]

The organisers highlighted as the dual inspirers for their project (published as a single massive volume in 1956), G.P. Marsh and the Russian geographer Ivanovich Woeikof. The latter had, like Marsh, made a series of clarion calls against thoughtless environmental degradation. The Chicago organisers were cognisant too of the significance of Nathaniel Shaler's essay entitled *Man and the earth* (1905) and quoted other lesser-known authorities as a way of giving

---

[39] Thomas (1984).
[40] McNeill (2000).
[41] In June 2005 an entire conference was devoted to debating 'Turning points in environmental history' at Bielefeld, Germany. The proceedings of the conference are in press.
[42] Thomas (1956).
[43] Thomas (1956), Introduction.

some credence to the undoubted timeliness of the conference.[44] It was opened, appropriately, with papers read by Carl Sauer and Clarence Glacken. For Sauer the meeting was to be a 'Marsh festival', while Glacken opened his address, on 'Changing ideas of the habitable world', by referring to Plato's remarks on the ruination of Attica by soil erosion.[45] The latter essay was a dry-run for the arguments produced in *Traces*, the book which would appear more than decade later in 1967.[46]

The papers presented to the Chicago meeting were largely written by white Anglo-Saxon men, but there was one very important exception, that given by E.K. Janaki Ammal, a distinguished woman geneticist and global plant geographer, a Professor of Botany at Madras and co-author of *The Chromosome atlas of cultivated plants*.[47] The latter was already a famous work and one which qualified her uniquely to comment on the evolution of global plant distributions, and human impacts upon them. But her contribution to the Thomas volume was to essay an environmental history of subsistence agriculture in India. Paying particular attention to the sharp difference between patriarchal, matriarchal and tribal (indigenous Indian) agronomies, Ammal can be said to have pioneered both indigenous and gendered environmental approaches to land-use history.[48]

Parallel developments were taking place in India itself in 1955–6, particularly under the aegis of the Ecole d'Extreme Orient. This institute, originally established in Phnom Penh and Saigon, was forced to move to Pondicherry in French India during the Vietnam War in early 1954. The first Director of the Institute in Pondicherry, Jean Filliozat, pioneered a major study of the forest and vegetation history of the whole of Southern India. This was a project in the fashion of Dudley Stamp's surveys, planned on an unprecedented scale and which in 2005 is still far from complete. But, combined with his *annaliste* training, the scope of the project led Filliozat to write some of the earliest 'ecologies historiques' of India. His first paper, written in 1956, concerned the environmental history

---

[44] Shaler (1905).

[45] Thomas (1956).

[46] Glacken, lecture to the University of Virginia Summer School, Emmanuel College, Cambridge, July 18 1981 entitled 'Writing a successor to *Traces on the Rhodian Shore*'; and pers.comm. Glacken.

[47] C.D. Darlington and E.K. Janaki Ammal, *The chromosome atlas of cultivated plants*, London, 1945

[48] Ammal was the Anglo-Indian grandchild of John Child Hannyngton I.C.S, British Resident at the courts of Travancore and Cochin, a Sanskritist and expert on Malayali culture. A relative, F. Hannyngton, was a well-known lepidopterist. Ammal was the first woman to gain a PhD in Botany in the United States, in 1938, and later became Professor of History at Madras and the first Director of the Central Botanical Institute, Lucknow. She was also the founder of ethnobotany in India.

of the Kallar country of South India.[49] As such it was a forerunner of what was to be a burgeoning school of environmental history in South Asia. Like Janaki Ammal, it was Filliozat's interest in ethnobotany that led Filliozat to think in terms of environmental history on large scale.

From hindsight what we can now see as the overwhelming influence of Glacken, a presence that was already disproportionate at Chicago, could not really have borne fruit without the development of strong national schools of environmental history. In the English-speaking world, these developed first in Britain, America, Australia and in India, although we may note that some British and American scholars, for example, can be identified as scholars of regions far from home. As we have seen, the first attempts at supra-regional or global environmental history had developed first among Classical scholars and archaeologists used to thinking at least in the dimensions of the known Classical and Mediterranean world.

In Britain, Hoskins wrote *The making of the English landscape* while at the Department of English Local History at Leicester University. This school concentrated on making use of the minutiae of local sources on environment and material culture. It also owed intellectual debts to Henry Darby (especially his 1936 work on *The draining of the fens*), the *annalistes* and to Fernand Braudel. Braudel, whose first major work on the Mediterranean appeared in 1949, shocked contemporaries with his holistic, globally referential and conspicuous attention to climate, environment, the sea and material culture.[50] However, like Glacken's work, the book was before its time and it was not for decades that comparable work would be done on the Mediterranean, by Claudio Vita-Finzi, John McNeill, Oliver Rackham, Peregrine Horden and A.T. Grove.[51] Instead Braudel was emulated elsewhere, due possibly to the sceptism of the French academic establishment towards his academic empiricism characterised as it was both by environmental determinism and its dismissal of grand theory.

This meant that the publication of the Braudelesque work of Hoskins acquired a still greater importance. Significantly Hoskins, although originally a medieval historian, collaborated closely with Dudley Stamp in producing *The commos lands of England and Wales*, another book with antecedents in colonial survey methods.[52] Hoskins's work helped to stimulate the growth of a whole field of agricultural and local history after 1953 and nurtured a group of scholars, many of them women, who carried Hoskins' methods (summarised in *Fieldwork in local history*, 1967)[53] to many parts of British provincial history, but

---

[49] Filliozat (1980).
[50] Braudel (1949).
[51] Horden and Purcell (2000); Grove and Rackham (2001); Vita-Finzi (1969).
[52] Hoskins and Stamp (1963).
[53] Hoskins (1967).

with a Braudelian rigour. Joan Thirsk, Margaret Spufford, Harold Fox and John Patten, among others, all owed an intellectual debt to the Leicester/Hoskins School as well as the methods of Herbert Finberg. One might note that Finberg, himself a follower of Frederic Seebohm's village-level approach in local history, was the founder of the multi-volume *Agrarian history of England and Wales*, much of which was in essence local environmental history.[54] Women, arguably, were less hesitant about crossing disciplinary boundaries and concentrating on the detail of material culture. Indeed, to date the disproportionate presence of women as authors of environmental history, just as they are preponderant in the environmental movement, has been conspicuous.[55] The Hoskins school also encouraged scholarship in English forest history and on hydraulic and water-management history. In the English Fenlands, for example, Jack Ravensdale wrote *Liable to Floods*, a highly creative Hoskinite development of the work of Henry Darby, examining the impact of seventeenth-century drainage on a Fen-edge village through many centuries.[56] Similarly, Victor Skipp, another Leicester product, published an innovative ecological history of the Forest of Arden.[57] There was an element of protest in the Leicester school approach, connecting both a rejection of the macroscopic concerns of metropolitan university history and the incorporation of a new landscape and environmental framework to history, that was implicitly local and global.

The Leicester school of local and regional history also gave inspiration to researchers entering the subject very much in the tradition of the geologist/ecologist Dudley Stamp, that is, directly from the natural sciences. Prominent among these was Oliver Rackham, a plant physiologist who happened to have received a Classical languages training at public school. Just as physiologist Joseph Needham became the foremost historian of Chinese science so Rackham became the leading exponent of English and European ecological and woodland history; what was originally his fungi and plant-collecting hobby became a profession strengthened by the rare ability to read medieval court rolls as if they were a daily newspaper. Rackham's work was as painstaking as Glacken's and his first book in ecological history was succeeded by a series of books on ancient woodlands and the countryside, all written much in the mode of Hoskins, but displaying with finesse the advantages of both a Classical and scientific training. Rackham seems at first to have been reluctant to admit to his intellectual heritage but he finally signposted it in his first major academic

---

[54] Finberg (1967 & 1972).
[55] See Carolyn Merchant (1980), cited as one of the classic texts in environmental history. The tradition of ecofeminist writing also bears this point see discussion of the tradition in Carolyn Merchant (1992), and M. Leach and Cathy Green (1997).
[56] Ravensdale (1977).
[57] Skipp (1978).

excursion outside Britain, in *The making of the Cretan landscape*, published in 1994.[58]

Other book titles also echoed Hoskins' work on the English landscape, notably Michael Williams' 1974 *The making of the South Australian landscape*.[59] The development of a French school of environmental history has always been rather separate from the Anglo-American tradition in sources and inspiration. However, the work of Le Roy Ladurie, especially his *Times of feast, Times of famine* made a great impression outside France and has had a global relevance.[60] Ladurie took a particular interest in the historical impact of climate change and climatic extreme events on local communities, and had been much influenced by his reading of British climate historians.

California and environmental history

The shift towards integrating new climatic knowledge into understandings of the dynamics of social and economic change had been prefigured by Glacken in *Traces*, in underlining the long intellectual history of theories on anthropogenic causes of climate change. Throughout most of the period between 1955 and 1967, Glacken continued to work away at *Traces on the Rhodian Shore*. When the book finally appeared it coincided with the simultaneous and explosive growth of both the anti-Vietnam war movement and the nascent popular environmental movement, many of whose influential advocates were, like Glacken, faculty members on University of California campuses. Meanwhile Carl Sauer, Glacken's patron, had published his own *The early Spanish main*, in 1966, with the University of California press.[61] The press had already been broken in to the task of publishing environmental history with the appearance in 1963 of Clifford Geertz's *Agricultural involution; the processes of ecological changes in Indonesia*. For both authors and publishers interdisciplinary research, as Geertz wrote, 'is always a gamble'.[62] But Geertz also left a clue to the growing interest that some historians were taking in the work of their more interdisciplinary colleagues. Quoting Marc Bloch, Geertz noted that 'just as the progress of a disease shows a doctor the secret life of the body, so to the historian the progress of a great calamity yields valuable information about the nature of a society so stricken'.[63] Undoubtedly, Geertz, Sauer and Glacken all wrote, at least in

---

[58] Rackham and Moody (1994).

[59] Williams (1974).

[60] Le Roy Ladurie (1988).

[61] Sauer (1966). Contemporary reviewers commented, 'History, written by a geographer, especially one of Professor Sauer's learning, is an enriching experience' (Cover sheet quoting *The Professional Geographer*)

[62] Geertz (1963), vii.

[63] Geertz (1963), vi.

part, from the standpoint of the growing post-war American imperial interest in Southeast Asia and South America and their personal experience of it coloured their writing. Their approach was intrinsically global. By contrast those earliest Americans actually to call themselves environmental historians, who were faculty at UC Santa Barbara, only had the barest acquaintance with non-American themes and indeed most California historians lived in ignorance of what their interdisciplinary colleagues at Berkeley were actually up to. Even less were they aware of the activities of Henry Bernstein, their pioneer compatriot at London University.

In 1967 Roderick Nash published *Wilderness and the American mind.*[64] The book was an immediate success with the California public but was a strikingly nationalistic and parochial product compared to the work of Glacken published in the same year. The book's success encouraged Nash to take his research further and to list a new course at UC Santa Barbara entitled 'Environmental History'. Nash seems himself to have re-coined the term while quite unaware of its longstanding use by Quaternary geologists, or indeed by Bernstein in London. As Nash wrote revealingly in 1972:

> I thought I was responding to the cries for environmental responsibility which reached a crescendo in the first months of that year. I also felt good about helping to make the university, and particularly the Department of History, more responsive to the problems of society. I was, at last, 'relevant'. Moreover, my previous work in American intellectual history, especially the research that led to *Wilderness and the American mind*, had familiarized me with broad patterns of interaction between Americans and their environment...but on the way back to my office, misgivings began. They grew into anxieties of major proportions during the next two weeks as 450 students enrolled in the inchoate course. What was I going to do with them? There were few places to turn for answers. To the best of my knowledge no similar course had ever been offered. Also lacking was the body of reading material.[65]

For a scholar teaching in California these remarks were startling. Nash clearly believed that it would be necessary to start from scratch, and appeared not to know of either Glacken or Sauer. 'Environmental history', he wrote, as if discovering something quite new, 'would refer to the past contact of man with his *total* habitat'. Casting around for an idiom or a comparison, Nash continued; 'the environmental historian, like the ecologist, would think in terms of wholes, of communities, of interrelationships, and of balances'. He should take as his first axiom John Muir's statement that 'when we try to pick out

---

[64] Nash (1967).
[65] Nash (1972).

anything by itself, we find it hitched to everything else in the universe'. Finally, he concluded, without any real justification (except perhaps in terms of a kind of proto-Deep Ecology) 'in a very real sense environmental history fitted into the framework of New Left history. This would indeed be history "from the bottom up", except that here the exploited element would be the biota and the land itself.' These remarks really gave the game away. Clearly Nash conceived of himself as the inventor of 'environmental history'. Nash went on to describe the contents of his course as it later emerged. It was remarkable for having absolutely no reference to any extra-North American material, with the exception of a reference to G.P. Marsh's *Man and nature*. In 1972 Nash contributed a chapter on 'The state of environmental history' to a volume on *The state of American history* edited by Herbert Bass.[66] Here, as in later article in the *Pacific Historical Review*, Nash appears to have equated environmental history purely with American history. This inward-looking bias was important since Nash and his colleagues went on to found the American Society for Environmental History, as a breakaway from the American Historical Association. Unsurprisingly the house journal of the Society, *The Environmental Review*, now named *Environmental History*, confined itself until very recently almost entirely to American issues.

This parochialism may have reflected a wider tendency to introspection after the Vietnam War. Subsequent moves towards a more global treatment of environmental history had to come from elsewhere, above all from ecologists, geographers, archaeologists and colonial or Asianist historians. Just as in the previous century, discourses of global environmental history after 1967 (i.e., after Glacken) developed in discrete and surprisingly identifiable areas of academe, above all in those concerned with a Classical training, experience in the Mediterranean region (or both) and among academic South and central Asianists. After 1967 another group of specialists also became particularly concerned with global change. These were historians of China and Japan, and historians of empire. In other words, they were scholars accustomed to think in the long time periods encompassing Classical history, extensive geographical areas and those accustomed to making geographical and temporal comparisons of environmental and social change. But overarching these incentives to writing global environmental history were an increasing understanding of the intrinsically globalised dynamics of climate history and the growing interest, in the decolonisation period, in all aspects of imperialism. Thus after Glacken, we see both a range of worldwide environmental treatises being produced and an ambitious series of major regional environmental histories of a kind that had not been attempted previously.

---

[66] Bass (1970).

Millennial and socio-ecological environmental history

Donald Hughes (at one time a history lecturer in Athens) continued to link his classical eastern Mediterranean preoccupations with the first professional historian's stab at writing an avowedly global environmental history published in 2001. From being the fantastical projection of the celluloid Space Odyssey, 2001 now became a millennial milestone in global environmental history.[67] What some have considered to be the 'closing' of the globalised world in terms of communication and the Internet had by now become the academic occasion for a whole series of treatments that took a retrospective look at the contribution of European and Chinese imperialisms in globalising resource and territorial control and transforming the world environment. The increasing scholarly interest taken in these matters by Americans seems to have paralleled the rapid growth in the United States imperial interests and ambitions after 1945, despite the quite separate introspective nature of environmental history in the United States itself. It may well have been this awareness of new superpower status, as well as the Cold war itself, that encouraged two men, Paul Colinvaux and Jared Diamond, both tropical ecologists, to attempt socio-ecological explanations of world history, respectively in the *Fates of Nations* (1980) and in *Guns, germs and steel* (1997).[68]

These books, both best-sellers, attempted to explain Western European 'supremacy' in terms of such concepts as 'the learned niche' (Colinvaux) and the rather less profound 'differences in environmental real estate' put forward rather inconclusively by Diamond as the reason for the differential economic trajectories of Europeans versus African or aboriginal societies. Neither of these authors attached primacy to the coalescence of networks of ideas in explaining technological differentiation as the McNeills did so much more satisfactorily in their later *Human web* of 2003.[69] Differences in intellectual training clearly counted in this instance. A more recent book by Diamond, *Collapses*, published in 2004, represents a populist attempt to explain societal collapse in terms of isolation, climatic anomalies and fossil fuel consumption was arguably somewhat hindered by its retreat into a mainly Americo-centric use of case-studies.[70] Nevertheless the sub-conscious influence of notions of millennial collapse and Social Darwinist superpower competition that underlies both of

---

[67] Hughes (2001); see also Radkau (2000) and Radkau (2003); Sörlin and Öckerman (1998). An excellent edited survey is de Fries and Goudsblom (2002). This book usefully assesses the environmental footprint of the Roman Empire.

[68] Colinvaux (1980); Diamond (1987).

[69] McNeill and Mc Neill (2003).

[70] Diamond (2005) The subtitle of this book is worthy of particular attention and, were Gunder Frank still alive, would no doubt have attracted an incisive response. However, one will not find Frank listed amongst the Diamond bibliographies.

these socio-ecological global history writings is quite transparent. This is characteristic, it must be said, of much socio-biological history written for the American market. But perhaps one can argue that there is another underlying flavour to the kind of judgemental global history preached by Colinvaux and Diamond; that of cultural superiority.

Neither author appears to have been aware of highly relevant earlier writings, such as those by Jan Christian Smuts whose 1926 visionary book, *Holism and evolution*, was itself heavily affected by the new plant ecology of Arthur Tansley, himself the author of one of the earliest environmental histories of Britain. Smuts had been much concerned with the relations between climate and human evolution. One question the unquestionably (although mildly) racist Smuts raised in a lecture in 1932: 'what has caused the immense difference between the European and the Bushman of today? We see the one the leading race of the world, while the other, though still living, has become a mere human fossil, verging to extinction', sounds eerily like many of the rhetorical questions underpinning the work of both Colinvaux and Diamond. Nor, it would seem, were they aware of that other Jungian disciple of Smuts, Laurens van der Post who, after much personal struggle, became a great early advocate of the natural sophistication of indigenous societies that had previously been scripted as 'primitive'. The root of the problem, it seems, is that ecological training could be used to argue for preferential race or caste classification or for quite the opposite, an egalitarian or cultural relativist view of human types and their varying ecological interactions and impacts. What purported to be 'objective' scientific environmental history proved, inadvertently, to be quite the opposite.[71]

Asianists and global environmental history

In contrast to the global views of the socio-ecologists, a refreshing and less culturally triumphalist view of the environmental impact of apparently 'successful' expansionist imperial societies has been taken by a distinctive group of historians of China, beginning with Lester Bilsky in 1980.[72] Bilsky, drawing mainly on the translated works of Mencius, Hsuntze and Chan-kuo Ts'e, pointed to the prolific evidence for local social collapse and famine-caused wars in the first millennium. Much more detailed evidence was published in the ensuing 20 years for what Elvin first called 'Three thousand years of unsustainable growth', evidence that was expanded further in Elvin's *The retreat of the elephants; an environmental history of China*.[73] Peter Purdue and Robert Marks,

---

[71] See Anker (2001).
[72] Bilsky (1980).
[73] Elvin (1993); Elvin (2004).

like Elvin, drew attention to the massive ecological transformation brought by the westward expansion of successive Chinese empires, findings which helped to place European colonial expansion in perspective and re-emphasise their comparable ecological impacts.[74] These kind of findings arose as a logical development of empirical and yet questioning approaches to longstanding assumptions in agrarian and economic history, questionings which had already started in Elvin's 1973 *The pattern of the Chinese past*, a book which already represented a controversial watershed in Chinese economic history, a field which had, in fact, almost entirely ignored environmental factors.[75]

A comparable shift towards an ecological questioning of conventional agrarian history developed among historians of South Asia in the early 1980s, and particularly in the minds of Richard Tucker and John Richards, a specialist in the monetary and agrarian history of Mughal India. Both men were convinced that ecological changes accompanying economic transition in the seventeenth to nineteenth centuries, while clearly large-scale, had never been properly quantified. Almost immediately Richards and Tucker, working in tandem, realised that their questioning could not be confined to South Asia but was equally requiring of answers researched in terms of global economic history in general, particularly with the advent of an era in which the connections between deforestation, carbon dioxide production and global warming were becoming major popular anxieties in what Teresa Brennan was already calling the 'Age of Paranoia'.[76] In two major edited works they therefore set out to review the global history of deforestation, especially in the tropics.[77] Stimulated by this gargantuan effort, the two scholars then set out in diverging directions. Richards decided on a project of Domesday Book proportions that aimed to chart the district-by-district ecological transformation of the whole of India and Southeast Asia in 1880–1980. This took place over a ten-year period under the aegis of the Duke Ecological History project involving ecologist Betsy Flint, and historians James Hagen and Edward Haynes.[78]

Meanwhile Richard Tucker started to gather material not this time on the impact of the Raj but on the new global raj of the United States for yet another appropriately millennial global history disturbingly named Insatiable Appetite: The United States and the Ecological Degradation of the Tropical World.[79] Like Tucker, Richards felt compelled to continue to investigate the ecological effects of globalisation, inspired partly by Janet Abu-Lughod's essay on 'pre-colonial'

[74] Marks (1997); Purdue (1987).
[75] Elvin (1973).
[76] Brennan (2003).
[77] Tucker and Richards (1983); Tucker and Richards (1988).
[78] Richards, Hagen and Haynes (1988).
[79] Tucker (2000).

globalisation. Taking up where Abu-Lughod had left off, Richards set out to survey the environmental history of the whole world between 1500 and 1800. But by ending his study at 1800, he joined scholars such as Glacken and Thomas who felt reluctant to consider the industrialising complexities of the nineteenth century, a century whose environmental history has, as yet, only been tackled in any effective way at all and then only regionally by Bonyhady and Raymond Williams, the latter as a study literary history. Nevertheless Richards' *The unending frontier* (2003) stands to date as the only work of its kind yet to rival in scale or authority Glacken's Traces on the Rhodian Shore.[80] It is true that attempts were made in 2001 and 2003 by Sing Chew and Michael Williams to essay world histories of degradation and deforestation over the last 5000 years.[81] However, these brave but unwieldy attempts are necessarily handicapped by large lacunae in understandings of the basic research on the environmental histories of large parts of the world, as well as major gaps in the history of ideas about the environment in the imperial context. By contrast, both Tucker and Richards had come to global environmental history as part of a realisation that the ecological impacts of empire, especially as forerunners of 'modern' globalisation, had been startlingly neglected. Nevertheless, in making the quick and logical intellectual leap from South Asia to world history, Richards and Tucker had ironically left the environmental history of South Asia itself largely undone with the exception of some very limited essays, regional studies and essay collections.[82]

Imperial historians, historians of science and global environmental history

The difficulty with writing a history of globalising colonial empires in terms of an expanding resource frontier was that Richards had necessarily to write snapshots of different parts of that frontier worldwide focussing on a smorgasbord of commodities and periods. This permitted a very effective broad-brush picture of what might have been happening in terms of actual material change and resource consumption in the early modern period. But Richards did not set out to explain the reasons, methods and motivation of the ecological

---

[80] Richards (2003).

[81] Williams (2003); Chew (2001).

[82] Guha and Gadgil (1992). This book and the article which prefigured it in the Calcutta journal *Economic and Political Weekly* were handicapped by their articulation of a highly controversial notion that the Indian caste system had a practical and evolutionary ecological basis. The caste system is without doubt one of the central reasons for structured inequality in the sub-continent and the authors' work, otherwise interesting and novel, was justly criticised over the apparent apologia for the caste system; for a review of other work in the field see the introduction to Grove, Damodaran and Sangwan (1998).

transformation in terms of the history of science, history of ideas or imperial organisation. Some of these ways of understanding the 'empires of nature', to use Mackenzie's term, were already well developed.[83] In particular, Roy Macleod, Deepak Kumar and Satpal Sangwan had, in their early excursions into the 'science of the Raj', started to unravel the complex matrix of British imperial science in India, all of them finding it was necessary to overturn Louis Pyenson and George Basalla's portrayals of colonial science as the handmaiden of metropolitan science. Other works such as John Mackenzie in *The empire of nature* (1988) and Richard Grove in *Green imperialism* (1995) found rich veins to follow in the history of imperial conservationism, some of them intersecting histories of hunting and natural history collecting.[84]

Henry Hobhouse in *Seeds of change; five plants that transformed mankind* (1985) and Alfred Crosby's 1986 *Ecological imperialism* (an enlargement of his 1972 *The Columbian exchange, the biological and cultural consequences of 1492*) focused on the effects of the introduction of crops and diseases by Europeans to the settler colonies and the West Indies.[85] Although illuminating, these books largely ignored the much longer pre-European histories of plant transfer and disease in Old World Africa, Asia and Oceania, many accounts of which had been written by scholars such as Thurstan Shaw, as well as many contributors to the journal *Economic Botany*. But Crosby and Hobhouse dramatised the issue. Lucille Brockway, in a precocious and pathbreaking book, *Science and colonial expansion*, explored the role of the Royal Botanic Gardens at Kew in promoting imperial botanic gardens and exchanges.[86] Then Ray Desmond (like John Richards originally a South Asianist) using expertise gathered as an archivist of the India Office library, published *The European discovery of the Indian flora* in 1992 and both he (in 1995) and Richard Drayton (in 2000) developed this theme in further closely related books on the history of Kew gardens and the impact of explorer-curators such as Joseph Hooker on imperial 'improvement' schemes.[87] Donal McCracken usefully supplemented these works in a global guide to the history of colonial botanic gardens.[88] All these scholars had been foreshadowed to some extent by Ian Burkill's 1965 treatise on chapters on the history of Botany in India.[89] Burkill, although unable to edit his manuscript properly through blindness, afforded remarkable insights into the significance of environmental thinking in British India. He pointed, for example, to the

---

[83] Kumar (1995).
[84] Mackenzie (1988); Grove (1995).
[85] Hobhouse (1985); Crosby (1986); Crosby (1972).
[86] Brockway (1979).
[87] Desmond (1992); Desmond (1995); Drayton (2000).
[88] McCracken (1997).
[89] Burkill (1965).

highly precocious advent of ecological thinking in botanical science before 1857, particularly by John Edgeworth and Ellerton Stocks, surgeon-botanists employed by the East India Company.

More modern themes in the history of ecological thinking in the imperial context were not taken up until Peder Anker did so in his *Imperial ecology* in 2001.[90] Anker argued that the structures of what appeared to be a globally referential 'science' were at least in part composed of a medley of holistic notions of the kind pursued by Jan Smuts, laced with notions of Freudian psychology in the work of Arthur Tansley, but above all based on military, racial and social nationalism. He showed too that, like Smuts, Tansley and Charles Elton (who, along with Frederic Clements, were the leaders of the new global and imperial ecology), all made connections between climate, ecology and 'climax' individuals and communities. The controversial racial echoes of Huntington's climatic social theories and his connections with eugenics were, once more, uncomfortably close. Deep and discriminatory fears about extreme climatic events, racial difference and identity run right through the histories of empires and formulations of global environmental and climate change.

An accumulation of knowledge about the chronology of extreme climate events and a more limited understanding of their dynamics might have been expected to reduce such fears by offering the prospect of a more predictable world. Indeed global climate histories are now the most rapidly developing aspect of global environmental history and, as we have seen, they have always formed a vital central aspect of the field. Instead, however, global climate history and its practitioners, on whom global climate modellers are entirely dependent, has in the last decade offered, in its presentation of global warming trends, a prospect almost as grim as that of the spread of modern epidemics of immune deficiency disease, an aspect of Global environmental history which we have touched on but little in this paper. A flurry of recent books on El Nino/Southern Oscillation events and their historical impacts has done little to lessen these anxieties, even though ENSO has no immediate connection to global warming but may instead be more directly connected to tectonic events and volcanism. Nevertheless, when social scientists' irrational phobias about 'environmental determinism' finally die down and the full history of the messy and complex interconnections between climate and human events becomes more sharply drawn, global environmental histories that are fully integrated with climatic chronologies will become a useful explanatory tool in alerting a global public to the vital task of stewardship of the atmospheric as well as earthly environments. Scholars and policy makers need to focus on the overlap between environmental concerns and human welfare and new vocabularies

---

[90] Anker (2001).

and frameworks need to be developed that will help us see nature and culture in a shared light. Then perhaps situations of the kind that led to the New Orleans disaster may instead become problems of adaptation rather than international calamity.

References

Afinson, J. (2003), *The river we have wrought; a history of the upper Mississippi*, University of Minnesota Press, Minneapolis.

Anker, P. (2001), *Imperial ecology*, Cambridge, Mass.

Barry, J.M. (1979), *Deep'n as it come; the 1927 Mississippi river flood*, Oxford University Press, New York.

Barry, J.M. (1997), *Rising tide: the great Mississippi flood of 1927 and how it changed America*, Simon and Schuster, New York.

Bass, H. (ed.) (1970), *The state of American history*, Chicago.

Bernstein, H.T. (1956), 'Steamboats on the Ganges, 1828–1840', PhD thesis No. ADDX1956, Yale University.

Bernstein, H.T. (1960), *Steamboats on the Ganges; an exploration in the history of India's modernization through science and technology*, Orient Longman, Calcutta, reprint 1987.

Bilsky, L. (ed.) (1980), *Historical ecology; essays on ecology and social change*, Kennikat Press, Port Washington.

Braudel, F. (1949), *The Mediterranean and the Mediterranean world in the age of Philip the Second*, Paris, trans.in Fontana edition, London (1973).

Brennan, T. (2003), *Globalisation and its terrors*, London.

Brockway, L. (1979), *Science and colonial expansion*, New York.

Burkill, I. (1965), *Chapters in the history of Indian botany*, Calcutta.

Carson, R. (1951), *The sea around us*, New York.

Carson, R. (1962), *Silent spring*, New York.

Chew, S. (2001), *World ecological degradation, accumulation, urbanisation and deforestation, 3000 BC– 2000 AD*, Chicago.

Colinvaux, P. (1980), *The fates of nations; a biological theory of history*, Simon and Schuster, New York.

Crosby, A. (1972), *The Columbian exchange; the biological and cultural consequences of 1492*, Westport, Conn.

Crosby, A. (1986), *Ecological imperialism; the biological expansion of Europe 900–1900*, New York.

Darby, H.C. (1952), *The Domesday geography of eastern England*, Cambridge.

Darlington, C.D. and Janaki Ammal, E.K. (1945), *The chromosome atlas of cultivated plants*, London.

Desmond, R. (1992), *The European discovery of the Indian flora*, Oxford.

Desmond, R. (1995), *Kew; the history of the Royal Botanic Gardens, Kew*, London.

Diamond, J. (1987), *Guns, germs and steel; the fates of human societies*, New York.

Diamond, J. (2005), *Collapse; how societies choose to fail or succeed*, Viking, New York.

Drayton, R. (2000), *Nature's Government; science, imperial Britain and the 'improvement' of the world*, New Haven.

Elvin, M. (1973), *The pattern of the Chinese past*, London.

Elvin, M. (1993), 'Three thousand years of unsustainable growth; China's environment from archaic times to the present', *East Asian History*, 6, 7–46.

Elvin, M. (2004), *The retreat of the elephants; an environmental history of China*, Yale University Press, New Haven.

Engel, C.C. (2004), 'Post-war syndromes: illustrating the impact of the social psyche on notions of risk, responsibility, reason and remedy', *Journal of the American Academy of Psychoanalysis and Dynamic Psychiatry*, 32, 321–334.

Filliozat, J. (1980), 'Ecologie historique en Inde du Sud; Le Pays des Kallars', *Revue des Etudes d'Extreme Orient*, 2, 22–46.

Finberg, H. (ed.) (1967 and 1972), *The agrarian history of England and Wales*, Vols. I and IV, Cambridge.

Foltz, R.C. (2003), 'Does nature have historical agency? World history, environmental history, and how historians can help save the planet', *The History Teacher*, 37, 9–28.

Forde, D. (1927), *Ancient mariners; the story of ships and sea routes*, London.

Forde, D. (1931), *Ethnography of the Yuma Indians*, Berkeley.

Forde, D. (1934), *Habitat, economy and society*, London.

Fries, B. de, and Goudsblom, J. (eds) (2002), *Mappae Mundi; humans and their habitats in a long-term socio-ecological perspective*, Amsterdam University Press, Amsterdam.

Geertz, C. (1963), *Agricultural involution; the processes of ecological changes in Indonesia*, Berkeley.

Glacken, C. (1955), *The Great Loochoo*, Berkeley.

Glacken, C. (1967), *Traces on the Rhodian Shore; nature and culture in western thought, from ancient times to the end of the eighteenth century*, University of California, Berkeley.

Gordon East, *The geography behind history*, T. Nelson Publishers, London, 1938, p. 11.

Grove, A.T. and Rackham, O. (2001), *The nature of Mediterranean Europe; an ecological history*, Yale University Press, London.

Grove, R. (1989), 'Scottish missionaries, evangelical discourses and the origins of conservation thinking in Southern Africa', *Journal of Southern African Studies*, January.

Grove, R. (1995), *Green imperialism. colonial expansion, tropical island Edens and the origins of environmentalism, 1600–1860*, Cambridge.

Grove, R., Damodaran, V. and Sangwan, S. (1998), *Nature and the Orient; the environmental history of South and Southeast Asia*, Oxford.

Guha, R. and Gadgil, M. (1992), *This fissured earth; towards an ecological history of India*, Oxford, India.

Hamilton, A.C. (1982), *The environmental history of East Africa*, London.

Hays, S. (1957), *Conservation and the gospel of efficiency*, Chicago.

Hobhouse, H. (1985), *Seeds of change; five plants that transformed the world*, London.

Horden, P. and Purcell, N. (2000), *The corrupting sea; a study of Mediterranean history*, London, Blackwell.

Hoskins, W.G. (1967), *Fieldwork in local history*, London.

Hoskins, W.G. and Stamp, L.D. (1963), *The common lands of England and Wales*, London.

Hough, F.B. (1878–9), *Report in forestry*, US Congress, Washington DC, 4 Vols.

Hughes, D. (2001a), *The environmental history of the world; humankind's changing role in the community of life*, Routledge, London.

Hughes, D. (2001b), 'Global dimensions of environmental history', *Pacific Historical Review*, 70, 91–101.

Kumar, D. (1995), *Science and the Raj, 1857–1905*.

Lacey, M. (1979), 'The mysteries of earth-making dissolve; a study of Washington's intellectual community and the origins of American environmentalism in the late nineteenth century', PhD thesis, George Washington University.

Leach, M. and Green, C. (1997), 'Gender and environmental history', *Environment and History*, 3/3, 343–370.

Lowenthal, D. (1958), *George Perkins Marsh; Versatile Vermonter*, Columbia University Press, New York.

MacKenzie, J.C. (1913), 'Climate in Burmese history', *Journal of the Burma Research Society*, 3, 40–6.

Mackenzie, J. (1988), *The empire of nature, hunting, conservation and British imperialism*, Manchester.

Marks, R.B. (1997), *Tigers, rice, silk and silt: environment and economy in late imperial South China*, New York.

Marco, G.J. (1987), *Silent spring revisited*, New York.

Mathur, A. and da Cunha, D. (2001), *Mississippi floods: designing a shifting landscape*, Yale University Press, New Haven, 2001;

McCracken, D. (1997), *Gardens of Empire; botanical institutions of the Victorian British Empire*, Leicester.

McNeill, J.M. (2000), *Something new under the sun; an environmental history of the twentieth century world*, Norton, New York.

McNeill, J. (2003), 'Observations on the nature and culture of environmental history', *History and Theory*, 42, 5–43.

McPhee, P. (1999), *Revolution and environment in southern France 1780–1830; peasants, lands and murder in the Corbieres*, Oxford.

Meiggs, R. (1982), *Trees and timber in the ancient Mediterranean world*, London.

Mellanby, K. (1967), *Pesticides and pollution*, Collins, London.

Merchant, C. (1980), *Death of nature, women, ecology and the scientific revolution*, Harper.

Merchant, C. (1992), *Radical ecology, search for a livable world*, London.

Montesquieu, Charles de Secondat, Baron de (1734[1965])), *Considerations on the causes of the greatness of the Romans and their decline*, Lowenthal, D., ed. and translated, New York.

Nash, R. (1967), *Wilderness and the American mind*, New York.

Nash, R. (1972), 'American environmental history; a new teaching frontier', *Pacific Historical Review*, 41, 362–372.

Nobles, G.H. (1997), *American frontiers: cultural encounters and continental conquest*, Hill & Wang, New York.

Petrow, R. (1968), *In the wake of the Torrey Canyon*, David Mackay, New York.

Ponting, C. (1991), *A green history of the world; the environment and the collapse of great civilizations*, London.

Powell, J.M. (1996), 'Historical geography and environmental history; an Australian interface', *Journal of Historical Geography*, 22, 253–273.

Purdue, P. (1987), *Exhausting the earth; state and peasant in Hunan, 1500–1850*, Cambridge, Mass.

Rackham, O. and Moody, J. (1994), *The making of the Cretan landscape*, Manchester University Press, Manchester.

Radkau, J. (2000), *Natur und Macht, Eine Weltgeschichte der Umwelt*, C.H. Beck Verlag, Munich.

Radkau, J. (2003), *Naturschutz und Nazionalsozialismus*, Campus Verlag, Frankfurt.

Ravensdale, J. (1977), *Liable to floods*, Cambridge.

Richards, J.F. (2003), *The unending frontier; an environmental history of the early modern world*, University of California, Berkeley.

Richards, J.F., Hagen, J.R., Haynes, E.S. (1988), 'Changing land use in Bihar, Punjab and Haryana, 1850–1980', *Modern Asian Studies*, 19, 679–732.

Sauer, C. (1966), *The early Spanish Main*, University of California Press, Berkeley.

Semple, E.C. (1911), *Influence of geographical environments on the basis of Ratzels anthropo-geography*, New York.

Semple, E.C. (1931), *The geography of the Mediterranean region in relation to its ancient history*, New York.

Shaler, N. (1905), *Man and the earth*, Fox and Duffield, New York.

Skipp, V. (1978), *Crisis and development: an ecological case study of the forest of Arden, 1570–1674*, London.

Sörlin, S. and Öckerman, A. (1998), *Jorden en o: En global miljohistoria, Bokforlaget natur och Kultur*, Stockholm.

Stamp, L.D. (1924a), 'Notes in the vegetation of Burma', *The Geographical Journal*, 43, 231–3.

Stamp, L.D. (1924b), 'The aerial survey of the Irrawaddy Delta forests', *Journal of Ecology*, 15, 262–76.

Stamp, L.D. (1930a), 'Burma; a survey of a monsoon country', *Geographical Review*, 20, 86–109.

Stamp, L.D. (1930b), *The vegetation of Burma; from an ecological standpoint*, University of Rangoon Research Publications, no 1. Rangoon.

Stamp, L.D. (1940), 'The IIrrawaddy river', *The Geographical Journal*, 95, 329–56.

Stamp, L.D. (1942), 'Siam before the war', *The Geographical Journal*, 99, 209–224.

Stamp, L.D. (1961), *The basic land resources of Burma*, Sarpay Beikam Press for the Burma Research Society, Fiftieth Anniversary Publication, Rangoon, 458–480.

Stamp, L.D. and Lord, L. (1923), 'The ecology of part of the riverine tract of Burma', *Journal of Ecology*, 2, 129–159.

Steers, J.A. (1940a), 'The coral cays of Jamaica', *Geographical Journal*, 95, 30–42.

Steers, J.A. (1940b), 'The Cays and Palisadoes, Port Royal, Jamaica', *Geographical Journal*, 96, 305–328.

Steers, J.A., Chapman, V.J., Colman, J. and Lofthouse, J.A. (1940c), 'Sand cays and mangroves in Jamaica', *Geographical Journal*, 96, 305–328.

Thomas, K. (1984), *Man and the natural world*, Oxford.

Thomas, W.L. (ed.) (1956), *Man's role in changing the face of the earth*, University of Chicago Press, Chicago, Illinois.

Trevelyan, G.M. (1929), *Must England's beauty perish? A plea on behalf of the National Trust for places of historic interest and natural beauty*, London.

Tucker, R. (2000), *Insatiable appetite; the United States and the ecological degradation of the tropical world*, Berkeley.

Tucker, R.P. and Richards, J.F. (eds.) (1983), *Global deforestation in the nineteenth century world economy*, Durham.

Tucker, R.P. and Richards, J.F. (eds.) (1988), *World deforestation in the twentieth century*, Durham.

Vita-Finzi, C. (1969), *The Mediterranean Valleys; geological changes in historic times*, Cambridge.

White, R. (1985), 'Environmental history; the development of a new historical field', *Pacific Historical Review*, 54, 297–335.

White, R. (2001), 'Environmental history; watching a field mature', *Pacific Historical Review*, 70, 103ff.

Williams, M. (1974), *The making of the South Australian landscape*, London.

Williams, M. (2003), *Deforesting the earth*, Chicago.

2
Separation, Proprietorship and Community in the History of Conservation

William M. Adams

On a beach on the remote Masoala peninsula in the northeast corner of Madagascar, a group of men and boys watch a motorboat disgorge a party of visiting conservation planners. The visitors come from Europe, North America, South America and Asia, and their purpose is to meet the people who live on the boundaries of the Masoala National Park. The village lies behind the thick-trunked trees that line the white sand shore in a band of cultivated fields growing rice, coffee, vanilla and cinnamon. Beyond the fields lies the forest, just a kilometre or so away. From the boat, the forest forms a vast sweep of green reaching unbroken from the beach to the distant watershed 1200 m above the sea, lit by bright sunshine and framed in dark grey-black rain clouds.

The Masoala National Park was created in 1997 to protect the extensive tracts of coastal and lowland rainforest that survived on the Masoala peninsula long after similar forest had been cleared elsewhere.[1] The park is the largest in Madagascar (211,000 ha)[2]. Its biological diversity is very high, and it contains numerous rare, endangered and endemic species such as red-ruffed and hairy-eared dwarf lemurs, the Madagascar serpent eagle, red owl, chameleons, geckos, butterflies and fish. It also incorporates three coral reef marine reserves in the Baie D'Antongil, where Humpback whales calve.

The establishment of Masoala National Park followed an almost textbook process of protected area (PA) planning.[3] This involved both Madagascan government organizations (Association National des Aires Protegees, ANGAP and (Direction des Eaux et Forets, DEF), and international NGOs (led by the Wildlife Conservation Society with other, including Care International). An ICDP (Integrated Conservation and Development Project) begun in 1990 proposed to set aside 3000 km[2] as a national park, with the rest of the peninsula

---

[1] Goodman and Benstead (2004).
[2] www.wcs.org/sw-around_the_globe/Africa/174291
[3] Kremen *et al.* (1999).

under community-based management.[4] The approach used in planning was to demarcate 'core' areas of forest for preservation, and to encourage sedentary sustainable agriculture in surrounding 'multiple use' zones that could meet human subsistence while ensuring protection of the core forests. Detailed survey from 1990 used GIS, satellite imagery and ground survey to create a species inventory, assess timber resources and map human settlement and resource use. Design criteria for the park sought to protect the full range of habitats and biodiversity, and it included corridors between the park and the adjacent PAs.

The survey showed that, while 97% of the forest was primary, uncleared, forest, the area was in no sense a wilderness, devoid of people. Fieldwork mapped 259 permanent villages containing 6500 people and 190 temporary villages. Most human occupation had no legal basis: few people had permits to reside or cut wood, although they did so.[5] To build support locally and nationally, the proposal was developed 'in consultation with people at local and national levels', and drew on the idea of integrated conservation and development.[6] In practice this meant that local people participated in the surveys of forest and village territories, and meetings held with local village members to collect feedback and to gauge local concerns. Conservation planners and government officials walked along the park's proposed borders, talking about specific points with community leaders. Community leaders were taken to other areas in Madagascar to show the impacts of unchecked logging and forest conversion for agriculture, and several communities were encouraged to grow formerly harvested rainforest products in pilot programmes.

People were removed from the park area, but only where 'local social norms' permitted it. Thus people who had invested heavily in the creation of irrigated rice paddies were not evicted, while those engaging in illegal 'slash and burn agriculture' near watersheds were moved to 'reintegrate' with permanent villages because they had not invested in the land, and their actions threatened downstream water supplies. The park's boundaries were adjusted to minimize conflicts between wildlife and the surrounding communities, particularly around major rural centres.

Conservation ideas

The Masoala National Park is in a sense typical of the dilemmas and challenges of contemporary conservation. The specific question of the displacement of

---

[4] Kremen *et al.* (1999).
[5] Kremen *et al.* (1999).
[6] Kremen *et al.* (1999), 1065.

people from PAs has become an important international policy issue.[7] After fierce debate at the Fifth World Parks Congress in Durban in 2003,[8] the 'Durban Accord' proposed a new paradigm for PAs, 'equitably integrating them with the interests of all affected people', such that they provide benefits 'beyond their boundaries on a map, beyond the boundaries of nation states, across societies, genders and generations'.[9] It called for an end to all involuntary resettlement and expulsions of indigenous peoples from their lands for PAs.

However, behind debates about population displacement lie more subtle and complex questions about the way conservation ideas are worked out in specific practical contexts. The concepts used by the park planners at Masoala are familiar within international conservation circles, but – despite the careful protocols of 'conservation planning'[10] – they are also strongly ideological. They reflect common science-based ideas about what non-human nature is like (the possibility of identifying categories of natural forest, the idea of biodiversity and rarity as ways to understand nature's variety and to prioritize conservation action). More seriously, they reflect strongly socialized ideas about human society: people are assumed to live in 'communities', and to share 'social norms'; agricultural practices are loosely referred to as 'slash and burn', and by this terminology are classified as destructive, illogical and unacceptable; investment in agriculture is taken to legitimize tenure, whereas other forms of land and resource use are not.

The work of the park planners at Masoala was careful, and expressed in the language of science, yet it raises key questions about the power of ideas in conservation to influence and direct change on the ground. Where do the conservation ideas applied in specific places and times come from? Why do some ideas come to be dominant over others? How do ideas change over time and vary in space?

There is a basic commonality to ideas about conservation, certainly in the twentieth century, during which such ideas moved from the marginal to the mainstream, from being minority patrician concerns to major elements in global political debate. But despite the rise of an international conservation movement in the second half of the twentieth century, and the development of standard approaches and methodologies, conservation remains a complex and highly diverse activity.[11] What I understand by conservation is that it comprises a set of ideas and practices that mediate the relationship between people and non-human nature. Conservation is therefore fundamentally a social

---

[7] Colchester (2004); Dowie (2005); Cernea (2006).
[8] Brosius (2004).
[9] World Conservation Union (2005), 220.
[10] Margules and Pressey (2000).
[11] Adams (2004).

phenomenon or social practice. It needs to be understood not as a neutral universal practice, but as something that is geographically diverse, historically changing and contested. As such, one might anticipate that there might be both geographical and historical specificity to ideas about conservation. Conservation movements, and the ideologies that underpin them and emerge from them, are likely to be highly diverse in space and time. Moreover, one might expect that these ideas, because variable in space and changing in time, will be contested, as they haven been, to some extent, at Masoala. Indeed, this proves to be the case, with marked differences in the ideas dominating different places at different times.[12]

The nature of the differences in ideas about conservation between different places and the way such ideas move and evolve in both space and time are fundamental to the political ecology of conservation. Some of these dynamics, and some contexts within which these changes take place, are well researched, others barely acknowledged in the research literature. A great deal has been written, for example, on the history of conservation in the United States and the United Kingdom. There has been some exploration of the development of ideas in different parts of the European colonial world and the development of ideas internationally in the Western-dominated twentieth century. The study of the development of thinking about conservation, or more broadly about idea of human relations with non-human nature, in many countries has barely begun.

In this chapter I want to explore the development of conservation ideas in the United States and the United Kingdom. The similarities and differences between British and American thinking about PAs and the separation between human and nature had considerable significance for the dissemination of ideas internationally. This chapter cannot begin to tease out the complexities of this story, or its many byways. However, it will sketch something of the two approaches and their influence in one continent, Africa.

Separating people and nature

Neumann describes four themes in the histories of national parks in Africa and the United States.[13] First, a denial or failure to recognize historic human occupation of land, and a human role in its ecology; second, a lack of empirically sound ecological justification for eviction of resident populations; third, that the way those managing PAs have dealt with people is closely related to other state policies of social control and spatial segregation, for example associated

---

[12] Brechin *et al.* (2003); Adams (2004); Brosius *et al.* (2005).
[13] Neumann (2004a).

with race or disease; fourth, the way parks have functioned as exclosures, curtailing access to common pool resources, with far-reaching implications.

These themes arise directly from the model of conservation adopted internationally following the Second World War, built around the protected area (PA) and the physical separation of spaces for nature and for intensive human activity. This partition of space reflected the prior and far more fundamental conceptual Enlightenment distinction between human and non-human, which also receives expression in Europe in distinctions between town and country, husbanded and wild, constructed and natural. Specifically as an appeal to the preservation or conservation of wild nature, the PA was derived from a mixture of North American and European experience, moderated by ideas in European colonial territories.[14]

In Britain itself there has always been an understanding that nature was not particularly pristine, and this distinction between natural and the human-influenced habitats was hard to draw in practice. While ideas of the majesty of mountain and hill informed ideas of landscape and of gentlemanly hunting in the Scottish Highlands,[15] elsewhere it was obvious that nature had been severely modified by human action. Away from mountain tops and sheer cliffs the countryside was an anthropogenic hybrid, its vegetation semi-natural, persisting in more or less disturbed patches, as the ecologist Arthur Tansley pointed out.[16] Some of those patches were diverse and beautiful, but very often that diversity derived from the way it had been managed in the past. The task for conservation, therefore, was in many instances to maintain or imitate past management practices, rather than stop all human interference. Conservation was a matter of influencing how people used the land, not stopping them from doing so. Diversity and beauty could flow from right management, just as uniformity and destruction would flow from ignorant or inappropriate management.

In thinking about PAs in the United States, the distinction between the realms of nature and humanity was thought of as more clear-cut. European ideas of wilderness were re-worked to underpin American ideas of nature, pristine and free of human influence of humankind evolution of thinking about wilderness from Europe to North America.[17] Yellowstone National Park was established by the US Federal Government in 1872 (although a park had been proposed by George Catlin in 1832), providing a model for central state protection for outstanding natural wonders and environments.[18] It was not the first

---

[14] Adams (2003); Adams (2004); Barton (2002).
[15] Toogood (2003).
[16] Sheail (1987).
[17] Schama (1995); Cronon (1995).
[18] Runte (1987).

experiment with state protection in the United States, however: Yosemite was established as a state park in 1864, although it only became a national park in 1890.[19] In the United States, as in Australia, wilderness became an important element in emergent national identity.[20] The US national parks model, epitomized by Yellowstone and Yosemite, was built on a conception of nature as something pristine, separate and separable from human-transformed lands. The concept excluded people as direct users of land and resources. It is this US model for PAs that has had most influence internationally.

The seductive and long-lived idea of such places as wholly wild or unmodified by human influence was, however, highly misleading, also based on a quite erroneous understanding of the human ecological transformation of precolonial America.[21] The idea of pristine nature that underpinned the creation of national Parks in the United States contributed directly to the removal of indigenous people, in a 'state-organized process of re-arranging the countryside, in which native peoples and nature were slotted into distinct categories and separated from one another'.[22] The idea of pristine nature contributed to some extent to the suppression of indigenous rights that was a wider feature of colonialism in North America, whereby indigenous people were actively suppressed by military action, marginalized by legal and bureaucratic power and physically removed from lands valued by settlers into small and economically marginal reservations. They were at the same time airbrushed from history, to fit the story America told about itself, of a frontier carved in the wilderness.

To the Europeans from the East who 'discovered' and 'explored' the rugged land of Yellowstone, or the meadows of Yosemite, it was indeed the apparent *lack* of human presence that made them seem so remarkable. Indian cultural sites were ignored and their role in creating the landscape was forgotten. Early tourists were shocked to encounter Indian bands, a feeling that was presumably mutual. However, the tourists' vision of these lands as parks was endorsed by state and federal government, and the Indians' view of theme as home did not. Indian heritage was effaced from parks, whose new place names featured 'natural' wonders. Economic use of natural resources in conserved lands by indigenous people became illegal. Thus Karl Jacoby describes how resource users were banned in the Grand Canyon National Park, Arizona (prior to 1919 the Grand Canyon Forest Reserve). The Havasupai hunting and farming economy was transformed, and the Havasupai people, formerly hunters and farmers, fell into near destitution as wage labourers, ironically dependent on the park

---

[19] Runte (1990).
[20] Dunlap (1999); Griffiths and Robins (1997).
[21] Whitney (1994).
[22] Jacoby (2001), 87.

authorities for subsistence, performing menial tasks associated with tourism and management of their former lands.[23]

The Yellowstone model inspired similar experiments elsewhere. In 1906 a delegation from the Society for the Preservation of the Wild Fauna of the Empire told the Secretary of State for the Colonies that it was 'the duty and the interest of Great Britain' to follow the US example in East Africa.[24] The model was copied in the British Dominions (Canada 1887, Australia 1891 and New Zealand 1894), in the Belgian Congo in 1925 and in South Africa in 1926.[25] Expansion in areas such as East Africa had to wait until after the Second World War,[26] but eventually the model was adopted there and in due course worldwide,[27] the basis of the global expansion of PAs. National Parks on the Yellowstone model became a global phenomenon after the Second World War (the years of 'conservation boom' in Africa).[28] The IUCN established a Provisional Committee on National Parks 1958, and in 1962 UN 'World List of National Parks and Equivalent Reserves'. The area of PAs globally doubled in the 1970s, 1980s, 1990s. By 2005, there were 100,000 PAs, covering 2 million km^2, or 12% of the world's land surface.[29]

Wild Africa

The emphasis of the US 'wilderness' park model on pristine nature and un-peopled wild land was echoed in the ideological framing of nature in other colonial jurisdictions, such as Africa. Thus the idea of Africa as an 'unspoiled Eden'[30] or 'a lost Eden in need of protection and preservation' was a potent element in colonial thinking about national parks in that continent, and especially thinking about people in African PAs.[31] Sir Alfred Sharpe, Commissioner of the Central African Protectorate, commented in 1905 that 'there seems to have been a general tendency, while rigidly restricting Europeans from shooting big game, to leave the native free to slaughter all he wishes without let or hindrance'.[32] White sportsmen could be persuaded to become conservationists

---

[23] Jacoby (2001).

[24] Rhys Williams to the Secretary of State for the Colonies, 9 June 1906, *Society for the Preservation of Wild Fauna of the Empire (hereafter SPWFE) Journal*, 14–19 (p. 15).

[25] Fitter with Scott (1974); McNamee (1993).

[26] Neumann (2002).

[27] Adams (2004); Jepson and Whittaker (2002).

[28] Neumann (2002).

[29] Chape *et al.* (2005).

[30] Anderson and Grove (1987), 4.

[31] Neumann (1998), 80; Adams (2004).

[32] Sir Alfred Sharpe (1905), Commissioner of the Central African Protectorate, *SPWFE Journal*, 2 (p. 18).

by the adoption of a sporting code,[33] but native hunters were a different matter. The dominant view of the sporting conservationists was neatly expressed by the prominent American conservationist William Hornaday, bracketing native hunters in the United States and Africa in formulating his 'sportsman's platform': 'an Indian or other native has no more right to kill game, or to subsist upon it all year round, than any white man in the same locality. The native has no God-given ownership of the game of any land, any more than its mineral resources; and he should be governed by the same laws as white men.'[34] African subsistence hunting was widely regarded by colonial administrators as haphazard, inefficient, wasteful and cruel. Moreover, it distracted rural people from gainful employment in cash crop production or wage labour. The idea that subsistence, let alone cultures, could properly depend on hunting was not countenanced.[35] G.W. Hingston, author of a report proposing national parks in East Africa, wrote in 1931, 'When natives hunt collectively, they then have the power to cause serious depletion through wholesale and indiscriminate methods employed.'[36]

With the Yellowstone model went the same experience of evictions and exclusions, for example in Australia, Russia and Canada[37] and widely in European colonial territories and elsewhere.[38] In Africa, game reserves and subsequently national parks were imagined as places for nature, not for people, just as they were in the United States: as Neumann comments of the creation of the Selous Game Reserve in Tanzania, 'what links the case of the Selous to the establishment of the first national parks in the US is the fact that it had to be created before it could be protected'.[39]

The idea of wilderness has enduring power: an article about the 1200 km 'mega transect' walked by American biologist John Fa in the Congo basin (following which the President of Gabon announced a total of 13 new national parks in 2002) was entitled 'Saving Africa's Eden'.[40] Reflecting the same language, pastoralists were evicted from the Mkomazi Game Reserve in Tanzania as recently as 1988 'making it "wilderness" for the first time, because of conservation planners' fears of the people, and their present and unknown future

---

[33] Mackenzie (1988); Adams (2009).

[34] W.T. Hornaday's Letter and Fifteen Cardinal Principles, *SPWFE Journal* 1909, Vol. 5, 56–8 (p. 57).

[35] Marks (1984).

[36] Hingston (1932), 404.

[37] Poirier and Ostergren (2002); Langton (2003); McNamee (1993).

[38] Homewood and Rodgers (1991); Neumann (1998); Ranger (1999); Colchester (1997); Colchester (2004).

[39] Neumann (2004a), 212.

[40] Quammen (2003).

impact'.[41] Ironically, as parks have spread, the eviction of people to create them have indeed created wilderness from inhabited lands.[42] While the history of displacement and land alienation is sometimes complex (e.g. in the Arusha National Park in Tanzania, created from land originally cleared for white settler farms and subsequently reclaimed by the state or purchased by conservation NGOs), the broad pattern of the removal of African land and right holders to make way for empty wildlife-filled lands is common.[43]

One result of the continued power of the concept of 'wild' nature and the urge to separate it from humanized landscapes has been the resurgence of calls for socially exclusive parks in international conservation in the 1990s.[44] This 'back to the barriers' movement[45] followed several decades during which conservation thinking was greatly influenced by the idea that conservation and development should be integrated through some kind of community-based conservation strategy.[46] Scepticism with such approaches have contributed to renewed support for the idea of strictly protected 'people-free parks'.[47]

Proprietorship: British models

The concept of PAs adopted internationally in the second half of the twentieth century as spaces set aside for nature, or rather places where humans could only engage with nature in very particular ways (as tourists, scientists or sporting hunters), was associated with a variety of different traditions. From Europe came the idea of a private exclusive royal or aristocratic hunting ground.[48] Access to such pleasure grounds was socially restricted, and the unlicensed killing of game (by rural people marked down as 'poachers') closely policed.[49] In Victorian Britain, the 'preservation of game' was an aristocratic obsession, with woods patrolled to preserve pheasants for the landowner's *battue*, and Scottish mountains, on estates cleared of tenants, were managed for deer stalking.[50]

From British aristocratic sport grew the tradition of big game hunting in the British colonial empire, where colonial administrator and army officer reveled

[41] Brockington and Homewood (1996), 104.
[42] Neumann (1996); Neumann (2001).
[43] Brockington (2002); Neumann (1996); Neumann (1998); Neumann (2001).
[44] Wilshusen *et al.* (2002).
[45] Hutton *et al.* (2005).
[46] Adams and Hulme (2001); Western *et al.* (1994); Hulme and Murphree (2001).
[47] Schwartzman *et al.* (2000); Terborgh (1999); Oates (1999); Kramer *et al.* (1997); Brandon *et al.* (1998); Struhsaker (1999).
[48] James (1981).
[49] Thompson (1975).
[50] Mackenzie (1988).

in the opportunity to stick pigs, hunt jackals with foxhounds, or shoot antelope, big cats and above all elephants.[51] It was from the ranks of these hunters, that 'penitent butchers'[52] emerged to argue for the conservation of game in the British Empire, most notably in the Society for the Preservation of the Wild Fauna of the Empire, founded in 1903.[53] In a sense, a cadre of aristocratic hunting conservationists were able to maintain and police in British colonies a version of the sportsman's playground that was the Victorian country estate long after that world had disappeared at home.[54]

The British tradition was for reserves that were exclusive through of private land ownership, where non-proprietors lacked rights of access and use. Developing this model in the colonial context, the colonial state substituted for the crown, the ultimate provider of aristocratic privilege and wealth. The Society for the Wild Fauna of the Empire was founded in 1903 following a meeting at the House of Commons to write a letter to Lord Cromer, Governor-General of the Sudan, to protest at a Colonial Office proposal to de-gazette game reserve in Sudan between the White and Blue Niles and the Sobat River.[55] The founder membership of the society was a distinguished and powerful role-call of politicians and aristocracy, colonial administrators, businessmen, hunters, scientists and naturalists. Between 1905 and 1909 the SPWFE took three delegations to the Secretary of State for the Colonies to argue for stronger hunting regulations and for game reserves. The government game reserve became the mainstay of the colonial conservation imagination in the British Empire, a resort for gentleman hunters, whether traveller or colonial servant.[56]

In Britain, the first nature reserves were the result of non-governmental action, and were conceived of as fitting into a privately owned countryside. There were attempts at large-scale protection, notably the opposition to the enclosure of common land around expanding cities for housing, for example by the work of the Commons, Open Spaces and Footpaths Society (established in 1865), or the work of patrician activists such as Edward North Buxton, who secured the preservation of Epping Forest in Essex by the Corporation of the City of London.[57] However, for the first half of the twentieth century, most British PAs were small. The areas set aside were nicely described by Sir Edwin Ray Lankester, formerly Director of the Natural History Museum, in a letter to *Nature* in 1914 announcing the establishment of the Society for the Promotion

---

[51] MacKenzie (1988); Adams (2004).

[52] Fitter and Scott (1974).

[53] Neumann (1996); Prendergast and Adams (2003).

[54] Neumann (1996).

[55] Prendergast and Adams (2003).

[56] Neumann (1996); Adams (2004); MacKenzie (1988).

[57] Buxton (1923); Sheail (1976).

of Nature Reserves (SPNR). In his words, the society proposed to secure from destruction 'the invaluable surviving haunts of nature'.[58] In Britain, PAs were first and foremost small areas of human-managed habitat set within a tapestry of developed and intensively managed land. The state had no great tracts of land to partition up and set aside, for although big estates existed, they were mostly held by crown or aristocracy. Conservationists had to buy their way into this landscape, or work to persuade landowners to preserve the best sites and manage their land in ways that favoured or respected nature.

Of course, the case for national parks, and a national responsibility for nature and nature reserves, was argued repeatedly, and from the later 1920s with increasing intensity.[59] The period of planning for post-war reconstruction in the early 1940s saw a significant sea-change in approach, with acceptance of the principle of government action. In the event, Britain adopted a strange double strategy, with separate provision for nature reserves and national parks under the National Parks and Access to the Countryside Act 1949. The Nature Conservancy[60] was established as a scientific organization with powers to acquire National Nature Reserves and to designate Sites of Special Scientific Interest (SSSIs), while the National Parks Commission was established with powers to designate national parks. However, the parks were managed by committees of their constituent County Councils, under the Town and Country Planning Act 1947. Land in national parks remained privately owned.[61] The 1949 Act made no provisions for national parks in Scotland or Northern Ireland. The first British National Parks were created in 1951, all in hill areas of long-proven recreational importance, the Peak District, Dartmoor, the Lake District and Snowdonia. Between 1951 and 1957, the National Parks Commission designated ten National Parks in England and Wales.[62] Legislation to allow national parks in Scotland was only passed in 2000, although much had been achieved by the Countryside Commission for Scotland in the meantime with a different concept, the National Scenic Area.[63] The first Scottish National Parks were declared in 2002 (Loch Lomond and the Trossachs, followed by the Cairngorms, both originally proposed in 1945[64]).

When national parks were eventually designated in Great Britain, they were therefore essentially planning designations, to protect natural beauty

---

[58] E. Ray Lankester (1914) 'Nature reserves'. *Nature*, March 12, 33–5.
[59] Sheail (1976).
[60] Not to be confused with the private US non-governmental organisation The Nature Conservancy, founded two years later, in 1951 (www.nature.org/aboutus/).
[61] Sheail (1991); Sheail (1998).
[62] Evans (1992).
[63] Moir (1991).
[64] Ramsey (1945).

and promote countryside recreation.[65] They contained lived-in landscapes that were dominated by agriculture, forestry (and increasingly tourism). They were mosaics of large and small landholdings, almost all of which were private although increasingly non-governmental organizations such as the National Trust have acquired critical areas to guarantee their protection.

The proprietorship of British national parks is therefore a public–private mix. Although from 1949 the Nature Conservancy (and its successors) had the power to acquire nature reserves, they had a very limited budget. Only a few smaller sites could be purchased – the rest had to be leased, and the largest reserves (and by far the dominant area) had to be secured under nature conservation agreements with private landowners.[66] Moreover, there were no real powers to protect SSSIs, which comprised the set of areas of greatest conservation value nationally: they could be identified, listed and studied, but their conservation interest survived at the whim of their owners. Here too, therefore, a hybrid model of conservation proprietorship prevailed. Government agencies and non-governmental organizations could but exhort landowners (principally farmers) to maintain conservation interests of their land (often in the face of clear economic pressures that would damage it), helped by an inadequate portfolio of small grants. In the face of development pressures (and especially agricultural improvement funded by windfall subsidies from the Common Agricultural Policy), this model became more and more obviously inadequate. From the 1980s, a new approach tried to bring about good management by paying incentives to landowners through management agreements and other arrangements.[67] The approach has had some success, although the acid observation that this was like keeping a balloon aloft by burning money[68] has proved prophetic: the United Kingdom's public–private partnership essentially involves the state renting natural diversity from private landowners.

Proprietorship: US models

In the United States, PAs were from the first created on land held by the state (at various scales from municipal through state to federal), not by private landholders. Indeed, Neumann argues that 'the wild area of national parks and reserves, as products of the creation of the modern nation state, are as much an expression of modernism as skyscrapers... an integral part of the practice of

---

[65] MacEwen and MacEwen (1982).
[66] Adams (1995).
[67] Adams (1995); Marren (2002).
[68] MacEwen and MacEwen (1982).

modern statecraft'. He suggests that in PAs, nature literally becomes a ward of the state.[69]

In the first US parks, that state control was enforced by direct military enforcement. The Yosemite Valley was cleared in 1852 by the army, who executed five Indians believed to have attacked prospectors, and scattered the remainder.[70] The US Army patrolled Yellowstone during the 'Sioux Wars' of 1876–7 and controlled the park from 1886 until 1918.[71] When the US Park Service was finally created, in 1916, it hired former soldiers as rangers, and copied military uniforms and drill.

In setting land aside for nature in this way, the state is in a sense simply turning it into another kind of resource. Birch describes wilderness preservation as simply 'another stanza in the same old imperialist song of Western civilization'.[72] He maintains that the very act of designating 'wilderness' reserves for wild nature constrains the 'otherness' of wild land. Such 'wilderness' exists at the whim of legislators and government policy, and it is 'managed', even if that management is hidden from human visitors. Wilderness parks are, just like mines, dams or cities, evidence of the power of Western civilization to 'manage, invade, declassify, abolish, desanctify the legal wildland entities it has created, and the creation of such entities on its terms does little to diminish this power'.[73]

The development of US national parks therefore needs to be understood in the context of Progressive Era rational state conservationism.[74] The establishment of the Bureau of Reclamation in 1902 and the US Forest Service in 1905 reflected a use-based conservation ethic that at one level was opposed to the aesthetic preservationism of the national parks movement (epitomized by John Muir and the Sierra Club, founded in 1892). However, it was overseen by Theodore Roosevelt, self-styled big-game hunter, 'frontiersman', and conservationist, and founder (in 1888) of the Boone and Crockett Club.[75] The idea that the developing state should set aside national parks to protect nature was perfectly compatible with the idea of mapping and partitioning virgin lands to maximize human benefit. Nature became a zone in a cartographic imagination of American internal colonialism.[76] This idea was one of the most powerful attractive features of the Yellowstone

[69] Neumann (2004a), 212.

[70] Runte (1990).

[71] Greene (1991); Jacoby (2001).

[72] Birch (1990), 4.

[73] Birch (1990), 22.

[74] Hays (1959).

[75] Jeffers (2003).

[76] Goetzmann (1966); Meine (2004).

model to the developing world as it emerged from colonialism from the 1950s.

This compatibility became increasingly obvious as the twentieth century progressed, as the global nature-based tourism grew. Tourism had always been central to the attraction of US national parks. Yosemite received its first tourists in 1855, and had a lodge by 1857. Parks were created where the emerging railway networks of the United States and Canada met the Rockies, at Glacier National Park (USA) 1910, Banff 1885 and Yoho 1886. The wonders of nature were harnessed to boost trans-continental travel (Yoho even importing Italian mountain guides to provide authentic gentile excitement). The coming of the motor car dramatically transformed accessibility. In 1916 more people travelled to Yellowstone by car than rail for the first time. In 1918 it was seven times more, and by the start of the 1950s, 99% visitors reached US Parks by car. Nature had become a consumer good, an icon of the urban good life.[77]

Similar developments occurred elsewhere. In South Africa, Kruger National Park, established in 1926 after years of campaigning by conservation advocates, became a hugely important place for the Afrikaans-speaking white community.[78] By 1930, there were 500 km of roads, and by the 1950s there was accommodation for 3000 people. By the 1930s, the British-based Society for the Preservation of the Fauna of the Empire was lobbying for national parks in East Africa on the basis that they would secure the development of an industry based on international tourism. In the 1950s the advent of inter-continental travel by jet airliner the length of Africa made mass tourism possible, and ultimately made the PAs of the developing world as much a mainstay of American and European urban consumers' lives as their domestic parks.

Community with nature

On the surface, there seems a fundamental distinction between US and British tradition in the way PAs were understood to separate people and nature. In the United States, national parks were a state-led institution to enclose and protect something imagined as wilderness. That model also provided the basis for British colonial conservation. In the United Kingdom itself, PAs were hybrids between state and private owners, and the nature they contained was also understood as a hybrid, deeply imbricated with human agency, its diversity and value as much a cultural as a natural product.

This distinction was reflected in the early development of ecological ideas about vegetation communities.[79] Despite considerable differences with regard

---

[77] Wilson (1992).
[78] Carruthers (1995).
[79] McIntosh (1983); Sheail (1987); Kingsland (2005).

to the idea of vegetation as organism or as a continuum of individuals, the ideas of American ecologists (Clements, Cowles and Gleason) all tended to approach the definition of plant communities as essentially the product of natural processes. In the United Kingdom, Arthur Tansley took a different view in his accounts of British vegetation in 1911 and 1939.[80] Tansley, living in a crowded island, interpreted its vegetation as semi-natural, adding the concept of deflected succession to an anthropogenic plagioclimax to the field of vegetation analysis. His revolutionary concept of the ecosystem gave a central place to human influence within the analysis of nature.[81]

If accepted as two broad schools, US-derived 'state protected wilderness' conservation and UK-derived 'mixed tenure semi-natural habitat' conservation, there can be no doubt which has been the most influential internationally, particularly in the tropical and former European colonial world. The US model for conservation, and especially its ideas of nature and PAs, has become the international model, just as the big US non-governmental organizations dominate the international movement with their power, wealth, scientific capacity and zeal.

However, there are other traditions. The first is the universal persistence of cultural dimensions to conservation thinking. This is particularly obvious in the United Kingdom, where the powerful (if problematic) imagery of the countryside as opposed to the town resonated through the twentieth-century conservation,[82] and enthusiasm for the conservation of cultural artefacts such as orchards,[83] or for recovery of the cultural associates of wild flowers.[84] However, the cultural dimensions of conservation are of wider importance. Thus Mulligan writes of the way white Australians put down roots in place, creating 'storied landscapes' natural (or semi-natural) places need to be understood in terms of relationships not blank spaces.[85]

Arguably, place-responsiveness seems to be a fundamental feature of human-response to nature.[86] Plumwood explores the possibility of a place-sensitive ethics of gardening, eschewing both the colonial rejection of indigenous species and the contemporary ideal of an 'all-native' garden in Australia.[87]

This argument is strongly made by US writers too, for example Meine laments that 'we lost the vision of conservation as a commitment binding people and places together across ideological and cultural divides, across landscapes, across

---

[80] Tansley (1911); Tansley (1939).
[81] Tansley (1935).
[82] Williams (1953); Bunce (1994).
[83] For example the work of Common Ground in establishing annual Apple Days and championing community orchards: www.commonground.org.uk/
[84] Mabey (1996).
[85] Mulligan (2003).
[86] Cameron (2003).
[87] Plumwood (2005).

generations'. Meine's inspiration here is Aldo Leopold, particularly his classic collection of essays, *A Sand County Almanac*, published posthumously, in 1949. *A Sand County Almanac* became one of the great books of conservation. A series of paperback editions from the 1960s brought it to a new generation of young readers and the heart of the growing environmental movement. The book became, and has remained, one of a small number of classics of conservation. In his resonant concept of the land ethic, Leopold observed that land is abused when regarded simply as a commodity, and respected (and conserved) when it is regarded as a community to which people belong. Meine argues against the 'simple sequestration' of conventional protectionist approaches to conservation, in favour of integration.[88]

The great idea of nineteenth- and twentieth-century conservation was to separate nature and human activity, to sequestrate the wild. However, even in US thinking, there are strong currents that recognize the dangers of such an approach. Arguably, the greatest challenge for the twenty-first century is therefore to recognize the inherent embeddedness of human agency in nature, and to build conservation strategies that acknowledge the conceptual inseparability of nature and society ecologically, economically and culturally, and move beyond attempts simply to separate them on the ground. Curt Meine describes the challenge as a 'journey to be native to this place'.[89]

That challenge lies behind the persistent enthusiasm of conservation planners for 'community-based' approaches. The hope is that these will anchor a concern for the maintenance of living diversity in the genuine aspirations and interests of local people.[90] That is certainly the expressed intention of the people who planned the Masoala National Park in Madagascar, with which this chapter began. That such a paradigm shift was attempted in the 1980s and 1990s shows how seriously the need to marry conservation to human need was taken. The limited success of such approaches is a marker of the magnitude of the challenge. The views of the farmers and fishermen on the shore in Madagascar, their families and others like them will be the acid test of conservation's real ability to imagine strategies than can integrate and not separate humanity and nature.

References

Adams, W.M. (1995) *Future Nature: A Vision for Conservation*, London: Earthscan.
Adams, W.M. (2009) 'Sportsman's shot, poacher's pot: histories of hunting and conservation', In Hutton, J., Dickson, B. and Adams, W.M. (eds) *Recreational Hunting and Conservation and Rural Livelihoods*, Oxford: Blackwell, 127–140.

---

[88] Meine (2004).
[89] Meine (2004), 209.
[90] Western *et al.* (1994); Adams and Hulme (2003); Brosius *et al.* (2005).

Adams, W.M. (2003) 'Nature and the colonial mind', In Adams, W.M. and Mulligan, M., (eds) *Decolonizing Nature: Strategies for Conservation in a Post-colonial Era*, London: Earthscan, 16–50.

Adams, W.M. (2004) *Against Extinction: The Story of Conservation*, London: Earthscan.

Adams, W.M. and Hulme, D. (2001) 'Conservation and communities: changing narratives, policies and practices in African conservation', In Hulme, D. and Murphree, M., (eds) *African Wildlife and Livelihoods: The Promise and Performance of Community Conservation*, London: James Currey, 9–23.

Anderson, D. and Grove, R. (1987) 'The scramble for Eden: past, present and future in African conservation', In Anderson, D. and Grove, R. (eds) *Conservation in Africa: People, Policies and Practice*, Cambridge: Cambridge University Press, 1–12.

Barton, G.A. (2002) *Empire Forestry and the Origins of Environmentalism*, Cambridge: Cambridge University Press.

Birch, T.H. (1990) 'The incarceration of wildness: wilderness areas as prisons'. *Environmental Ethics*, 12(1), 3–26.

Brandon, K., Redford, K.H. and Sanderson, S.E. (1998) *Parks in Peril: People, Politics and Protected Areas*, Washington: Island Press, for the Nature Conservancy.

Brechin, S.R., Wilhusen, P.R., Fortwangler, C.L., and West, P.C. (eds) (2003) *Contested Nature: Promoting International Biodiversity with Social Justice in the Twenty-first Century*, New York: State University of New York Press.

Brockington, D. (2002) *Fortress Conservation: The Preservation of the Mkomazi Game Reserve, Tanzania*, Oxford: James Currey.

Brockington, D. and Homewood, K. (1996) 'Wildlife, pastoralists and science: Debates concerning Mkomazi Game Reserve, Tanzania', pp. 91–104. In M. Leach and R. Mearns (eds) *The Lie of the Land: Challenging Received Wisdom in African Environmental Change*, Oxford: James Currey.

Brosius, J.P., Tsing, A.L. and Zerner, C. (2005) *Communities and Conservation: Histories and Politics of Community-based Natural Resource Management*, Walnut Creek Ca: Altamira Press.

Brosius, P. (2004) 'Indigenous peoples and protected areas at the World Parks Congress', *Conservation Biology*, 18, 609–612.

Bunce, M. (1994) *The Countryside Ideal: Anglo-American Images of Landscape*, London: Routledge.

Buxton, E.N. (1923) *Epping Forest*, London: E. Stanford.

Cameron, J. (2003) 'Responding to place in a post-colonial era', pp. 172–196. In W.M. Adams and M. Mulligan (eds) *Decolonising Nature: Strategies for Conservation in a Post-colonial Era*, London: Earthscan.

Carruthers, J. (1995) *The Kruger National Park: A Social and Political History*, Natal University Press.

Cernea, M.M. (2006) 'Population displacement inside protected areas: A redefinition of concepts in conservation politics'. *Policy Matters* 14, 8–26.

Chape, S., Harrison, J., Spalding, M. and Lysenko, I. (2005) 'Measuring the extent and effectiveness of protected areas as an indicator for meeting global biodiversity target'. *Philosophical Transactions of the Royal Society* B. 360, 443–455.

Colchester, M. (1994) *Salvaging Nature: Indigenous Peoples, Protected Areas and Biodiversity Conservation*, Montevideo: World Rainforest Movement.

Colchester, M. (1997) 'Salvaging nature: Indigenous peoples and protected Areas', In Ghimire, K. and Pimbert M. (eds) *Social Change and Conservation: Environmental Politics and Impacts of National Parks and Protected Areas*, London: Earthscan, 97–130.

Colchester, M. (2004) 'Conservation policy and indigenous peoples'. *Cultural Survival Quarterly*, 28(1), 17–22.

Cronon, W. (1995) 'The trouble with wilderness', pp. 69–90. In W. Cronon (ed.) *Uncommon Ground: Toward Reinventing Nature*, New York: W.W. Norton and Co.

Curt Meine, C. (2004) *Correction Lines: Essays on Land, Leopold and Conservation*, Washington: Island Press.

Dowie, M. (2005) 'Conservation refugees: When protecting nature means kicking people out'. *Orion* November/December 2005 (http://www.oriononline.org/pages/om/05-6om/Dowie.html).

Dunlap, T.R. (1999) *Nature and the English Diaspora: Environment and History in the United States, Canada, Australia and New Zealand*, Cambridge: Cambridge University Press.

Evans, D. (1992) *A History of Nature Conservation in Britain*, London: Routledge.

Fitter, R. and Scott, P. (1974) *The Penitent Butchers: 75 Years of Wildlife Conservation*, London: Fauna Preservation Society.

Goetzmann, W.H. (1966) *Exploration and Empire: The Explorer and the Scientist in the Winning of the American West*, New York: Alfred A. Knopf.

Goodman, S.M. and Benstead, J.P. (eds) (2004) *The Natural History of Madagascar*, University of Chicago Press.

Greene, J.A. (1991) *Yellowstone Command: Colonel Nelson A. Miles and the Great Sioux War 1876–1877*, Lincoln: University of Nebraska Press.

Griffiths, T. and Robins, L. (eds) (2007) *Ecology and Empire: Environmental History of Settler Societies*, Keele: Keele University Press.

Hays, S.P. (1959) *Conservation and the Gospel of Efficiency: The Progressive Conservation Movement 1890–1920*, Cambridge, Mass: Harvard University Press.

Hingston, R.W.G. (1932) 'Proposed British National Parks for Africa'. *Geographical Journal*, 76, 1–24.

Homewood, K.M. and Rodgers, W.A. (1991) *Maasailand Ecology: Pastoralist Development and Wildlife Conservation in Ngorogoro, Tanzania*, Cambridge: Cambridge University Press.

Hulme, D. and Murphree, M. (eds) (2001) *African Wildlife and Livelihoods: The Promise and Performance of Community Conservation*, Oxford: James Currey.

Hutton, J., Adams, W.M., and Murombedzi, J.C. (2005) 'Back to the barriers? Changing narratives in biodiversity conservation'. *Forum for Development Studies*, 32(2), 341–370.

Jacoby, K. (2001) *Crimes Against Nature: Squatters, Poachers, Thieves, and the Hidden History of American Conservation*, Berkeley: California University Press.

James , N.D.G. (1981) *A History of English Forestry*, Oxford: Basil Blackwell.

Jeffers, H.P. (2003) *Roosevelt the Explorer: Teddy Roosevelt's Amazing Adventures as a Naturalist, Conservationist, and Explorer*, Lanham NY: Taylor Trade Publishing.

Jepson, P. and Whittaker, R.J. (2002) 'Histories of protected areas: internationalisation of conservationist values and their adoption in the Netherlands Indies (Indonesia)'. *Environment and History*, 8, 129–172.

Kingsland, S.E. (2005) *The Evolution of American Ecology, 1890–2000*, Baltimore: Johns Hopkins University Press.

Kramer, R.A., Schaik, C.P. van and Johnson, J. (eds) (1997) *The Last Stand: Protected Areas and the Defense of Tropical Biodiversity*, New York: Oxford University Press.

Kremen, C., Razafimahatratra, V., Guillery, R.P., Rakotomalala, J., Weiss, A. and Ratsisompatrarivo, J.-S. (1999) 'Designing the Masoala National Park in Madagascar Based on Biological and Socioeconomic Data'. *Conservation Biology*, 13(5), 1055–1068.

Langton, M. (2003) 'The "wild", the market and the native: indigenous people face new forms of global colonization', pp. 79–107. In Adams, W.M. and Mulligan, M. (eds) *Decolonizing Nature: Strategies for Conservation in a Post-colonial Era*, London: Earthscan.

Mabey, R. (1996) *Flora Britannica*, London: Chatto and Windus.

MacEwen, A. and MacEwen, E. (1982) *National Parks: Conservation or Cosmetics?* London: George Allen & Unwin.

MacKenzie, J. (1988) *The Empire of Nature: Hunting, Conservation and British Imperialism*, Manchester University Press.

Margules, C.R. and Pressey, R.L. (2000) 'Systematic conservation planning'. *Nature*, 405, 243–253.

Marks, S.A. (1984) *The Imperial Lion: Human Dimensions of Wildlife Management in Central Africa*, Boulder, CO: Westview Press.

Marren, P. (2002) *Nature Conservation: A Review of the Conservation of Wildlife in Britain, 1950–2001*, London: HarperCollins. (New Naturalist Series 91).

McIntosh, R.P. (1985) *The Background of Ecology: Concept and Theory*, Cambridge: Cambridge University Press.

McNamee, K. (1993) 'From wild places to endangered spaces: a history of Canadian national Parks', pp. 17–44. In Dearden, P. and Rollins, R. (eds) *Parks and Protected Areas in Canada: Planning and Management*, Toronto, Oxford University Press.

Meine, C. (2004) *Correction Lines: Essays on Land, Leopold and Conservation*, Washington: Island Press.

Moir, J. (1991) 'National Parks north of the border', *Planning Outlook*, 34, 61–7.

Mulligan, M. (2003) 'Feet to the ground in storied landscapes: disrupting the colonial legacy with a poetic politics', pp. 268–289. In Adams, W.M. and Mulligan, M. (eds) *Decolonising Nature: Strategies for Conservation in a Post-colonial Era*, London: Earthscan.

Neumann, R.P. (1996) 'Dukes, Earls and ersatz Edens: aristocratic nature preservationists in colonial Africa'. *Environment and Planning D: Society and Space*, 14, 79–98.

Neumann, R.P. (1998) *Imposing Wilderness: Struggles Over Livelihood and Nature Preservation in Africa*, Berkeley: University of California Press.

Neumann, R.P. (2001) 'Africa's "last wilderness": reordering space for political and economic control in colonial Tanzania'. *Africa*, 71, 641–665.

Neumann, R.P. (2002) 'The postwar conservation boom in British colonial Africa'. *Environmental History*, 7, 22–47.

Neumann, R.P. (2004a) 'Nature-state-territory: towards a critical theorization of conservation enclosures', In Peet, R. and Watts, M. (eds) *Liberation Ecologies: Environment, Development, Social Movements*, London: Routledge, 195–217.

Oates, J. (1999) *Myth and Reality in the Rain Forest: How Conservation Strategies Are Failing in West Africa*, Berkeley: University of California Press.

Plumwood, V. (2005) 'Decolonising Australian gardens: gardening and the ethics of place'. *Australian Humanities Review*, vol. 36, July 2005, special issue on 'Desert Gardens: Waterless Lands and the Problems of Adaptation', http://www.lib.latrobe.edu.au/AHR/archive/Issue-July-2005/09Plumwood.html

Poirier, R. and Ostergren, D. (2002) 'Evicting people from nature: indigenous land rights and national parks in Australia, Russia and the United States'. *Natural Resources Journal*, 42, 331–51.

Prendergast, D.K. and Adams, W.M. (2003) 'Colonial wildlife conservation and the origins of the Society for the Preservation of the Wild Fauna of the Empire (1903–1914)'. *Oryx*, 37, 251–260.

Quammen, D. (2003) 'Saving Africa's Eden'. *National Geographic*, 204(3), 48–75.

Ramsey, Sir J. Douglas (1945) *National Parks: A Scottish Survey; A Report by the Scottish National Parks Survey Committee*, HMSO London.

Ranger, T. (1999) *Voices from the Rocks: Nature, Culture and History in the Matopos Hills of Zimbabwe*, Oxford: James Currey.

Runte, A. (1987) *National Parks: The American Experience*, Lincoln: University of Nebraska Press.

Runte, A. (1990) *Yosemite: The Embattled Wilderness*, Lincoln: University of Nebraska Press.

Schama, S. (1995) *Landscape and Memory*, London: HarperCollins.

Schwartzman, S., Nepstad, D. and Moreira, A. (2000) 'Arguing Tropical Forest Conservation: People versus Parks', *Conservation Biology*, 14(5), pp. 1370–1374.

Sheail, J. (1976) *Nature in Trust: The History of Nature Conservation in Great Britain*, Glasgow: Blackie.

Sheail, J. (1987) *Seventy-five Years in Ecology: The British Ecological Society*, Oxford: Blackwell Scientific.

Sheail, J. (1991) *Rural Conservation in Inter-war Britain*, Oxford: Clarendon Press.

Sheail, J. (1998) *Nature Conservation in Britain: The Formative Years*, London: The Stationery Office.

Society for the Preservation of Wild Fauna of the Empire Journal (1906–9) vol. 2–5.

Struhsaker, T.T. (1999) *Ecology of an African Rain Forest: Logging in Kibale and the Conflict between Conservation and Exploitation*, Gainesville: University Press of Florida.

Tansley, A.G. (1911) *Types of British Vegetation*, Cambridge University Press.

Tansley, A.G. (1935) 'The use and abuse of vegetational concepts and terms'. *Ecology*, 16, 284–307.

Tansley, A.G. (1939) *The British Isles and Their Vegetation*, Cambridge University Press.

Terborgh, J. (1999) *Requiem for Nature*, Washington DC: Island Press.

Thompson, E.P. (1975) *Whigs and Hunters: The origin of the Black Act*, London: Allen Lane.

Toogood, M. (2003) 'Decolonizing highland conservation', pp. 152–71. In Adams, W. and Mulligan, M. (eds) *Decolonizing Nature: Strategies for Conservation in a Post-colonial Era*, London: Earthscan.

Western, D., Wright, M. and Strumm, S. (eds) (1994) *Natural Connections: Perspectives in Community-based Conservation*, Washington DC: Island Press.

Whitney, G.G. (1994) *From Coastal Wilderness to Fruited Plain: A History of Environmental Change in Temperate North America from 1500 to the Present*, Cambridge: Cambridge University Press.

Williams, R. (1953) *The Country and the City*, London: Chatto and Windus.

Wilshusen, P.R., Brechin, S.R., Fortwangler, C.L. and West, P.C. (2002) 'Reinventing a square wheel: Critique of a resurgent "Protection Paradigm" in international biodiversity conservation'. *Society and Natural Resources* 15, 17–40.

Wilson, A. (1992) *The Culture of Nature: North American Landscape from Disney to Exxon Valdez*, Oxford: Blackwell.

World Conservation Union – IUCN 2005: *Benefits Beyond Boundaries. Proceedings of the Vth World Parks Congress*, Cambridge UK, World Conservation Union.

3

The Environmental History of Pre-industrial Agriculture in Europe

Paul Warde

Agricultural history and environmental history are inextricably linked. This is not simply because agriculture directly exercises a profound influence on the surface and climate of our planet, but also because until very recently indeed the great majority of humans were agriculturalists. For millennia the passage of their lives as rural consumers and producers gave the human role in changing the face of the Earth a distinctly bucolic tint, even when these populations exercised their power indirectly, through demand for the products of cities, manufactories and mines. Yet while the *land* has been a perennial concern of environmental historians over recent decades, their interest in *agriculture*, most especially perhaps in Europe, has frequently been rather incidental, though with notable exceptions.[1] Cultivation has frequently been viewed simply as another means by which humans inflict change, and pre-eminently damage, on the 'environment'. Conversely, agricultural historians, while producing extensive writings on what we might consider 'the environment', have tended to consider it as either the backdrop to human endeavour, or written narratives by which nature's fecundity is increasingly harnessed and enhanced by human ingenuity.

How did this happen? Raise the issue of 'the environment' with anyone, and they will almost instinctively think of wild, open and supposedly untrammelled hillsides; of otters or barn owls but not pigs or cows; of hedgerows and not the tractor-furrowed field, although the hedge is just as much the outcome of agricultural management as the furrow. These inculcated instincts express a hugely influential strand in thinking that has shaped the ideals and interests of many who now practice environmental history, as well as 'popular' attitudes, one that Donald Worster has characterised as an 'arcadian' approach to ecology.[2]

---

[1] Cronon (1984), Cronon (1991), Pfister (1984). Indeed, in this volume Damodaran and Grove describe much of the *Agrarian History of England and Wales* as essentially environmental history.
[2] Worster (1994).

Some historians, such as the early garden historian of the 1830s, J.C. Loudon, or in more recent times Keith Thomas, have attributed the rise of the arcadian ideal to a reaction against the early enclosure and 'geometrification' of the English countryside. Once the tilled field and the linear quickset hedge came to dominate the landscape, the search for beauty in the land shifted away from the patterned Classical garden to the apparently naturalistic sweep of the 'English' country estate, and the untamed solitudes of the mountain. Indeed, historians of later nineteenth- and twentieth-century England have written extensively on the links between national identity, an 'idyllic' view of the countryside, and the rise of industrial and urban society, though with little reference to an explicitly 'environmental history', or, indeed, agriculture.[3]

What of those who wrote about agriculture itself? How did they imagine their art: a 'Restauration of Nature', as Timothy Nourse thought in the 1690's, where the uncultivated common was the 'very abstract of degenerated Nature', and cultivation 'restor'd [the land] to its Primitive Beauty in the State of Paradise?'[4] Or did agronomists too come to imagine the product of their endeavour in purely utilitarian terms, happily conceding aesthetic primacy to the romantic? This chapter will contend that the modern understanding of the 'environment', that owes so much to arcadian ideas, is not simply the result of responses to agricultural revolution, to urbanisation and to industrialisation. It is part and parcel of a broader process, and the way in which we think about human activity. In regard to the land, this is especially true of agriculture, and the task of agriculture as formulated by the 'classic' agronomists of the Enlightenment. In other words, the delineating of subsequent environmental understanding was not simply reactive, but inscribed into the way in which the most influential writers on land management conceptualised their task. By examining how this understanding arose, a process by which the environment has come to be seen as the 'other' of agriculture we can not only provide an environmental perspective on agrarian history. We may also reflect on the concept of *environment* itself – a task undertaken all too infrequently by 'environmental historians'.

It is true, of course, that cultivators have always seen that agriculture is dependent on the beneficence of what we now think of as the 'natural environment'.[5] However, what I will argue is that writings on agricultural practice, and indeed the historic collective regulation of agricultural practice, have been concerned above all with defining a world of action for those who cultivated the land

---

[3] Thomas (1983), 261, 263–4; a classic on urban-country relations is Williams (1973); for a useful historiographical discussion, see Burchardt (2007). This last piece only mentions the 'environment' once, in the very last paragraph.

[4] Nourse (1706), 2–3, 99.

[5] On the hazards of this expression, see Ingold (2000), Ch. 1.

and husbanded the animals. In demarcating this realm for human action, and establishing parameters for that action, agriculturalists and agronomists simultaneously demarcated an 'environment' beyond the scope of their influence and that thus was not a central object of their attention. The legacy of that demarcation, that itself has shifted over time, has in turn strongly imprinted our notion of what 'the environment' is. Indeed, in the introduction to this volume it is argued that our modern, politicised concept of the environment has arisen from a gradual dawning consciousness of the historicity of this exclusion and problems that it generates.

This chapter is divided into three parts. The first will examine 'traditional' agricultural practice in the centuries before the nineteenth. The focus will be on central Europe, which happens to be the region with which I am most familiar, but much of what is discussed is equally applicable to other lowland regions of western Europe. Through these centuries agriculture was a tightly regulated enterprise, subject to the collectively managed supervision of lords (or more directly, their stewards and bailiffs), or village and municipal authorities.[6] From the late medieval period onwards numerous ordinances, field orders, cadastres and accounts survive that allow us to reconstruct agricultural practice and how collective controls were exercised. In interpreting these sources, historians frequently adopt what might be called an 'ecological' perspective, seeing the rules by which peasant communities and manors operated as seeking to ensure the sustainability of the agricultural system. Obviously, it is a sound ecological principle that for any material process to be sustainable you have to put in at least as much as you take out, and we may presume that pre-industrial farmers wanted to ensure that they had at least as much food to eat the next year as they had this year. The (relatively speaking) slow pace of economic change prior to the nineteenth century copper-fastens an intuition that before the Industrial Revolution farming fostered sustainable values and an ethic of care for the environment. How could it not have been so?

Yet I will argue that this is a misreading of the sources; not because cultivators had no concept of what we think of as 'sustainability', but because the rule-making was fundamentally concerned with *human action* and its consequences (above all, in fact, how the actions of one person had consequences for another). These actions did indeed have implications for 'sustainability', but the rules themselves were not intended as blueprints for sustainable living.

The second section examines the 'new husbandry' and 'experimental' agriculture that began to develop in northern Europe from the 1760s. Coeval with the widespread abolition of collective controls on agriculture (the 'enclosure' movement and the dissolution of the feudal regime), the

---

[6] For example, Hopcroft (1999), de Moor, Shaw-Taylor and Warde (2002); Troßbach and Zimmermann (2006).

agronomists of this period showed little interest in problems of neighbourly relations, and sought ways of maximising a sustainable profit from their land. Their concern remained what action the wise farmer should take to achieve this aim, and the core of their ideas focused on the recycling of nutrients within the farm by the use of manure. They stressed that the cultivator himself could determine his fortunes, and what we would now term the 'environment' received only cursory consideration and a limited role in determining agricultural fortunes.

The third part examines the heritage of the 'new husbandry' in agricultural and environmental history. Enlightenment agronomists shaped not just subsequent agricultural practice, but our perceptions of what had gone before. During the nineteenth century, increased understanding of soil chemistry and biological processes led to a gradual reappraisal of the role of recycling and agricultural practice, coupled with the introduction of artificial fertilisers. The recognition of feedback effects between human action and environmental impacts has since blurred the previous lines of demarcation, but agricultural historians, even those who have taken a consciously 'ecological' turn in recent decades, have however largely remained wedded to parameters for study laid down by the 'new husbandry'. Thus this form of history has not been especially 'environmental' in its thinking; and the gap that is left in fact provides environmental historians with a task that should be central to the enterprise, but is poorly understood: the history of 'environing', the study of how all human activity demarcates and generates an 'environment', an outside that haunts the space in which people choose to act.[7]

The traditional agricultural economy

European agriculture was diverse indeed, but the most prevalent form of collective management was known as the 'three-field' system, and I will take this as an exemplar. For centuries the bulk of the population of lowland western Europe lived in compact 'nucleated' villages that were surrounded by 'open' fields subject to collective regulation of a three-course crop rotation, generally of two main grain crops and a fallow field. The cultivated plots were 'privately' held by a farming family, usually as tenants, though often interspersed with large estate farms run by lords and wealthy landowners. But the agricultural calendar was governed by a greater institution, such as the village commune or the court of the manorial lord who owned the soil from which the tenant farmers eked out a living. This body would determine when plots were open to the use of all for grazing, when the harvest began, and the timing of other agricultural

---

[7] These observations are indebted in part to the work of Luhmann (1995) and Hastrup (1990).

tasks. Such rules effectively determined which crops could be grown, as the calendar was geared towards the production of favoured varieties. If, for example, everybody's plots had to be open to general grazing by the village's herds and flocks at a certain time, then obviously to grow a crop upon this land that would mature later than this date would expose it to being eaten by the animals. Similarly, the payment of rents in kind often determined the cropping regime.

In the 'three-field system', a 'winter crop' such as rye or wheat was sown in the autumn on one great, 'open' field. On the second 'open' field, a crop such as barley or oats would go into the ground in spring, while the third field was left fallow, that is, uncropped and open both for grazing and to be prepared for a future winter-sown crop. After the harvest in August, each field would shift on to the next phase in the cycle: spring-sown following the winter crop, fallow following the spring-sown crop, and an autumnal sowing of the winter crop succeeding the fallow. Each of the three great fields would be divided into the numerous smaller plots held by the cultivators. In turn, each 'farm' was an agglomeration of these small plots, with the cultivators usually trying to ensure that their land was scattered across the three fields to provide a guarantee that they always brought in a balance of the desired crops. The fielden landscape was thus a mosaic of many long, narrow strips, so shaped to facilitate ploughing, with each farm's land intertwined with the plots of the neighbours.

Beyond the fields, open pasture and woodland was usually 'common', meaning it was subject to collective rights of grazing, and fuel-collection (such as of wood, furze and bracken). When not cultivated, nearly all land was subject to common grazing rights. It is important however that we understand the nature of the 'commons'. They were not a free-for-all, or 'open access' as the technical term has it. Common rights were vested in the village or manorial government and were regulated by them. Grazing, for example, often had to be conducted in one collective herd overseen by a communally employed herdsman. It was rare that *every* household in the village had the right to exploit these open common lands. More frequently access was restricted in some way, such as being enjoyed only by those who were tenants of land, and not the landless labouring classes.[8]

The three-field system effectively meant that changes in land-use, at least in theory, required collective or official approval. In many places, not only the use of land, but the employment of labour was subject to restrictions. Villagers were required to make ploughs, the horses and oxen to haul them, and human labour available in a timely fashion for important tasks, establishing reciprocal relationships between smallholders (who had plenty of hands ready to work) and wealthier farmers (who had the capital: the livestock and equipment).[9]

---

[8] de Moor, Shaw-Taylor, and Warde (2002); Rheinheimer (1999), 154–6.
[9] Warde (2006), 52–94.

Grazing was subject to communal oversight, and the number of animals that each household could hold was also restricted, thus limiting the overall grazing pressure on community resources. Movement of 'biomass', that is, all the matter derived from living organisms, within and without the community was restricted. Frequently hay, straw, wood and manure could not be removed from the village jurisdiction, or was offered for sale to locals first. Landlords also laid down very similar restrictions on their tenant farms.

The collective guarantee

The life of the cultivator was to a large degree orientated around the passage of the seasons and the life of plants. Daily life ran around these patterns with an intimacy hard to imagine for most of us today. Up until the 1780s most agricultural writers laid constant stress on how the weather dictated work patterns. Sow your spring crop when you hear the frog's chorus, suggested the German pastor Martin Grosser in 1593.[10] Most ground had different uses at different times; crops had their cycles of germination, maturity and decay; grass grew predominantly in late spring and summer, and so on.

In this world, property rights were generally thought to pertain to the *use of a material for a specified purpose*, rather than simply being a claim to something – such as a piece of land – from which others can be excluded. As the product of the land altered over the year, so did the property claims to any given space. The particular property claim to exclusive use of products from the soil of a defined space, that with which we are most familiar today, was a very important right but still only one among many.

As any agricultural enterprise will fail if it does not have access to certain resources – seed, fuel, fertiliser, fodder, ground – denial of access to material to an enterprise that needed it could actually be considered an infringement of rights, of one's '*necessitas domestica*'.[11] Thus it was possible one person might own a patch of soil, but through the collective system of management another could claim the right to the grass, leaves or fruit that grew upon it. Of course, this did not mean that you could demand from your neighbour anything you considered useful, but that communities created a series of collective guarantees that resources would be made available when and where they were essential to the functioning of a household. This was the kernel of the whole system of collective regulation. The rules were indeed concerned obliquely with sustainability, but not of an eco- or agricultural system, but the *household*. And they did not *guarantee* sustainability, as the history of households makes very clear. Rather, they sought to ensure that the action of *others* did not reduce your ability to

---

[10] Grosser (1589[1965]), 24.
[11] On this idea, see Blickle (1992); and Warde (2006), 335–41.

provide for yourself. This intertwining of household fortunes in a system of management could also, as the early 'improving' writers recognised, provide disincentives to changes in agricultural practice. Among the reasons listed for a reluctance to 'improve' or enclose common land by English agricultural writer Walter Blith in 1649 were the fears of the farmers that the cost they bore for the changes would not bring proportional benefit, or that they were not in so good a position as their neighbours to exploit new-found freedoms.[12] A collective guarantee had a flip side of mutual suspicion among 'envious neighbours', as John Worlidge put in 1669.[13] Thus for much of the year peasants had free disposal over their assets, but at other times they were subject to tight control.

Many of the rules pertaining to common land seem designed to prevent 'free riding', that is taking excessive benefit from a resource without bearing the cost of its maintenance, leading to degradation of that resource.[14] In ecological terms, 'free riding' could effect traditional agriculture through overgrazing land or clearing wood for short-term gain without putting anything 'back in', leading to a decline in output. In economic terms, it might mean a tenant running down a capital asset (such as the soil), leaving the land useless to future tenants. Collective rule-making hence sought to prevent *over-use* and excess consumption.

However, the traditional agricultural economy was just as concerned with not *under-using* resources. As was first clearly recognised, at least in learned and formalised discourse, by the soil chemist Justus von Liebig in the 1840s, the successful elements of an agricultural economy work only in combination. Applying ever-greater quantities of one input (such as an ammonium fertiliser) will do no good at all if there are not other inputs that permit the uptake of the fertiliser by the plant, the right kind of growth, and successful harvesting.[15] And so it is with any factor of production: capital, labour, land, energy, nutrients. Thus many of the rules of collective agriculture sought to ensure that land was made available when most useful, hay was cut at a time when it had the maximum nutritious effect, that labour would be on tap for the harvest so it would not spoil on the stalk or vine, and that those who could not afford draught animals and ploughing gear would have it made available to plough their smallholdings. This too was part of the collective guarantee.

What these rules sought above all else was *predictability*. Of course, the outcome of agricultural endeavour was (and remains) unpredictable. But in the realm of human action, one could at least limit consequences arising from the caprice of one's neighbours, whether the regulation was concerned with the

---

[12] Blith (1652), 84.
[13] Worlidge (1669), 10.
[14] The classic statement of this problem is Hardin (1968).
[15] Liebig (1862), 38.

boundaries and access rights of plots, controlling marauding geese, allotting wood-cutting rights or requiring that ready hands or ploughs, or draught teams had to be made available for timely use. The by-laws controlling pre-industrial village life followed a pattern that later agriculturalists (and perhaps most economic enterprises) would follow. They sought to *simplify* the level of caprice that they faced, and relegated those areas too complex to control to the 'environs'.

The agronomist writers of this age did not argue that the cultivator should accept the hand that the fates dealt them. From the sixteenth-century hand-books of advice and estate management were published in northern Europe, partly in imitation of classical and Mediterranean predecessors. Early and enduringly popular forms of the genre took the form of calendars advising on what agricultural tasks to perform throughout the year. In mid-seventeenth-century England a more ambitious literature of 'improvement' was established which developed much longer thematic discussions of good practice, mixing practical example, report on authorial experiment, and the (occasionally dis-counted) wisdom of the ancients.[16] These writers did not imagine that they could control the natural world, and placed their advice within a wider frame-work of cause and effect in which agriculture unfolded. For the collections of letters and discourses edited by republican Samuel Hartlib in the 1650s, this context was divine: 'It is the Lord that maketh barren places fruitfull', so a husbandman must, 'walk as becommeth a Christian, in all Sobriety, Righteous-nesse and Godlinesse; not to trust his confidence in his own labours, and good Husbandry; but on the Lord that hath made all things.'[17] The great synthesis of previous works provided by John Worlidge in 1669 devoted two chapters to, 'The common and known external Injuries, Inconveniences, Enemies and Diseases incident to, and usually afflicting the Husbandman', and prognos-tications of the weather and pests. His work consistently gave space to the 'uncertain Dispositions of an Over-ruling Providence', and asserted that the task of the husbandman was to ascertain the 'natural Temper of his Land', and what 'agreeth with the nature of his land'.[18] Some 30 years later Timothy Nourse would agree, the farmer, 'after he has cast his Business to the best Method his Reason can propose, must still depend upon Providence, as to the event, here being so many Accidents which may traverse his Designs, and such as can be never provided against, nor foreseen'.[19]

As a result, these authors were often very particular in their observations, stressing the different practices to be employed for the melioration of particular soils and aspects of the land; the balance of dressings applied the soil, quality of

---

[16] Thirsk (1984), 533–89; Abel (1962), 201–88; Ambrosoli (1997).
[17] Hartlib (1659), 81.
[18] Worlidge (1669), 179, 35.
[19] Nourse (1706), 37.

tilth and amount and type of dung were to be carefully assessed. The cultivator (the 'husbandman') was the principle agent in, and subject of their exhortations, but this did not entail a profound division between human activity and the action of 'providence'. Indeed, it was incumbent on the good farmer to prepare for contingency and the specific challenges of their land. The husbandman might be subject to many forces beyond his control, but there was no clear delineation between the agricultural world and an 'environment'.

The 'new husbandry'

With some justice, though also with some hyperbole, the agricultural writers of the late eighteenth century waxed large in their self-importance. Arthur Young, still considered the greatest authority of the agriculture of the era, noted brusquely in his *Rural Oeconomy* of 1773 that he did not include the usual encomia for earlier authors because not a single one had written anything of much use to practical men such as himself.[20] In good English fashion Young insisted that only the empirical observations gathered by years of careful experimentation – as he himself undertook – could provide sure knowledge of good agricultural practice. In fact this was rather unfair to predecessors such as Walter Blith who had been farmers themselves and advocates of the virtues of Baconian experimentation, and many writers included a self-justificatory comment early in their texts on the years of experience they had accumulated.[21]

Young's German equivalent was Albrecht Thaer, a Hanoverian doctor turned farmer and founder of the Prussian agricultural academy. Thaer stressed that 'the science of agriculture rests on experience', as opposed to 'simple tradition', although he was following most eighteenth-century agronomists in so arguing. Indeed, some would assert that tradition was itself based on, and evidence for, positive experience.[22] But while Thaer was more generous than Young to earlier authors, and even the 'mere growers of potatoes...and petty gardeners' who surrounded him, it was important to Thaer's reputation to insist that the combination of 'accident and necessity' in his own experimental conduct had brought him success in discovering efficacious new crop rotations; and not the chance perusal of English authors such as Young, of whom he was in fact a conscious imitator.[23] Experiment was all.

---

[20] Young (1773), 2–3.
[21] Indeed elsewhere Young was a little more forgiving of his predecessors, but not much. Blith (1652), 111, 150–1; Worlidge (1669), 63; Nourse (1706), 25; Ellis (1773), Preface, iii; Thirsk (1984), 536–7, 561–7; Ambrosoli (1997), 262–398; Young (1770), viii–xvi.
[22] Thaer (1844), 2–3.
[23] Thaer (1844), 3, 232–3; indeed, Thaer was not the first to claim grounding in practical experience of farm management as the trope is given prominence in authors from the sixteenth century onwards, notably von Eckhardt in 1754. Abel (1962), 202.

At the heart of experimental agriculture was a new theory of the soil, that instead of emphasising how the cultivator had to adapt to the edaphic conditions prevailing in his fields (categorised variously by colour, or from 'heaviness' to 'lightness', or in earlier versions by humoral theory resting on dryness, heat, coldness and moisture[24]), the quality of the soil was in the gift of the farmer. Soil was part of a system of action. Agriculturalists have always known that the style of tillage affects the tilth, but previous writers had attempted simply to ameliorate problems arising from cultivated the soils that they inherited and were subject to, with dressings of lime, marl or clay to encourage alkalinity, or the introduction of manures to improve the nutrient (or 'nitrous juices') level or texture. 'It is very necessary in marling lands to finde out the true proportion, how much on every Acre, that you add not too much, nor too little.'[25] In contrast the 'practical farmers' sought to bend the soil to their will.

At the same time, the abolition of systems of collective constraint through the enclosure movement, that waxed most prominently and early in England but that would embrace most of western Europe in the nineteenth century, removed the issue of the consequences of neighbourly action from the agronomist's purview.[26] The system of agricultural action became the farm unit, the environment denoted as 'natural' or 'market' forces.

Over time the works of Young, Thaer and their contemporaries have become historians' standard tools of reference for both 'traditional' agricultural practice and the new age of improvement. These writers provided the first really comprehensive accounts of how a traditional agricultural system is supposed to have functioned. The practical farmers themselves drew a clear line between what had gone before and what should come to pass under their influence. By the time another highly influential 'classic', the *Chemistry in its application to agriculture and physiology* of Liebig was printed in numerous, ever-updated editions in the mid-nineteenth century, he could clearly divide the schema of his volume into agriculture as it existed 'before' and 'after' 1840, roughly when the first edition of his work appeared.[27]

Sustainability and improvement

Key to the new agronomy was the principle of maintaining a balance between the extent of meadow and pasture, and the arable. The main aim of production was high yields of cereals, the main foodstuff of the pre-industrial world. Each harvest of cereal crops diminished soil fertility as the nutrients in the crop

---

[24] Blith (1652), 140; Evelyn (1676); Rohr (1726), 131–2.
[25] Hartlib (1659), 11; Worlidge (1669), 62.
[26] Warde (2003).
[27] Liebig (1862), Vol.1.

were removed and consumed by hungry mouths elsewhere, instead of rotting, like most dead plant matter, back into the soil from whence they had sprung. According to Young the primary manner in which fertility could be replenished was copious additions of manure, an idea taken up a few years after Young's first book by Adam Smith in *The Wealth of Nations*.[28] Of course the properties of many types of manure and alkalis had been well known and applied for many centuries, but now took on a novel centrality.[29] Some manure could come from feeding livestock the straw of the previous harvest, but obviously this still involved net loss of nutrients (or 'succulent juices') in the recycling process. The answer was to pasture livestock on meadows, transferring biomass onto the animal fields, and logically to optimise output one needed the correct ratio of meadow to arable to sustain continued high yields.

Establishing this ratio, and all the 'proportions' derived from this, was the stated aim of Young at the very beginning of his *Rural Oeconomy*, and 'if any of the proportions...are broken, the whole chain is affected...so much does one part of a well managed farm depend on the other'.[30] Even if the danger of getting the proportions wrong was not apparent in the short-term, it would eventually cause soil exhaustion. The idea of balancing livestock numbers and tillage because of a desire for the appropriate supply of manure was again not novel by any means. Previous writers such as the German cameralist and administrator von Justi had considered this to be a remedy for lower than optimum supplies of dung. Indeed, cameralist Professor Johann Friedrich von Pfeiffer already made crude calculations as to intensity of manuring, numbers of animals and meadow required, and doubtless many farmers had done so in their heads for centuries.[31] It was however pioneering to understand cultivation as a systematic enterprise where manure was the critical vehicle for recycling nutrients within a closed system. Earlier authors such as John Evelyn had thought farmers' concern for the application of manure, and especially dung, had been too great.[32] Indeed, while the importance of manure was almost universally acknowledged, predecessors to Young frequently argued that weeding and the quality of the tilth were just as important, and indeed that the act of aerating the soil in turning in manure was one of the primary benefits of the practice.[33] But according to Young, 'It may be laid down almost as a maxim, that there is no farming without manure, and that in plenty too.'[34]

---

[28] Smith (1776), I.11.197.
[29] Abel (1962), 166–8.
[30] Young (1773) 7, 12–13,
[31] Achilles (1991), 8–9; Abel (1962), 283, 306–7.
[32] Evelyn (1676), 63; Worlidge (1669), 8.
[33] The transference of weed seeds via manure was considered a major problem. Nourse (1706), 29–30; Ambrosoli (1997), 325, 348–56.
[34] Young (1773), 8–9, 30, 61, 157.

Thus though the idea of these relationships was hardly novel, the new husbandry elevated these insights in a systematic model that, through experiment
and calculation, could guarantee success. Young's approach put the onus on
human action. As much as experiment, farming was to become a matter of
accounting. Low yields were the consequence of insufficient manuring, and
insufficient manuring was the product of an ill-proportioned husbandry. Therefore, poor yields or uncultivated waste were not fundamentally the product of
local edaphic conditions or inclement weather, but bad husbandry, pure and
simple.[35] This went in the face of much earlier writing, that while waxing
lyrical about the possibilities of improvement, had been more ready to concede, 'that most of the barren an unimproved lands in England are so, either
because of Drought, or the want of Water or Moisture, or that they are poi-
soned or glutted with too much.'[36] Thus at the very beginnings of modern
agronomy, Young's model of the farm as ecosystem was writing what we would
come to call 'the environment' out of the story as an agent in agricultural
change.

Albrecht Thaer conducted the bulk of his 'experiments', and published his
classic work, some 40 years after Young began to write. The starting point of
his arguments remained similar to Young's: '...the produce generally depends
more on the quantity of the manure than on the nature of the soil', and poor
land was commensurately a product not of poor soil but 'want of manure'.
Naturally this required a correct proportioning of tillage and livestock, or more
precisely, tillage and fodder.[37] Now that fodder crops such as legumes were
gaining a greater place in arable rotations, this was no longer an issue that
focussed so exclusively upon meadow or pasture.[38] Working from the principle that it was manure that recycled the necessary 'nutritive juices' into the
soil for crop growth, Thaer developed a form of accounting for this process
based on 'degrees' of fertility and developed a model of the ideal proportions
to be established between fodder inputs, animal numbers and agricultural
output.[39]

Young and Thaer, the two most influential writers on agriculture of their
age in Britain and Germany, thus established early in their careers a relatively
'closed' system-like model of nutrient flows within the farm, mediated via

---

[35] Young (1773), 25. Though the idea that lack of capital and idleness were the causes
of low yield can be found, for example, in the mid-seventeenth-century work of Walter
Blith, who however only provided a rather unsystematic list of possible improvements to
cultivation. Blith (1652), Ch. 2.

[36] Worlidge (1669), 189.

[37] Thaer (1844), 24, 130, 142, 161.

[38] Thaer (1844), 27.

[39] Thaer (1844), 138–48, 162–5, 170–84.

manure. While they remained keen observers of wider farming practice and the importance of producing a good tilth and combating weeds, they argued that only in this way could yields be varied and long-term sustainability achieved. The inference of subsequent historians has been that the long-lived traditional systems of cultivation must have operated along similar lines, but with inadequate production of fodder to permit large numbers of livestock and win high yields. Yet this is the despite the fact that the Enlightenment proponents of the 'new husbandry' did not consider such practices to be widespread at all. Young especially lambasted farmers for being 'bad managers', their 'neglect' and 'execrable' practices where 'every means of exhausting the soil are taken, but none of replenishing it'.[40] The model of the new proselytisers that, as we shall see, would be taken by successors to be the apotheosis of a long-established pre-industrial agricultural system yet had been considered by these self-same 'practical farmers' to be a revolutionary break with the past.

The 'Sewer question' and Liebig's ecological history

Thaer had understood well that plant growth must be the consequence of some kind of chemical action, and in combination with the chemist Einhof attempted a series of experiments aiming to demonstrate how an understanding of physio-chemical processes was essential to the agriculturalist, though with rather little success.[41] The active fertile agent could be specified with no more exactitude than 'nutritive juices'.[42] Early nineteenth-century chemists developed an 'organicist' theory of plant growth, arguing that it was living matter in the humus of the topsoil that was in turn absorbed by and that fed growing plants, a life-force inhering in organic matter. Manure, of course, was the agent of transmission in this recycling.[43]

This viewpoint would be overturned by the theories of Justus von Liebig who from the 1830s onward laid the foundations of modern soil science by establishing that the basis of plant and soil chemistry was the circulation of elements and compounds (nitrogen, phosphates and so forth). Yet much of the writing on pre-industrial agricultural history veered off along another course, retaining the circulation theories of the 'experimental farmers' of the Enlightenment, but combining these with economic models that gave scant attention to subsequent developments in agronomy. As G.P.H. Chorley could write in 1981, 'When [historians] discuss [the problem of nutrient supply], they do so...in terms that most often suggest that Liebig, Gilbert, and their successors in the

---

[40] Young (1773), 9, 17, 59.
[41] Thaer (1844), 6–7, 132–3, 271–335.
[42] Thaer (1844), 220–4, 259.
[43] Thaer (1844), 336, Liebig (1862), 13–14, 137.

development of agricultural science had not existed.'[44] The only impacts of the environment in much of this writing are 'exogenous', that is, thunderbolt-like blasts of natural power from skies that farmers hoped would be forever blue. Agricultural history had become largely a narrative of choices made by the farmers, where it was the application of sound economy on the basis of the right kind of technical knowledge that determined whether they could turn a profit.

Justus von Liebig published the first edition of his *Chemistry* in 1840, responding in part to a request to write up an address to the British Association for the Advancement of Science in Liverpool in 1837. He dedicated the work to the great natural historian of the age, Alexander von Humboldt. In this and subsequent editions Liebig largely dismissed the experimental agriculture of the age, and its apparent argument that agricultural success simply lay in he hands of the farmer and his wisdom in the application of manure. He criticised especially the notion that observations gathered from just one farm could become a universal basis for agricultural practice. This dismissal of course mirrored the rhetoric of the experimental agronomists themselves, and much the same claim was made by the contemporary soil scientist and garden historian J.C. Loudon.[45] Opposed to the earlier 'organicist' perspective, Liebig argued that the fundamental roots of plant physiology were not some life-force pertaining to organic matter, but the complex interactions of inorganic trace elements contained in the soil and atmosphere.[46] Liebig could not explain how this actually occurred, as the bacterial action that decomposed matter and made elements assimilable by plants was not discovered until the 1880s. The basic insight, however, would prove to be essentially correct. Equally, the understanding of the chemical underpinnings of agricultural production also led Liebig to the conclusion that simple recycling of organic matter was insufficient to maintain soil fertility because each stage of processing involved irretrievable loss of elements, and thus led to long-term decline in yields.[47]

Liebig did not dismiss the recycling of manure, indeed he insisted on making it as efficient as possible. But he argued that the particular contribution this could make would vary greatly according to soil type; that inevitable wastage meant that additional supplies of elements had to be obtained from the atmosphere and processes of weathering of the land, and that in the face of a rising European population, alternative sources of nutrients had to be found if food

---

[44] Chorley (1981), 71; see also Ditt et al. (2001), 5; for a recent exception, see Allen (2004).
[45] Mårald (2002), 67; Liebig (1862), viii, 1–6, 136; Winiwarter (2002), 226.
[46] It should be noted that Liebig's use of 'organic' and 'inorganic' in chemistry has quite a different meaning to that employed in the discipline today.
[47] Liebig (1862), 11–15.

supplies could be kept in step with demand.[48] Liebig was deeply engaged with the social and economic implications of his studies, and developed a model of historical change that saw history's obsession with the role of war as mere froth in contrast with the essential role of the sod. Partly due to some wildly inaccurate population estimates, he saw the path of civilisations ever since the Greeks as being a story of slow, relentless overexploitation of the earth and consequent decadence and decline. 'A people arises and develops in proportion to the fertility of the land, and with its exhaustion they disappear.' Only China and Japan represented an exception of steady growth because of their extreme success in recycling, basically through cutting wastage with the removal of livestock husbandry and directly applying human faecal matter to the land.[49] Europe, and especially England, was only able to sustain current population expansion through a vampiric dependency on imports, especially of finite stocks of guano, but including the gruesome excavation of bones from battlefields and ossories.

What was worse, in conditions of rapid urbanisation this huge input of elements was not being recycled within the system, but due to the construction of the water-closet and sewerage systems, was then discharged uselessly into the sea. Liebig saw this as no less than a process of 'self-annihilation' (*Selbstvernichtung*) that would speed society to a sterile doom. The entire future of civilisation rested upon the resolution of this 'sewer question' (*Kloakenfrage*) and the retrieval of city dwellers' waste for the farmers' fields.[50]

The economic turn: von Thünen

Liebig provided an avenue towards an ecological history of agriculture, as well as the foundations for modern agronomy, with a deeply felt sensibility as to the interconnectedness of physical and social processes in circulating chains of causation. Agricultural historians would however take a largely divergent tack, one mapped out by another German near contemporary: Johann Heinrich von Thünen.

Von Thünen had long experience as an estate manager in eastern Germany, and it is not surprising that his classic work of the 1820s, 'The Isolated State' was above all an exercise in accounting and determining what kind of farming would turn a profit. Of course, he was still a cultivator, and argued that, 'The beast is to be seen as the indispensable machine by which hay and straw are turned into manure'; and he engaged in the same kind of crop and

---

[48] Liebig (1862), 27. The idea that different qualities of land demanded different types and amounts of manure was a prevalent view in earlier centuries.
[49] Liebig (1862), 110–1.
[50] Liebig (1862), 128–9, 153.

manure accounting as his predecessors.[51] He also absorbed, with some critical considerations, some of the insights of the new soil chemists into his work, although this still formalised biological relationships into equations with rather vague elements such as the 'Richness' and 'Quality' of the land.[52]

Von Thünen's major and lasting contribution was to analyse how agriculture in a market economy had to adapt to what the market could bear. Imagining a land that was an 'isolated state' with invariant environmental conditions and one urban major centre, von Thünen modelled what could be profitably produced at any given point in space. Von Thünen envisaged this landscape as a set of concentric circles displaying different land-uses laid outwards from the urban centre, each a function of price, transport costs and costs of production (wages, capital and rents). This model was highly influential for two reasons. It was intuitively pleasing and certainly did provide an accurate description of land-use zoning both to be found in the relatively undifferentiated environments of northern European cities. London imported cattle from Scotland as Amsterdam imported them from Denmark, but both got their fresh vegetables and milk from their immediate hinterlands. Secondly, von Thünen laid out a model where agricultural change could be modelled using long-term price series, which was about the one type of chronologically sustained, place-specific data relatively easily available to historians. Thus political economy in many parts of Europe would turn away from a direct interest in the processes of agriculture or the environment, a consequence at which Liebig could only express his bewilderment.[53]

Agricultural history has taken many twists and turns since. But the paradigmatic approaches remain those centred on two concepts. One is that of the price conjuncture, interpreted in Wilhelm Abel's classic 1960s history of western Europe as a function of long waves in population trends. For Abel, the limiting factor for agricultural productivity was the industry and wisdom of the cultivator. The fundamental transition away from a traditional economy exemplified in the writings of Young and Thaer came with the more favourable price conjunctures driven by the eighteenth- and nineteenth-century demographic transition, combined with the abolition of traditional agrarian institutions encouraging investment that allowed the West to 'overcome the crisis of supply'.[54] Abel's study of German agricultural history also paid homage to the classics in that he provided discrete

---

[51] Indeed, von Thaer also conceptualised the ultimate aim of the farmer as the maintenance of the highest possible sustainable profit. Von Thünen (1842), 243, 49–56, 83–7, 123.

[52] Von Thünen (1842), 56, 71–80.

[53] Liebig (1862), 135.

[54] Abel (1980); Abel (1962), 250, 342–5.

chapters, breaking the course of a general narrative, to provide accounting models of farm production at times employing data directly from von
Thünen.[55]

The second strand, to be found in the writings of the great Dutch agricultural historian B.H. Slicher van Bath, or historians of medieval England Postan
and Titow, retained the focus on manure. Slicher van Bath declared early in his
still classic *Agrarian History* that, 'In agriculture man is up against nature ... he
has to fight.' Soil and water supply were central determinants of what success
nature would reluctantly concede to the farmer. However, the environment
had no history: the 'E' for 'Environment' in Slicher van Bath's flow diagrams of
the agricultural system is a 'constant quantity'. Along with market forces, the
environment was an 'external factor' to the business of farming. In turn the
key to farming success was the manure that maintained soil fertility through
nutrient recycling. 'The farm had to be kept harmoniously balanced', above
all that proportion between pasture and arable that determined the number
of cattle.[56] If farms were almost by definition sustainable recycling machines,
and environmental change negligible, and technical change allowing productivity improvements a function of price incentives driven in turn by population
change, then no more work is really required to explain agricultural change. All
that remains is to fill in the details of how farmers would have responded to the
trends indicated in the price series.[57]

These classic works of the mid-twentieth century had not of course lost sight
of the role of the environment. Few agricultural historians would dissent from
the four factors influencing the intensity of cultivation listed by Christian Langenthal in the mid-nineteenth century: soil, climate, institutional burdens on
the cultivator, and transport links.[58] Yet even as the environment as 'agent' re-
emerged in the agricultural history of the later twentieth century, it remained
a force beyond the realm of action, outside of the box. The pre-eminence of
manure recycling so forcefully and pleasingly asserted by the Enlightenment
agronomists was retained.[59]

Thus the most ambitious and unsurpassed opus of the 1980s, Christian
Pfister's two-volume work on the climate and agricultural history of

---

[55] This also drew on the approach of Christian Langenthal's volumes published between
1847 and 1856, although Langenthal simply glossed the contents of important writers
on agricultural themes such as Johann Coler, and also some estate accounts. Langenthal
(1847–56); Abel (1962), 100–4, 250.

[56] Slicher van Bath (1963), 7–10, 14, 18–22; Postan (1972); Titow (1972); See Stone
(2005), 15.

[57] Slicher van Bath (1963), 14, 195–310; for a more recent version of these arguments, see
Kopsidis and Fertig (2004).

[58] Abel (1962), 167.

[59] Slicher van Bath (1963), 10–11.

Switzerland, harked back to Swiss agricultural writers of the 1760s to make a case that 'the level of yields essentially depended on the frequency and intensity of manuring'. According to Pfister, climatic deterioration combined with population pressure in the late sixteenth century led to reduced yields of fodder, and excessive expansion of tillage, causing livestock numbers and yields to plunge, reducing manure output, and in turn depressing human nutrition, raising mortality and lowering fertility. Pfister had put the environment back in, but the problem for farmers remained a 'manuring hole', an insufficient supply to maintain yields which could secure food supplies. Only 'a great revolution in agriculture', to cite the agronomist Jeremias Gotthelf as he looked back over the previous half century from 1850, through new crops and better recycling techniques, was able to close this hole.[60] Pfister's work was, however, representative of a new wave of studies that attempted to re-introduce ecological thinking to agrarian studies, frequently through the use of static models of how systems functioned at a given point in time.[61]

Escaping from the 'manuring hole': an environmental history of agriculture

By the 1990s, 'economic' and 'ecological' approaches had brought immense gains to our understanding of agricultural change. Earlier, through periods when 'sustainability' had not been a widely recognised issue, agricultural history persistently highlighted the brute fact that in the past, and for much of the contemporary rural populace, some kind of ethic of sustainability, or perhaps less pejoratively, of durability, is a basic requisite for any society or household. Both more 'economic' and more 'ecological' approaches, sometimes combined in the same works, usually saw the basic unit of cultivation as the self-sustaining farm. Of course this 'farm-system' was buffeted by the exogenous impact of the weather, disease and market forces that determined its long-term fortunes, but the farm itself persisted on the same basic, cyclical model, its internal variance dependent on the stimulus from outside. For the pre-industrial period both the 'economic' and the 'ecological' approaches of historians have envisaged an agriculture that was locked into a low-yield but sustainable system of management based on by-laws and the three-field system. The new husbandry and new techniques allowed the abolition of the traditional order, leaping to another higher-yielding one through a revolutionary package of innovations.

Yet this is not how writers such as Young and Thaer saw the situation. They judged their peers to be largely indifferent or ignorant of issues of sustainability,

---

[60] Pfister (1985), 27–8, 112–15.
[61] Pfister (1985), 18–19; Bayliss-Smith (1982); Shiel (2006); Winiwarter and Sonnlechner (2001); Winiwarter (2001); Tello et al. (2006); Bork et al. (1998).

and to be consistently tempted into error by the prospect of short-term profit. Indeed, they, along with Liebig, viewed traditional practice as persistently dissipating the riches of the soil: not as an optimisation of resources in a world of limited economic incentives, but rather a drawn-out slide into exhaustion. Of course, we should not assume that they were correct. Yet given the fact that even during the nineteenth century the basic processes that made the land fertile were only poorly understood, it remains unclear how the responses to price incentives that historians have placed at the heart of economic explanations of change could repeatedly have led to ecologically wise habits of husbandry; or why a supposedly ecologically wise husbandry could be sensitive to the prompts of price incentives. Can the economic and ecological approaches thus really be melded together?

Like most history, agricultural history has been a history of human actions, of people seeking to shape their own possible fortunes. What is too often missed by historians who desire to understand how humans have grappled with the natural world is that the documentary record left to us is also largely concerned with the consequences of human action. We should therefore pay very close attention to the audience that these records addressed. Pre-industrial regulators did not expect to act directly to strongly influence agricultural production, but they did think that their neighbours' actions could affect their fortunes. Hence by-laws and manorial records were preoccupied with neighbourly relations rather than ecological management.

Early writers on husbandry thought there was great scope for agricultural improvement, but they recognised that cultivators were as much acted upon as acting on. They also recognised that the world of action was intimately associated with knowledge of local conditions and saw it as part of a wider chain of consequence; there could be no clear-cut division between the agricultural world and other worlds (or environments). Their relations were too intimate, too pregnant with implication to be distinguished.

In contrast advocates of the 'New husbandry' argued that agricultural fortunes could be controlled by a process of closure and systemisation. The 'outside', the environment, appeared in two forms: the market (which previously had been tightly regulated) and 'nature', essentially seen as off-farm phenomena or pests, and not including the topsoil, the latter being largely a function of farm practice. Neither market nor nature was permitted to upset the functioning of the well-oiled farm machine. Of course, the *output* of the farm could be adapted to market conditions, but it did not fundamentally intervene in the agricultural process itself, which was governed by the decisions of the farmer. The 'environment' was placed outside of this world of endeavour, and its historical role could only be passive.

Of course, we must remember that the Enlightenment thinkers had resonance at the time and among their historians precisely because their methods brought about success, although late twentieth-century writers have called into question the centrality of manure in achieving yield improvements.[62] Yet what was perhaps at least as revolutionary as their results was the discourse that envisaged the farm as a closed entity in which the accomplished experimental farmer could manage that perfect circulation of materials; the farm to which the soil and air, but also the market, were environments. Indeed, the market has gradually taken on the mantle of not just an exterior force, but a law of nature itself. This discourse suggested that success was predicated upon a once-and-for-all change of system, a relatively rapid process that agricultural historians have spent much of their efforts ever since trying to locate precisely in time. Indeed, perhaps ironically some more ecological approaches come closer to another tradition in agrarian history, that focused on institutional limits to growth in the pre-industrial countryside that were swept away by enclosure (in England) and the dismantling of the feudal system of property rights in central Europe. In both cases, agricultural change is a 'step' change.[63] The Enlightenment shift in discourse did express, and indeed help bring about, a transformation of property rights that much of the agricultural land of Europe underwent in the eighteenth and especially nineteenth centuries, from collective to largely privatised systems of agriculture. Yet it is a much more doubtful proposition that the wise application of the lessons of 'practical agriculture' in the appropriate private property regime were responsible for a systematic shift from a low- to a high-yielding regime in European agriculture, rather than the incremental effects of numerous changes over a long period of time.

An environmental history that takes as its subject the very process of 'environing' in each society can seek to get to the heart of what governed action, to the heart of the world in which those acting dwelt. In defining those areas where one could successfully act (the neighbourhood of agriculturalists and their households, or the farm unit), and those areas where the consequences of actions were too complex to unravel or simply axiomatic or unfathomable (the environment), individuals and societies came to effectively determine what they themselves might be. 'Being' in society or as a person is a process of simplification, of placing much of what happens into the 'outside', to better control what seems controllable.[64] Thus pre-industrial agricultural regulators were largely concerned with isolating and controlling the caprice of fellow humanity, and in doing so regulation became an area for the contestation of power,

---

[62] Allen (2004); Chorley (1981), 80–2.

[63] For some recent discussions of these issues, that more generally (although not exclusively) seek to realign rural history with social and cultural approaches than environmental history, see Bruckmüller et al. (2004).

[64] This formulation owes much to Luhmann (1995); in praise of modelling as an analytical strategy in agricultural history, see Winiwarter (2002b), 104.

where lords and village institutions became key determinants of the agricultural order. This should be understood as a 'social system', or a set of rules intended to govern social action, rather than a system of ecological management.

By the mid-eighteenth century new models of a farm system were emerging out of early agronomy texts, and the experimental farmers would devise fully blown systems of control, where the practices of the farm simply had to be adjusted as was appropriate to 'environmental' inputs in the form of market prices and knowledge of the (largely invariant) soil and climate. In fact this model was quite misleading because of the continual importance of off-farm inputs such as seed.[65] By the twentieth century and with the development of artificial fertilisers this model might seem to have been justified in retrospect, though in fact the new dependency on science and the chemical industry bound farming into far greater systems of resource use over which the farmer had very little control.

The articulation in the sources of 'the environment' that goes hand-in-hand with that of delineating the agricultural world of action also helps us comprehend what these discourses were supposed to *do*, and understand that the rules of village communities and manorial courts, or the prescriptions of 'practical farmers' such as Young, did not constitute 'environmental regulation' as we think of it today, which represents an actual intervention in 'natural' processes to maintain ecological systems we think desirable. Rather, they sought to demarcate areas of control, and an 'environment' external to those areas; and then sought to govern human behaviour within those areas thought to be controllable. In tracing this process over time, environmental history can also discover that as a discipline it does not lie at the margin, in the environs of what really matters, but that it is inscribed in the active heart of what people thought they were capable of doing.

References

Abel, W. (1962), *Geschichte der deutschen Landwirtschaft vom frühen Mittelalter bis zum 19. Jahrhundert*, Stuttgart.
Abel, W. (1980), *Agricultural Fluctuations in Europe from the 13th to the 20th C*, London.
Achilles, W. (1991), *Landwirtschaft in der frühen Neuzeit*, Munich.
Allen, R.C. (2004), *The Nitrogen Hypothesis and the English Agricultural Revolution; a Biological Analysis*, Unpublished paper.
Ambrosoli, M. (1997), *The Wild and the Sown: Botany and Agriculture in Western Europe: 1350–1850*, Cambridge.
Bayliss-Smith, T. (1982), *Ecology of Agricultural Systems*, Cambridge.
Blickle, R. (1992), 'From subsistence to property: traces of a fundamental change in early modern Bavaria', *Central European History*, 25, 377–86.

---

[65] Ambrosoli (1997), 408–14.

Blith, W. (1652), *The English Improver Improved or the Survey of Husbandry Surveyed Discovering the Improueableness of All Lands*, London, 3rd ed.

Bork, H.-R., Bork, H., Dalchow, C., Faust, B., Piorr, H.-P. and Schatz, T. (1998), *Landschaftsentwicklung in Mitteleuropa. Wirkungen des Menschen auf Landschaften*, Gotha.

Bruckmüller, E., Langthaler, E. and Redl, J. (ed.) (2004), *Agrargeschichte schreiben*, Jahrbuch für Geschichte des ländlichen Raumes, Innsbruck.

Burchardt, J. (2007), 'Agricultural history, rural history, or countryside history?' *Historical Journal*, 50, 465–81.

Chorley, G.P.H. (1981), 'The agricultural revolution in northern Europe, 1750–1880: nitrogen, legumes, and crop productivity', *Economic History Review*, 34, 71–93.

Cronon, W. (1984), *Changes in the Land: Indians, Colonist and the Ecology of New England*, New York.

Cronon, W. (1991), *Nature's Metropolis. Chicago and the Great West*, Chicago.

Ditt, K., Gudermann, R. and Rüße, N. (eds) (2001), *Agrarmodernisierung und ökologische Folgen. Westfalen von 18 bis zum 20. Jahrhundert*, Paderborn.

Ellis, W. (1733), *Chiltern and Vale Farming Explained, According to the Latest Improvements. Necessary for all Landlords and Tenants of Either Ploughed-grass, or Wood-grounds*, London.

Evelyn, J. (1676), *A Philosophical Discourse on Earth Relating to the Culture and Improvement of Vegetation, and the Propagation of Plants, etc.*, London.

Grosser (1589 [1965]), *Anleitung zu der Landwirtschaft*, edited by Schröder-Lembke, G., Stuttgart.

Hardin, G. (1968), 'The tragedy of the commons', *Science*, 162, 1243–8.

Hartlib, S. (1659), *The Compleat Husband-man: or, a Discourse of the Whole Art of Husbandry; both Forraign and Domestick*, London.

Hastrup, K. (1990), *Nature and Policy in Iceland 1400–1800 : An Anthropological Analysis of History and Mentality*, Oxford, 1990.

Hopcroft, R.L. (1999), *Regions, Institutions, and Agrarian Change in European History*, Ann Arbor.

Ingold, T. (2000), *The Perception of the Environment. Essyas in Livelihood, Dwelling and Skill*, London.

Kopsidis, M. and Fertig, G. (2004), 'Agrarwachstum und bäuerliche Ökonomie 1648–1800. Neue Ansätze zwischen Entwicklungstheorie, historicher Anthropologie und Demographie', *Zeitschrift für Agrargeschichte und Agrarsoziologie*, 52, 11–22.

Langenthal, C.E. (1847–56) *Geschichte der teutschen Landwirthschaft*, Jena.

Liebig, J. von (1862), *Die Chemie in inhrer Anwendung auf Agricultur und Physiologie (7. Auflage). Teil I: Der chemische Proceß der Ernährung der Vegetabilien*, Braunschweig.

Luhmann, N. (1995), *Social Systems*, Stanford.

Mårald, E. (2002), 'Everything circulate: agricultural chemistry and recycling theories in the second half of the nineteenth century', *Environment and History*, 8, 65–84.

Moor, M. de, Shaw-Taylor, L. and Warde, P. (eds) (2002), *The Management of Common Land in North West Europe, c.1500–1850*, Turnhout.

Nourse, T. (1706), *Campania Fœlix. Or, a Discourse of the Benefits and Improvements of Husbandry: Containing Directions for All Manner of Tillage, Pasturage, and Plantation*, London, 2nd ed.

Pfister, C. (1984), *Bevölkerung, Klima und Agrarmodernisierung 1525–1860. Das Klima der Schweiz von 1525–1860 und seine Bedeutung in der Geschichte von Bevölkerung und Landwirtschaft*, Bern.

Postan, M.M. (1972), *The Medieval Economy and Society: An Economic History of Britain in the Middle Ages*, London.

Rheinheimer, M. (1999), *Die Dorfordnungen im Herzogtum Schleswig: Dorf und Obrigkeit in der frühen Neuzeit. Bd.1.*, Stuttgart.

Rohr, J.B. von (1726), *Compendieuse Haushaltungs Bibliothek*, Leipzig.

Shiel, R.S. (2006), 'Nutrient flows in pre-modern agriculture in Europe', in McNeill, J. and Winiwarter, W. (eds), *Soils and Societies. Perspectives From Environmental History*, Strond, 216–42.

Slicher van Bath, B.H. (1963), *The Agrarian History of Western Europe 500–1850*, London.

Smith, A. (1776[1904]), *An Inquiry into the Nature and Causes of the Wealth of Nations*, London.

Stone, D. (2005), *Decision-making in Medieval Agriculture*, Oxford.

Tello, E., Garrabou, R. and Cussó, X. (2006), 'Energy balance and land use: the making of an agrarian landscape from the vantage point of social metabolism (the Catalan Vallès County in 1860/1870)', in Agnoletti, M. (ed.), *The Conservation of Cultural Landscapes*, Wallingford.

Thaer, A.D. (1844), *The Principles of Agriculture*, trans by Shaw, W. and Johnson, C.W., London.

Thirsk, J. (1984), 'Agricultural innovations and their diffusion', in Thirsk, J. (ed.), *The Agrarian History of England and Wales*. Vol. V., Cambridge, 533–89.

Thomas, K. (1983), *Man and the Natural World: Changing Attitudes in England 1500–1800*, London.

Thünen, J.H., von (1842), *Der isolierte Staat in Beziehung auf Landwirtschaft und Nationalökonomie*, Rostock.

Titow, J.Z. (1972), *Winchester Yields: A Study in Medieval Agricultural Productivity*, Cambridge.

Troßbach, W. and Zimmermann, C. (2006), *Die Geschichte des Dorfes von den Anfängen im Frankenreich zur bundesdeutschen Gegenwart*, Stuttgart.

Warde, P. (2003), 'La Gestion de Terre Commune dans le Nord-ouest de l'Europe' in Demelas', M.-D. and Vivier, N. (eds), *La propriété collective 1750–1914*, Rennes.

Warde, P. (2006), 'Fears of wood shortage and the reality of the woodlands in Europe, c.1450–1850', *History Workshop Journal*, 62, 28–57.

Williams, R. (1973), *The Country and the City*, St. Albans.

Winiwarter, V. and Sonnlechner, C. (2001), *Der soziale Metabolismus der vorindustriellen Landwirtschaft in Europa*, Stuttgart.

Winiwarter, V. (2001), 'Landschaft, Natur und ländliche Gesellschaft im Umbruch. Eine umwelthistorische Perspektive zur Agrarmodernisierung', in Ditt, K., Gudermann, R. and Rüße, N. (eds), *Agrarmodernisierung und ökologische Folgen. Westfalen von 18 bis zum 20. Jahrhundert*, Paderborn, 733–67.

Winiwarter, V. (2002), 'Landwirtschftliches Wissen vom Boden. Zur Geschichte der Konzepte eines praktischen Umgangs mit der Erde', in Erdekunst- und Ausstellungshaue der Bundesrepublik Deutschland (ed.), *Wissenschaftliche Radkation*, Cologne, 221–32.

Winiwarter (2002b), 'Der umwelthistorische Beitrag zur Diskussion um nachhaltige Agrar-entwicklung', *Gaia*, 2, 104–12.

Worlidge, J. (1669) *Systema Agriculturae the Mystery of Husbandry Discovered*, London.

Worster, D. (1994), *Nature's Economy: A History of Ecological Ideas*, Cambridge, 2nd edition.

Young, A. (1770), *A Course of Experimental Agriculture*, London.

Young, A. (1773), *Rural Oeconomy, or, Essays on the Practical Parts of Husbandry*, London.

4
The Global Warming That Did Not Happen: Historicizing Glaciology and Climate Change

Sverker Sörlin

The iconography of climate change has for some time used glaciers as a main symbol and indicator. Glaciers grow or melt and they therefore represent a change in climate, although the connections are much more complex than the visual culture of the media sometimes suggests. In this chapter we will focus on glaciers during a period when they had not yet been established as indicators of climate change and environmental degradation. This is a history of field experiences in the 20th century, but also of ambitions to introduce glaciers and glaciology as an item on the agenda of science politics and polar diplomacy.

Popular history of the climate issue tends to reiterate old style history of science, presenting 'forerunners', 'early warners', and 'forgotten' but 'rediscovered' papers, all lined up along the path of enlightenment ending up with common knowledge.[1] However, on closer inspection the current view on these matters is comparatively recent as is also the role of climate as an historical factor, and as a key explanation of social change. Climate in this role has a fairly saddening prehistory, too, in the sense that climate causality tended to go along with environmental determinism. This was a widespread view in theories, and justifications, of imperialism, racism, and European superiority, such as they appeared for example in Benjamin Kidd's *The Control of the Tropics* (1898). They appeared in an extreme version in the speculative writings of Yale geographer-ecologist Ellsworth Huntington in the first quarter of the 20th century. Not only did Huntington provide the magisterial *Civilization and Climate* (1st edition 1915) but he also published a range of papers and books that focused more squarely on the actual changes of climate in different regions, notably Asia, and

---

[1] There is by now a small avalanche of books presenting the genealogy of climate change, some indeed very good. See, for example, Fleming (1998); Weart (2003), especially its constantly updated hypertext version on http://www.aip. org/history/climate/index.html; Flannery (2005), although only too rarely they dwell on the more complex science politics of the issue. See also the brief state of the art summary in Oreskes (2004), 1686.

how they had affected mankind.[2] For Huntington anthropology and human behaviour came first, and climate was more of an explanatory factor. However, his works were cited in the literature as part of the emerging science of historical climate change, and Huntington in turn was extremely eager to pick up on new, often speculative, ideas of climate change, notably the idea of sunspots, or 'solar cyclones', affecting terrestrial conditions, ideas that he propagated in his books *Climatic Changes* (1922) and *Earth and Sun* (1923).[3]

In the early decades of the 20th century there had been a growing interest in climate in general, not just glaciers, although they were an important part of the story. After the groundbreaking work by Agassiz and Charpentier in the 19th century, it had become an established fact that not only had the world seen one but several ice ages, four altogether it was believed at the time of the First World War, and the reasons for those were hotly debated, as were the reasons for postglacial climate change.[4] Among those who did work intensely on the issue were a group of geophysicists in Sweden, including the leading chemist Svante Arrhenius, who in the 1890s came up with the modern idea of a global warming based on CO_2 emissions, the so-called 'greenhouse effect'.[5]

How was the knowledge of historical climate change established in this period and how did the knowledge pass from observation and hypothesis into a more established and consistent scientific view? In fact, this process is not the same as the establishment of the idea of human climate forcing. While the latter has become gradually accepted worldwide in the last quarter of the 20th century the fact that historical, that is postglacial, climate change has occurred at all is much older and was considered common knowledge already between the two world wars and even earlier. It was also quite quickly received as an explanatory factor behind social change and demographical trends, whereas the reverse idea, that humans can affect climate, was harder to imagine and proved much harder to defend.

One quite remarkable observation is that climate was perceived as a factor behind historical and social change so prevalently even in a world where the nature, timescales and directions, not to speak of the causes, of a changing climate were yet so vaguely understood. A good example is the career of the hypothesis presented by Swedish oceanographer Otto Pettersson in 1914 that in all likelihood the waters around Greenland had been much warmer in the early Norse period in the 10th century. His own research on tidal variations as the cause of climate change had led him into the issue of variability of ocean temperature, but his reasons to believe in a mild Medieval Arctic were derived

---

[2] Huntington (1910); Huntington (1919). On Asia, see especially Huntington (1907).
[3] Huntington (1922) and (1923); Fleming (1998), ch. 8.
[4] Brooks (1926, 1949).
[5] Crawford (1996).

from the sagas and tales of Norse sailors who had travelled far up the west coast of Greenland.[6]

By the 1920s, when British meteorologist C.E.P. Brooks wrote his influential *Climate through the Ages*, the interest in postglacial climate was still at an early phase, but the number of theories of relevance was already large indeed. In the 1949 edition he sums them up towards the end of his book, a full dozen of possible explanations, ranging from elevation, landmass to astronomica, solar radiation and changes of atmospheric composition, suggested by a total of several dozen scientists. Carbon dioxide hypotheses were presented by John Tyndall in 1859, Arrhenius in 1896, Thomas Crowder Chamberlin in 1897, and then seminally by Guy Stewart Callendar in 1938, unknown to Brooks in 1926.[7]

We will return to the fate of Callendar. At this point, however, we should note the absence of environmental effects that were calculated from human action. Huntington and a handful of fellow climate determinists aside, the creativity among scientists in establishing changes in the climate by and large disregarded the opportunity to look at humans as agents of climate change. This may not be surprising in the early years, but when the CO_2 hypotheses were out there, and especially when Callendar's challenge became known, it is more surprising that human-induced climate change was not considered. I will argue that glaciology in the 20th century provided a case of non-acceptance of systematic human-induced climate change and also a case of science politics as it occurred before the political and popular breakthrough of environmentalism. Once environmentalism became more widely acknowledged, which was only in the 1960s and 1970s, the patterns of interpretation could much more easily include climate forcing by society. This is notwithstanding alternative explanations, in particular the enormous importance of computing and modelling power in the 1960s and onwards.[8] The important point is that field observations and melting glaciers are not enough: the interpreting narrative must also be in place. Before 1960 this was not the case.

Field and trust

Glaciers told stories, if not of deep time, so at least of a time that was deep enough to make it relevant for a whole range of scientific disciplines.[9] Apart from a discipline in its own right, glaciology soon became an auxiliary to other sciences, including history and archaeology, providing temperature records

---

[6] Pettersson (1914). See also Svansson (2006).
[7] Arrhenius (1896); Chamberlin (1897); Brooks (1926); Brooks (1949), 384–86. Fleming (1998), chapters 6 and 7.
[8] Bolin (2007), ch. 3; Bocking (2004), 112; Akera (2007).
[9] Rudwick (1992); Macdougall (2004).

and informed opinion on historical climate. In 1939, in the midst of a quick upsurge of interest in historical climate records, the term 'the Little Ice Age' was coined by glacial geologist Francois Matthes.[10] Glaciers were, however, not the only way that post-glacial change could be recorded. In Sweden quaternary geologists, pollen-analysts and geoscientists measuring time-layers of clay accompanied dendrochronologists in their quest to establish a better understanding of the past. This was of particular concern to the glacier-dotted northern nation of Sweden, which – as geologist Gerard De Geer said in a popular book (1890) – 'was born out of ice and water'. Post-glacial natural conditions were basic to any story of nationhood, and these stories were indeed reflected in the nation's identity politics of the period.[11]

Glaciology's concern with trust in the mid-war period can be seen in a wider historical and sociological context. In the 'history of truth' the geological sciences occupy a special position.[12] Just as had been the case with experimental science in the Renaissance both the place and the social position of knowledge workers were important. Geological field science enhanced its credibility, and was turned into a science of 'gentleman geologists', when amateur collecting was assimilated with aristocratic hobbies of horse riding and emerging out of doors.[13] In the course of the 19th century we can see the gradual differentiation of geophysical research, which evolved with distinctly different relationships to credibility. For a long time this differentiation was hardly visible; the dividing line was rather between the geophysical field sciences and the laboratory sciences.[14] The tension between the field and the laboratory is perhaps most clearly illustrated by the fact that the Nobel Prize in physics excluded geophysics altogether.[15] The science of glacier motion, for example, could rather be attached with lingering features of Romanticism and a 'heroic' and 'muscular' culture of fieldwork, as was the case with Tyndall, who was besides his successful career as scientist and public speaker also an ardent mountaineer, with significant experience of Alpine glaciers.[16] Clear demarcations between the different lines of geophysical field research become visible in Scandinavia in the early 20th century, for example in aurora research and in meteorological research, and in physical geography and glaciology.

The case of climate change served as an early and important test case, and a forceful driver, for the development of a modernized field science of glaciology and meteorology in the circumpolar north. Climate change was a knowledge

---

[10] Matthes (1939); Clague and Slaymaker (2000); Fagan (2000).
[11] Nordlund (2001); Sörlin (2002).
[12] Shapin (1994).
[13] Porter (1978).
[14] Kohler (2002).
[15] Friedman (2001).
[16] Hevly (1996).

claim on such a large scale, with such complexity, and with such far ranging social and economic ramifications, that it inevitably invited objections, at the same time as it was hard to prove. Such claims therefore forced the development of methods, precision practices, that could make the complex claims credible. In essence, there is an old, deep-seated relation between the growth of a culture of precision in Arctic geophysical field research and the long-standing controversies of climate change. When, in the middle of the 20th century, computer-based modelling of climate change appears as a reality, and revolutionizes meteorology, yet another demarcation line appears which rearranges the relative strengths of the claims that could be made by different scientific disciplines and research schools. It is also visible in the narratives, and the foundation histories, produced by these research communities.[17]

Hans W:son Ahlmann and Stockholm glaciology

Following on a rich tradition of Swedish earth science a radically new line of that work emerged in physical geography in the mid-war period under the leadership of Hans W:son Ahlmann.[18] Born in 1889 Ahlmann was a man of many talents and interests, among those both literary and artistic, and with a broadranging interest in geography stretching from field work in Libya to a major study of the modern Stockholm region. In the 1910s he started working on the mass balance of the Jotunheimen glaciers in southern Norway. He continued his work in the 1920s on the Kårsa glacier near the scientific station in Abisko in Swedish Lapland before assuming the Stockholm chair of geography in 1929.

An unforeseen event in 1930 was to have strong effects on his career. In that year he became a member of the committee which took charge of the newly discovered remnants of the ill-fated Andrée expedition, led by engineer Salomon August Andrée, who set out in a hot-air balloon to cross the Arctic Sea in 1897 but failed and had since then been missing. The return of the corpses to Stockholm in October 1930 was a solemn event that could be read as the kiss of death for an already long overdue era of patriotism in Swedish polar research.[19] Ahlmann co-edited the expedition report.[20] At the same time, and undoubtedly inspired by the work in the committee, he was looking for a northern extension of his glaciological work in Scandinavia which had started when the quest for a better understanding of post-glacial geography was fervent in Sweden and did not lack nationalist overtones.[21] Like many other Swedish geoscientists

---

[17] Sörlin (2009) explores the narrative theme further.
[18] Biographical details on Ahlmann in Hoppe (1990); and Sörlin (1998); in English, see a short article, 'Hans W:son Ahlmann', *Ice: News Bulletin of The British Glaciological Society*, vol. 9, January (1962).
[19] Sörlin (1999).
[20] *The Andrée diaries* (1931).
[21] Nordlund (2001).

Ahlmann had sometimes made patriotic claims of the calling of glaciers and northern science.[22] Polar science, he said on one occasion, had contributed to the best science ever made by Swedes, just like, he claimed, English polar science had 'developed the English kind of man, that the world has to give his honour'.[23]

Albeit respectful to the works of earlier generations of scientists, he would increasingly argue that there were other reasons than national identity politics that could serve as drivers of scientific energy. He articulated a new, consciously modernist, rhetoric of polar science, stressing international cooperation and stringent method. He also started gradually to relate his glaciology to the much wider phenomenon of climate change, although his conceptions of that change were still only very vague in the early to mid-1930s. He saw signs of change, but his impulse – and indeed the impulse of his entire research tradition – was to ask for more field data, and he was reluctant to frame his empirical observations as a general, let alone a man-made, climate change encompassing the entire planet.[24]

In the 1930s Ahlmann carefully collected his data during long field sessions, often lasting for weeks and months on distant Arctic glaciers. He went on two expeditions to Spitsbergen, 1931 and 1934, followed by an expedition to Iceland (Vátnajökull) in 1936 and to north-east Greenland in 1939.[25] The shifting rationale of glaciology in Ahlmann's work says a lot of how the entire issue of rapid post-glacial climate change became a scientific, although not yet a political, issue in this period.

Ahlmann was a dedicated data collector and he had a clear sense of the importance of long-time series produced under reliable conditions. On that note he sought to establish regular observation points. Towards the end of the Second World War he initiated work, with the help of a range of followers that had by now grown out of his popular research programme and at the Tarfala glacier complex in Swedish Lapland he and his closest colleagues founded a research station that has since then been in uninterrupted use by members of Stockholm University's Department of Physical Geography. The station became the ultimate material manifestation of Ahlmann's constant interest in the institutionalization of precision and data gathering in the field although realistically he had previously had to make do with temporary camps. The Tarfala station, with its surrounding *observation landscape* with instruments and

---

[22] Ahlmann (1921), 1–2.
[23] Ahlmann (1931).
[24] Sörlin (1996); Sörlin (2002), 115. It should also be recalled that in 1930, Norwegian–born Swedish metreologist Tor Bergeron published an important article that heralded a more dynamic view of postglacial climate, 'Richtlinien einer dynamischen Klimatologie', *Meteorologische Zeitschrift* 47(1930).
[25] Ahlmann and Thorarinnsson (1937–1943); Ahlmann (1936 [1938]); Ahlmann (1939).

glaciers became a *microgeography of authority*,[26] a site for training new recruits in Stockholm glaciology and a *sine qua non* for building and preserving the Ahlmann school.[27]

Climate embetterment

Even if Ahlmann did not considerably change his empirical work style during the 1920s and 1930s, his own contextualization of what he was doing changed radically. As he found overwhelming evidence of shrinking glaciers his thoughts were gradually led into the long-term patterns and also the causes of phenomena. In 1941 he published in *Ymer*, the journal of the Swedish geographical and anthropological society, for the first time an entirely new interpretation of the data he had assembled, stressing this time climate fluctuation.[28] On the basis of his observations of major retractions of the northern Greenland glaciers he suggested that the reasons might be 'meteorological factors'. He also reiterated his modernist position on polar field science: 'The scientist of today is no explorer as in the old days – today it is the exact science that comes to the fore – everything is geared towards maximum precision and completeness.'[29] There was no way that issues of this magnitude could be verified without careful research.

In a series of lectures delivered in the Geographical Auditorium at Oslo University in November 1940 Ahlmann claimed that both his own research, from his first studies in Norway, published in 1919, to the latest results from north east Greenland 1939–40, and glaciological work throughout the world, now universally demonstrated the same pattern: a gradual retreat of glaciers indicating a process of *polar* warming, that is, *not* global warming.[30] In the third and last of his lectures Ahlmann discussed possible causes of deglaciation, stressing particularly the role of strong winds, which could be the reason why glaciers retreated relatively more closer to the coasts.[31]

---

[26] For a brief contextualization, see my contribution to an IPY 2007–2008 research project on the history, culture and governance of field stations: http://museum.archanth.cam.ac.uk/fieldstation/members/sorlin/.

[27] Holmlund and Jansson (2002) indicates the importance of the station for the continuity of monitoring and research programs.

[28] Ahlmann (1941).

[29] 'Snebreene rundt Norske-havet går voldsomt tilbake', *Aftenposten* [Oslo] 25 November 1940.

[30] Ahlmann (1919), 1–2.

[31] The Norwegian press reported on the news: 'Breene i Norge går med underskudd', *Bergens Tidende* 29 November 1940; 'Norges og Öst-Grönlands breer går tilbake med femti procent', *Morgenbladet* 29 November 1940; 'Vindhastigheten har stor indflydelse på breenes tilbakegang', *Aftenposten* 30 November 1940.

At this time no single theory could explain what Ahlmann would call the 'embetterment of climate [*klimatförbättringen*] during the years 1920–1940, one of the most remarkable phenomena that has occurred in my field [of research] for a very long time'. The knowledge was vague, patchy, sometimes contradictory. Observational data seemed solid enough, but the time records were limited, only a few decades. Ahlmann was not even sure that the warming was an ongoing process; it could soon reverse, he said, which indeed it did later in the 1940s and 1950s. The milder climate in cold areas, such as the Arctic and the North Atlantic, was not accompanied by any similar climate change elsewhere, he maintained, for example in the tropics, but in a lecture in 1943 he assumed that polar melting occurred in Antarctica as well, which could explain the observed rise in sea levels since the 19th century.[32] Thus was born what we may term a counter-narrative to the Arrhenius' narrative of greenhouse effects, disregarding human climate forcing. There was climate change, but not global warming, and the change was a 'climate embetterment', a phrase often used by Ahlmann, not a sign of disaster. The change was part of natural fluctuations, although they were as yet not well known.

The science politics of climate

There was also a political dimension to Ahlmann's research programme. Ahlmann, ripe with results when the war started, had been forced to wait for the war to end in order to get along with his quest for certainty on the issue of polar warming. As soon as the powder had dried he started his diplomatic moves to unite Nordic efforts and instigate a wider international cooperation, including a visit to the Soviet Union. On his return to Stockholm he gave an interview in the Stockholm daily *Dagens Nyheter* where he stressed the generous and overwhelming reception – Stalin attended the final banquet himself – and also the qualities of Soviet science, especially in the fields of physics, chemistry and physical chemistry. Soviet scientists, claimed Ahlmann, had excellent resources and new facilities for geophysical research were to be installed near Leningrad. The best Soviet scientific institutions were on the same level as the American, he said. Botanist Eric Hultén added that, in Soviet Union scientific institutions, 'everybody looks happy'.[33] These enthusiastic statements by the

---

[32] Lecture in the Swedish Association for Anthropology and Geography 31 March 1943. A translation to English by K.A. Gleim, 'Ice and Sea in the Arctic' (mimeo), has found its way to the Vilhjalmur Stefansson collection at Dartmouth College, Stef MSS 242(1). See also ' "Vi måste känna både det som förenar och det som skiljer oss": Intervju om nordiskt samarbete och geografiska rön', *Hufvudstadsbladet* [Helsinki], 27 March 1943 [interview with Ahlmann].
[33] 'Våra vetenskapsmän imponerade i Ryssland: Sovjets plan hämtade alla gäster' [Our scientists impressed in Russia: Soviet airplane picked up all guests], *Dagens Nyheter* 3 July

Swedish scientists reflected the generally favourable view of the Soviet Union that was prevalent in Sweden after the war. But Ahlmann went further than most scientists would. In the following years he frequently spoke and published on the political, economic and scientific development in the Soviet Union, constantly in a hopeful and sympathetic way; he even polemicized with Arthur Koestler's *The Yogi and the Commissar* (1945).

After the visit to Leningrad and Moscow in June and July 1945 he went to Norway for vacations and business and chose the opportunity to tell the Norwegian press that he had hopes for Nordic cooperation in polar research. 'The most urgent', he told a Bergen newspaper, 'is continued research into what is called the current polar warming or the embetterment of climate in higher latitudes, both in the Arctic and in the Antarctic.' Such expeditions are costly, he said, and demand long-term planning. In fact, he now also mentioned the Soviet Union as a possible player in polar research, probably inspired by his recent meetings there, but he also stressed cooperation between several West European countries.

Already before the war ended Ahlmann had in both scientific circles and in public media voiced the need for East-West cooperation and underlined the role of Sweden as a bridge-builder between the western and the eastern blocs. In November 1945 Ahlmann was invited to England to lecture at the Royal Geographical Society in London, at Cambridge, Oxford, and at the newly formed British Glaciological Society. He used the occasion to inquire about possibilities of having the British join an expedition to the Antarctic. If one driving force for his diplomatic activities was scientific it was now complemented by his peace-making ideals and, significantly, his affection for Norway and Nordic cooperation; he had indeed been walking in the first line of anti-Nazi protest marches in Uppsala during the war years.[34]

It is, however, quite important to regard Ahlmann's internationalist agenda in the light of his changing scientific position. When he had started his northern extension of glaciology the motivation had been to continue the tradition from Otto Torell and Gerard De Geer who had both of them been convinced that the answers to the secrets of Scandinavia's climate, soil, and indeed its history, were to be found in the Ice Age past. That was the rationale of Arctic research. Norden's past lay in the ice, therefore was also her future a product of

---

1945. 'Galabankett och segerparad hos Stalin toppunkten vid professorsvisiten' [Gala banquet and victory parade with Stalin peak of professorial visit], *Stockholms-Tidningen* 3 July 1945.

[34] Transcript of taped recording with Professor Hans W:son Ahlmann, KVA Archives, MS Secretary's Archives 25a:1, p. 8.

the Ice Age. Formulations of this abound in Ahlmann's 1931 popular article on 'The Value and Legitimacy of Polar Research':

> [T]he investigation of these inland ices [Arctic and Antarctic glaciers] is a necessary task especially for science in those countries to which the Ice Age and its consequences have played a decisive role... [- - -] It has been increasingly clear, that certain parts of the polar regions, like Greenland, exert a considerable influence on those factors that determine climate, weather and winds in such areas as Scandinavia...[35]

If we go to the 1940s, and especially the post war years, there is very little of that tone left. International cooperation, of course, did not benefit from patriotic statements. But it was not just because of tactical deliberations that Ahlmann made his post-war moves. When he realized that one of the prime reasons to study glaciers was to assess and determine climate fluctuations the Scandinavian scope crept out of focus. Changes in climate, he now reasoned, might affect the entire planet. He now had to count on the possibility that he was dealing with global geophysical and environmental phenomena.

After the war he also started another major data collecting project. With the help of his Norwegian colleague and friend Harald Ulrik Sverdrup,[36] and with colleagues in the United Kingdom, he launched the idea of a Norwegian-British-Swedish Antarctic expedition, which he had conceived during the war years and which indeed took place from 1949 through 1952. In fact, Ahlmann had been in London, Oxford and Cambridge in November 1945 to solicit support for his plans.[37] By then he was already a respected name in the Royal Geographical Society, in the Scott Polar Research Institute, and in international glaciological and meteorological circles.[38] He did not join the expedition himself, however.

---

[35] Ahlmann (1931).

[36] Sverdrup had just returned to his native Norway to head the newly established Norwegian Polar Institute after serving for more than a decade as the director of Scripps Oceanographic Institute, La Jolla, California. He had also contributed to the Andree-volume in 1930.

[37] The Swedish press reported on the plans and on increasing British contacts: 'Svensk-brittiskt samarbete i ny Antarktisexpedition', *Svenska Dagbladet* 1 November 1945, 'Englandsintresse för Sverige', *Svenska Dagbladet* 24 November 1945, 'Celebra brittiska kulturgäster hit', *Stockholms-Tidningen* 24 November 1945. The following years saw continued contacts between Ahlmann and the Scott Polar Research Institute, where he was invited to lecture in May 1948. In correspondence he and the Institute's Director W.L.S. Fleming discussed, among other matters, plans for the Antarctic expedition. See, for example, Ahlmann to Fleming 4 September 1947 and 25 February 1948. University of Cambridge, Scott Polar Research Institute, Archives, File 92: Ahlmann.

[38] He was invited to write the foreword to the very first issue of *The Journal of Glaciology*, published in London in January 1947, a period when Ahlmann's British contacts were

Instead he was named ambassador to Oslo, the Norwegian capital, where he served from 1950 through 1956.

Human climate forcing?

Around 1950 Ahlmann's fame reached its peak. His contribution to the more and more widespread notion of an ongoing, yet enigmatic, process of warming, made him occasionally a figure in the mass media, both in Scandinavia, Europe, and the United States. The panorama of consequences of warming was by and large optimistic; the British, for example, could look forward to exposing their white legs a bit more in the summer. But there was also cause for concern, the entire British Isles might be threatened by the deluge to be released by melting polar ice caps.[39]

All along, Ahlmann was sceptical to the idea of human climate forcing. His main explanations lay within the realm of meteorology and geophysics. His favourite theory, which he defended all the way up to his Bowman lecture in New York 1952, *Glacier Variations and Climate Fluctuations*, was that warm tropical air had excessively in recent decades flowed towards the poles and caused an unusual increase of warm wind melting off glaciers, in accordance with, for example, his own results in the Icelandic Vatnajökull studies.[40] Despite his programmatic openness Ahlmann did not take much interest in theories that originated in other disciplines or fields of knowledge and could broaden his views.

There were clearly other theories around, including ones that gave a significant role to anthropogenic effects.[41] Guy Stewart Callendar, originally a British steam power engineer and an autodidact in meteorology, is the person most commonly associated with the renaissance of Arrhenius' theory of a greenhouse effect caused by combustion of fossil fuels. According to the current historiography of this particular environmental problem, however, Callendar's work was not taken seriously until much later. Indeed, he was never to exert any major

---

particularly active. See Ahlmann (1947). The journal intended to replace the *Zeitschrift für Gletscherkunde*, that had ceased publication during the war. In 1962 Ahlmann was named Honorary Member of the International Glaciological Society; *Ice* 1962 (Issue 9), pp. 8–9, and *Ice* 1972:1, p. 16. He was also President of the International Geographical Union from 1956 to 1960.

[39] This coverage of Ahlmann's work was by no means the only one. The Ahlmann collection in the KVA Archives (vols. 34–37 in particular) contains dozens of articles from newspapers around the world that uniformly give the picture of popularity because of spectacular results.

[40] The Isaiah Bowman Memorial lecture given to the American Geographical Society in New York in 1952 was published the following year as a booklet with a foreword by W.O. Field. Ahlmann (1953).

[41] Weart (2003); Goudie (1992), ch. 7.

role in the genealogy of the discovery, and the broader acknowledgement of his work is of a fairly recent date.[42] Human-forced climate change as a serious hypothesis was, instead, mainly an outcome of computerized models and Callendar's recognition became quite belated, and that of the quintessential 'disbelieved forerunner'.[43]

A counterpoint – the British meteorological debate

The British Meteorological community was well positioned to take a stand on the greenhouse effect and put Callendar's hypothesis in a perspective on his own home turf. The British response corroborates the general pattern. In the years preceding Callendar's seminal 1938 article there had been a slight but growing interest in post-glacial climate change, for example marked by a series of articles by C.E.P. Brooks in the *Quarterly Journal of Royal Meteorological Society*, and in his above-mentioned book *Climate Through the Ages*, first published in 1926, where he proposed his idea that the Arctic was in fact ice-free during the hypsithermal – from approximately 8,000 to 3,000 years ago – but did not give much credibility to ideas of contemporary global warming.

In those discussions there had been few references made to the greenhouse effect, probably because they just seemed like unrealistic fantasies. In a popularizing book in 1906 Svante Arrhenius wrote, 'we may hope to enjoy ages with more equable and better climates, especially as regards the colder regions of the Earth, ages when the Earth will bring forth much more abundant crops than at present, for the benefit of rapidly propagating mankind'.[44] Arrhenius's colleague and friend, meteorologist Nils Ekholm, who had worked in the Arctic and barely escaped being on Andrée's ill-fated balloon (he was part of an aborted attempt to fly in 1896 and declined an invitation to join the fatal 1897 journey), speculated already in 1901 over a future where humans would regulate the global climate – using burning of fossil fuels as a gas pedal and the planting of CO_2-eating plants as the brake...[45]

Ideas as fabulous as these did not win acclaim in the Royal Meteorological Society, nor did they impress on Ahlmann, on the contrary. The new results on historical climate change indeed opened for the opposite to speculation: they made it possible to establish a distance both to climatic determinism, at least the outrageous variety propounded by Ellsworth Huntington, which

---

[42] Fleming (2007). Callendar's most seminal paper on the issue was 'The Artificial Production of Carbon Dioxide and Its Influence on Climate', *Quarterly Journal of Royal Meteorological Society* 64 (1938), although he wrote numerous articles through to the early 1960's commenting on different aspects, several of them published in the same journal. See also Callendar (1949).

[43] Paterson (1992). Jones and Henderson-Sellers (1990), esp. 5–6.

[44] Arrhenius (1906).

[45] Ekholm (1901).

quickly grew out of fashion, and to the older tradition of belief in human forcing of micro-climates that had been strong, for example in debates on European forestry, lake management, moor ditching, and other forms of landscape change, that geographers had long been sensitive to but of which they had no real ability to judge the full climatic ramifications.[46] In issues of climate history and climate change, science was taking over and precision was crucial to make the argument trustworthy.

The Scandinavian data seemed to impress the British meteorologists, not least because of the rich empirical work and the sceptical non-speculative basic approach of Ahlmann and his co-workers in the field. In contrast stood – which was sometimes pointed out explicitly in the *Quarterly Journal* – 'German' science, always leaning to the intuitive and sweepingly Romantic. A particular friend of the Ahlmann tradition was Gordon Manley, who after the war assumed the position of President of the Royal Meteorological Society. He advised caution against primary causes, or grand theories such as the theory of solar orbits advanced by Milutin Milankovitch. He was also sceptical towards human agency in general. In a review article in 1944 he cited Ahlmann's 1940 article in [the Swedish *Geografiska Annaler*] *Annals of Geography* and claimed that this 'eminent Swedish authority distrusts all theories involving variations in solar radiation... In his [i.e. Ahlmann's] view discussion is impracticable until we have acquired a thorough knowledge of the connexion between climatological factors and the regime of the glaciers.' In the same article, Callendar was brushed away in a sentence, and his theory was not even presented.[47]

Critics soon found, however, that Manley and other cool sceptics did not give causal explanations, they just pushed the line of defence backwards. What, after all, caused changes in winds and circulation? It moved causes of global effects to the tropics – not the first time, in epidemiology, in climate determinism, or other fields – but was it more than just a way of telling about (possible) causal relations? Ahlmann, as we recall, had done just the same. It is perhaps not surprising that Manley expressed the virtues represented by his Swedish colleagues. Nor is it sensational that Fred Hoyle, the creative and controversial astronomer, in a 1949 discussion in the British Meteorological Society supported theories giving a major role to the sun and the 'change of the solar constant', an idea he had presented already in 1939. His approach was Olympic, and classically *Ignorabimus* – we shall never know, he claimed, what might have been the impact of the sun from its galactic passage through clouds of interstellar gas

---

[46] An early and classical example is George Perkins Marsh, *Man and Nature* (1864). Russian and European examples are given in Stehr and von Storch (2000). For Sweden, see Sörlin (1989). On Huntington and the link to ecology and human ecology, see Cittadino (1993), and Kingsland (2005).

[47] Manley (1944), 198.

is a matter of a 'million years', and thus unspeakable on the human timescale, but he admitted that we could detect change over periods of a thousand years.[48]

Hoyle spoke in the context of a major discussion on 'Post-Glacial Climatic Change' in the Meteorological Society held in 1949. Many different points of view came to the fore from a wide range of specialists, none of which addressed human climate forcing. Each of them represented an almost paradigmatic blueprint of their own different disciplinary research agendas. The biologist spoke on pollen analysis, the geophysicist spoke on various atmospheric particles and problems of measurement. There was, however, one peculiar exception to the rule, geographer W.G.V. Balchin, then at King's College, London, who had sent a written statement where he drew attention to possible man made changes in climate. He was modest: 'It would be interesting to estimate the relative amounts of man-made and natural pollution in an effort to assess the extent to which the human addition has contributed, if at all, to climate change.'[49] Still, around 1950, that was a minority view, whereas Ahlmannian geophysical explanations enjoyed widespread support. C.E.P. Brooks, the old guard authority on historical climate, despised greenhouse warming ideas altogether, claiming they were a thing of the past.[50] Swedish glaciology and British meteorology united on a counter-narrative that would soon grow into an embryo of climate scepticism.

How the details in the British debate played out after the early 1950s is beyond the scope of this chapter; it has been presented here as a counterpoint with strong links, both cognitive, institutional and personal, with the Scandinavian situation. In Sweden, however, it is evident that the Ahlmann tradition was maintained through a strong network of influential and well-positioned geoscientists and physical geographers, among those Ahlmann's own students and the team that carried out the British-Norwegian-Swedish Antarctic expedition, and made the Tarfala station their empirical platform of what turned out, more and more, to be an international bulwark of climate scepticism.[51]

---

[48] Hoyle (1949), 163.

[49] Hoyle (1949), 182. Balchin had already since the 1930s taken an interest in landscape change and possible man made effects on climate, see, for example, Lewis and Balchin (1940).

[50] Brooks (1951).

[51] Among the faculty were members who would come out later as notable climate change sceptics such as Professor Wibjörn Karlén. See, for example, Karlén (2001), 6. A generally sceptical attitude was prevalent in the department, although with time and as evidence of climate change grew, there was a softening of the opinion, particularly since the 1990's. Interview with Professor Per Holmlund, Department of Meteorology, Stockholm University, 1 September 2005. The same absence of interest in human factors is evident in the *Festschrift* for Ahlmann's 60th birthday in November 1949, Kirwan, Mannerfelt, Rossby and Schytt (1949), which involved several of his previous students and collaborators.

When the greenhouse effect came on the agenda again in the last quarter of the 20th century the Stockholm school of physical geography stood firmly with its long-held beliefs in long timescales and a predominance of geophysical explanations, and a basic 'we don't know yet' attitude. They had their glaciers, 'the only, or at least the most reliable, evidence of the history of climate', as Ahlmann had claimed in his 1947 special foreword to the first issue of *The Journal of Glaciology*. But, as he added, 'As yet we know very little' about their past and their variations: 'Before these questions are satisfactorily settled, the glaciers cannot be utilized as the climatographical registrars as they actually are.'[52]

The *longue durée* of ideas

I have argued that there are *longue durées* of disciplinary ideas, and I would like to suggest that these ideas are particularly strong when they are bound to field experiences, that is when the knowledge is of what we may call a bodily nature and has involved investment in physical work, installations and instrumentation of a fixed and long-term character. The field scientist has, somehow, physically appropriated the reality that he is at the same time claiming. The *field* is a strong and powerful carrier of the data and experiences and research. In that sense, the field contains boundary objects – sites, research stations, equipment, and other field monuments – that serve the function of uniting the group of scientists that share in the experience.[53]

In the case of Swedish glaciology and its provision of theories and explanations of polar warming the prime builder of these boundary objects was Ahlmann himself. He also constructed a modernist narrative of Swedish polar research in order to frame and legitimize his ambition. He made a specific point of not re-visiting the sites that had served as monuments of the older, heroic period of Arctic field science. Instead, he visited seemingly insignificant sites, of value, he claimed, for their scientific purpose only. The glacier alone would not have sufficed as a boundary object, it was the *instrumental glacier*, replete with apparatus and observing scientists, that could take on that status. His repeated praise of precision and masses of data reinforces the status of *measurement*. We should also bear in mind that Gordon Manley indeed used the expression 'instrumental period' to denote the period of climate history when there had been adequate instruments for measurement.[54]

The field 'monuments', and the monitoring instruments in particular, served also as symbolic markers of the professional. The amateur, who carries an

---

[52] Ahlmann (1947), 3.
[53] The classical study is Star and Griesemer (1989).
[54] Manley (1949), 163.

important role in the history of meteorology, would not be able to collect data in the same way as the scientist. Amateurs might be good friends of the discipline and the profession, but they must not be mixed up with real scientists. In fact, Ahlmann made at least one explicit, published comment on Callendar. Quoted by Gerald Seligman in 1944 he says that Callendar in a 1942 article has misrepresented his view. That is all. He did not discuss Callendar's hypothesis, but we could be quite certain that he shared his British colleagues' scepticism.[55] As much as this was founded in scepticism towards the amateur, it was also a praise of the production of knowledge in the field.[56]

In particular, caution was warranted when there was a risk for attic, basement or desk-top science. Gone were, since long, the days of the gentleman scientist. In that respect Callendar was the archetypical adversary of the geophysical scientist, who was working in teams, on projects, and following the kinds of detached aloof programs that Ahlmann did. Callendar was not a trained geoscientist – and he was working on calculations done by others, lacking deeper field experience of his own. He was not part of the discipline; in particular he was not part of its observing network and instrumentation. He was an outsider.

Callendar's theory of 1938 was quite simple and straightforward, regarding the atmosphere as a box which he divided into layers through which heat coming in had problems coming out, and where circulation downwards of cooler air in higher layers was turning more difficult.[57] The very simplicity of his idea made it less readily acceptable. It was already clear from contemporary geophysics and atmospheric chemistry, and perhaps most notably from meteorologists' studies of air currents, that such calculations could be imprecise at best, possibly useless. Why waste time and effort on what must be just speculation? So, interestingly, the way the discipline of glaciology acted vis à vis Callendar and the climate forcing hypothesis limited its possibilities to test it. Glaciologists were not impressed and, accordingly, continued their earlier lines of research.

---

[55] Seligman (1944), 22. The small note is a sign of reticence. We could note that Seligman was then since 1936 the founding president of the Association for the Study of Snow and Ice, which in 1946 changed its name to The British Glaciological Society. He was also on the first editorial board of *The Journal of Glaciology*, which invited Ahlmann to write the foreword of the first issue of the journal. Seligman and Ahlmann had entertained a collegiate and friendly correspondence since 1934. Indeed, Seligman had written Ahlmann to ask his opinion, and Seligman could be confident that his Swedish colleague sustained his view. KVA Archives, Ahlmann collection, vols. 11–19. Callendar was equally aware of Ahlmann's work, see for example Callendar (1942), where he refers to recent articles by both Ahlmann and Thorarinsson.
[56] Cf. Kohler (2002). Fleming (1998), p. 118, plays down Callendar's amateur status, although his examples of other scientists' use of his work date mainly from the 1950s and is among geophysicists rather than among glaciologists.
[57] Callendar (1938).

As we know, the climate forcing hypotheses ultimately became very successful. There were a number of factors to explain this – computers became ever more powerful; growing masses of historical data sustained it; it was supported politically; and, not least, it dovetailed with an environmental agenda in the 1960s and 1970s, which created a new general framework of interpretation.[58] But there were other impulses. An early one came from scientists who disliked deterministic interpretations of the world. Man-made climate change demonstrated that no physicist could predict the weather one year ahead, claimed Arthur Eddington, diametrically opposed to the deterministic Huntingtonian *Weltanschauung*, although still many of his colleagues would 'not yet' be convinced. They would probably claim that if we just knew everything about volcanoes erupting and local variations we would know everything. But that is not enough, says Eddington. As long as the human mind guides human action which can alter the chain of events on earth, physical determinism is impossible. Therefore,

> We must penetrate into the recesses of the human mind. A local strike, a great war, may directly change the conditions of the atmosphere; a lighted match idly thrown away may cause deforestation which will change the rainfall and climate. There can be no fully deterministic control of inorganic phenomena unless the determinism governs mind itself.[59]

A truly congenial argument for an Eddingtonian universe, perhaps. And, somehow, also part of the story why the role of the human mind, and thereby human society, could become such a powerful explanatory factor in the latter half of the 20th century. Humans had a choice, therefore they also had responsibility. Although it is by far not clear how Eddington's position would have fitted with then current ideas of climate change (in fact they did not; he was rather alone in his climate voluntarism), the modelling version of which stays squarely within a rationalistic, predictability-focused paradigm which he much distrusted.

General conclusion

Glaciological research, as performed by Ahlmann and his colleagues, suggested a theory of global warming but he did not provide one, he even opposed such a theory when he had occasion to sustain it. What can we learn from that? What we should further explore is how polar and global warming was socially

---

[58] I have argued for the environmental explanation elsewhere, for example in Sörlin (1996).
[59] Eddington (1929), 310–11.

constructed as an *environmental* problem, and not only treated as a geophysical and climatic phenomenon. It is evident that this process was fuelled by the general rise of environmentalism in the 1960s. Callendar was a forerunner, in a sense doomed not to be listened to, and Ahlmann's repertoire of possible explanations was primarily determined by his scientific training and his disciplinary background. The result was that his glaciological research, that we may now assume was indicative of the greenhouse effect, was not interpreted as such at the time but was used instead as a vehicle of mobilizing resources for international cooperation and large-scale polar research involving Sweden.

We should also penetrate further into the cognitive patterns of disciplines and how they relate to methodology. Field scientists were trained to put enormous weight on field data. The role of precision was crucial. Massive gathering of data was necessary to sustain any claim in such a complex project as making projections of climate change, historical and future. Data was strongly linked to explanation, in this case of glacier reduction and climate fluctuation. It is beyond the scope of this paper, but in the following decades glaciology and related geoscience disciplines developed an empirical friction to climate modelling, claiming experience from the field and down-to-earth time series as ice cool, and rock hard, evidence speaking against lofty mathematics and projections of earth systems modelling scientists. The tension between these scientific approaches is still with us, and we now also know that they carry political weight.

Not until much later – one very early case in point is the International Geophysical Year 1957/58 and its aftermath – was there any earnest attempt to coordinate research programmes with the greenhouse theory hypothesis and with atmospheric science and climate modelling. Thus, in this strand of glaciology, method and tradition limited the realm of possible empirical platforms in explaining the phenomena, which in essence meant that glaciology lost the initiative it had on climate change in the 1930s and 1940s to meteorology and other areas of geophysics in the late 1940s and 1950s.[60] Negating greenhouse effects already in the 1930s and 1940s, made glaciology path-dependent on its own counter-narrative, which later turned some of its practitioners into climate scepticism. Its many field observations of climate change became, in retrospect, signs of a global warming that never happened.

---

[60] Bruno Latour, among others, has pointed to the importance of being part of the practical work of observation as a defining feature of belonging to a discipline in the social sense; Latour (1987). Similar, but widened notions of disciplinarity are expressed by Timothy Lenoir (1997), 45–74, citing Michel Foucault and a range of STS-scholars. I acknowledge these important studies and theoretical forerunners, while at the same time noting that my argument puts more emphasis on the instrumentation and the research program as a demarcation line not only of discipline but of defining truth.

References

Ahlmann, H. (1919), 'Geomorphological studies in Norway', *Geografiska Annaler* [Annals of Geography], 1–2.

Ahlmann, H. (1921), 'The economical geography of Swedish Norrland', *Geografiska Annaler* [Annals of Geography], 1–2.

Ahlmann, H. (1931), 'Polarforskningens värde och berättigande', [The value and legitimacy of polar research], *Ord&Bild*.

Ahlmann, H. (1936), *Land of Ice and Fire: A Journey to the Great Iceland Glacier*, Engl. transl., London: Routledge & Kegan Paul, 1938.

Ahlmann, H. (1939), 'Färden till Nordostgrönland sommaren 1939', *Ymer*.

Ahlmann, H. (1941), 'Den nutida klimatfluktuationen: Det varmare vädret i Norge och på Svalbard', [The current climate fluctuation: The warmer weather in Norway and Svalbard], *Ymer*.

Ahlmann, H. (1947), 'Foreword', *The Journal of Glaciology* 1:1, 3–4.

Ahlmann, H. (1953), *Glacier Variations and Climate Fluctuations*, Bowman Memorial Series 3, New York: The American Geographical Society.

Ahlmann, H. and Thorarinnsson, S. (1937–43), 'Vatnajökull: Scientific results of the Swedish-Icelandic investigations 1936–37', 6 parts, Stockholm: Institute of Geography.

Akera, A. (2007), *Calculating a Natural World: Scientists, Engineers, and Computers during the Rise of U.S. Cold War Research*, Cambridge, MA and London: MIT Press.

The Andrée diaries being the diaries and records of S.A. Andrée, Nils Strindberg and Knut Fraenkel written during their balloon expedition to the North Pole in 1897 and discovered on White Island in 1930, together with a complete record of the expedition and discovery (1931), transl. Adams-Ray, E., London, 1931.

Arrhenius, S. (1896), 'On the Influence of Carbonic Acid in the Air upon the Temperature of the Ground', *Philosophical Magazine* (1896), 237–76.

Arrhenius, S. (1906), *Världarnas utveckling*, transl. *Worlds in the Making: The Evolution of the Universe*, New York: Harper & Brothers, 1908.

Bergeron, T. (1930), 'Richtlinien einer dynamischen Klimatologie', *Meteorologische Zeitschrift* 47(Berlin), 246–62.

Bocking, S. (2004), *Nature's Experts: Science, Politics, and the Environment*, New Brunswick, NJ and London: Rutgers University Press.

Bolin, B. (2007), *A History of the Science and Politics of Climate Change*, Cambridge: Cambridge University Press.

Brooks, C.E.P. (1926), *Climate through the Ages*, London: Ernest Benn.

Brooks, C.E.P. (1949), *Climate through the Ages*, 2nd revised ed. London: Ernest Benn.

Brooks, C.E.P. (1951), 'Geological and Historical Aspects of Climate Change', Malone, T.F. (ed.), *Compendium of Meteorology*, Boston, MA: American Meteorological Society.

Callendar, G.S. (1938), 'The Artificial Production of Carbon Dioxide and Its Influence on Climate', *Quarterly Journal of Royal Meteorological Society*, 64, 223–40.

Callendar, G.S. (1942), 'Air Temperature and the Growth of Glaciers', *Quarterly Journal of Royal Meteorological Society*, 68.

Callendar, G.S. (1949), 'Can CO_2 Influence Climate?', *Weather*, 4, 310–14.

Chamberlin, T.C. (1897), 'A Group of Hypotheses Bearing on Climatic Changes', *Journal of Geology* 5, 653–83.

Cittadino, E. (1993), 'The Failed Promise of Human Ecology', In Shortland, M. (ed.), *Science and Nature: Essays in the History of the Environmental Sciences*, Stanford-in-the-Vale: British Society for the History of Science.

Clague, J.J. and Slaymaker, O. (2000), 'Canadian Geomorphology 2000: Introduction', *Geomorphology*, 32, 203–11.

Crawford, E. (1996), *Arrhenius: From Ionic Theory to the Greenhouse Effect*, Canton, MA: Science History Publications.

Eddington, A. (1929), *The Nature of the Physical World*, Gifford Lectures 1927, Cambridge: Cambridge University Press.

Ekholm, N. (1901), 'On the Variations of the Climate of the Geological and Historical Past and Their Causes', *Quarterly Journal of Royal Meteorological Society*, 61.

Fagan, B. (2000), *The Little Ice Age: How Climate Made History 1300–1850*, New York: Basic Books.

Flannery, T. (2005), *The Weather Makers: The History and Future Impact of Climate Change*, Melbourne: Text.

Fleming, J.R. (1998), *Historical Perspectives on Climate Change*, Oxford and New York: Oxford University Press.

Fleming, J.R. (2007), *The Callendar Effect: The Life and Work of Guy Stewart Callendar (1898–1964)*, Boston: American Meteorological Society.

Friedman, R.M. (2001), *The Politics of Excellence: Behind the Nobel Prize in Science*, New York: Times Books.

Goudie, A. (1992), *Environmental Problems: Contemporary Problems in Geography*, 3rd ed., Oxford: Clarendon.

Hevly, B. (1996), 'The Heroic Science of Glacier Motion', *Osiris*, 2nd Series, Vol. 11, *Science in the Field* 66–86.

Holmlund, P. and Jansson, P. (2002), *Glaciological Research at Tarfala Research Station*, Stockholm: Stockholm University and Swedish Research Council.

Hoppe, G. (1990), 'Till 100-årsdagen av Hans W:son Ahlmanns födelse: En tillbakablick', *Ymer*.

Hoyle, F. (1949), 'External Sources of Climatic Variation', *Quarterly Journal of Royal Meteorological Society* 75, 161–3.

Huntington, E. (1907), *The Pulse of Asia: A Journal in Central Asia Illustrating the Geographic Basis of History*, Boston: Houghton Mifflin.

Huntington, E. (1910), 'The Burial of Olympia', *Geographical Journal* 36, 657–86. (London).

Huntington, E. (1915), *Civilization and Climate*, New Haven, CT: Yale University Press.

Huntington, E. (1919), *World Power and Evolution*, New Haven, CT: Yale University Press.

Huntington, E. (1923), *Earth and Sun: An Hypothesis of Weather and Sun Spots*, New Haven, CT: Yale University Press.

Huntington, E. and Visher, S.S. (1922), *Climatic Changes: Their Nature and Causes*, New Haven, CT: Yale University Press.

Jones, M.D.H. and Henderson-Sellers, A. (1990), 'History of the Greenhouse Effect', *Progress in Physical Geography* 14, 1–14.

Karlén, W. (2001), 'Global temperature forced by solar irradiation and greenhouse gases?' *Ambio* 30, 6.

Kingsland, S. (2005), *The Evolution of American Ecology, 1890–2000*, Baltimore, MD: The Johns Hopkins University Press.

Kirwan, I.P., Mannerfelt, C.M., Rossby, C.G. and Schytt, V. (eds) (1949), *Glaciers and Climate: Geophysical and Geomorphological Essays Dedicated to Hans W:son Ahlmann*, Stockholm: Generalstabens Litografiska Anstalt.

Kohler, R.E. (2002), *Landscapes and Labscapes: Exploring the Lab-Field Border in Biology*, Chicago, IL and London: The University of Chicago Press.

Latour, B. (1987), *Science in Action: How to Follow Scientists and Engineers through Society*, Cambridge, MA: Harvard University Press.

Lenoir, T. (1997), *Instituting Science: The Cultural Production of Scientific Disciplines*, Stanford, CA: Stanford University Press.

Lewis, W.V. and Balchin, W.G.V. (1940), 'Past Sea-levels at Dungeness', *The Geographical Journal* 96, 257–85.

Macdougall, D. (2004), *Frozen Earth: The Once and Future Story of Ice Ages*, Berkeley, CA: University of California Press.

Manley, G. (1944), 'Some Recent Contributions to the Study of Climatic Change', *Quarterly Journal of Royal Meteorological Society* 70, 197–219.

Manley, G. (1949), ['Comment',] in 'Post-glacial Climatic Change: Discussion', *Quarterly Journal of Royal Meteorological Society* 75, 163.

Marsh, G.P. (1864[2003]), *Man and Nature*, republished Seattle, WA and London: University of Washington Press.

Matthes, F.E. (1939), *Report of the Committee on Glaciers*, Transactions of the American Geophysical Union, 20.

Nordlund, C. (2001), ' "On Going Up in the World": Nation, Region and the Land Elevation Debate in Sweden', *Annals of Science*, 58, 1, 17–50.

Oreskes, N. (2004), 'Beyond the Ivory Tower: The Scientific Consensus on Climate Change', *Science*, 306: 5702, 1686.

Paterson, M. (1992), 'Global Warming', In Thomas, C. (ed.), *The Environment in International Relations*, London: Royal Institute of International Affairs, 155–98.

Pettersson, O. (1914), 'Climatic Variations in Historic and Prehistoric Time', *Svenska Hydrogrografisk-Biologiska Kommittén*, Stockholm: Royal Swedish Academy of Science, Skrifter 5.

Porter, R. (1978), 'Gentlemen and Geology: The Emergence of a Scientific Career, 1660–1920', *Historical Journal*, 21, 809–36.

Rudwick, M.J.S. (1992), *Scenes from Deep Time: Early Pictorial Representations of the Prehistoric World*, Chicago, IL and London: University of Chicago Press.

Seligman, G. (1944), 'Glacier Fluctuations', *Quarterly Journal of Royal Meteorological Society* 70, 22.

Shapin, S. (1994), *A Social History of Truth: Civility and Science in Seventeenth-Century England*, Chicago, IL and London: The University of Chicago Press.

Sörlin, S. (1989), *Land of the Future: Norrland and the North in Sweden and European Consciousness*, Center for Arctic Cultural Research, Umeå University, Miscellaneous Publications 8, Umeå.

Sörlin, S. (1996), 'Hans W:son Ahlmann, Arctic Research and Polar Warming: From a National to an International Scientific Agenda, 1929–1952', In *Mundus librorum: Essays on Books and the History of Learning*, Helsinki: Publications of the Helsinki University Library 62.

Sörlin, S. (1998), 'Den stora och den lilla världen: Kring en biografisk studie av Hans W:son Ahlmann, geograf och vetenskapsdiplomat', [The big world and the small: On a biographical study of Hans W:son Ahlmann, geographer and science diplomat], In Baudou, E. (ed.), *Forskarbiografin: Föredrag vid ett symposium i Stockholm 12–13 maj 1997*, Stockholm: Almqvist & Wiksell International, 129–42.

Sörlin, S. (1999), 'The Burial of an Era: The Home-coming of Andrée as a National Event', In Wråkberg , U. (ed.), *The Centennial of S.A. Andrée's North Pole Expedition*, Stockholm: Royal Swedish Academy of Sciences.

Sörlin, S. (2002), 'Rituals and Resources of Natural History: The North and the Arctic in Swedish Scientific Nationalism', In Bravo, M.T. and Sörlin, S. (eds), *Narrating the Arctic: A Cultural History of Nordic Scientific Practices*, Canton, MA: Science History Publications.

Sörlin, S. (2009), 'Narratives and Counter-Narratives of Climate Change: North Atlantic Glaciology and Meteorology, c.1930 to 1955', *Journal of Historical Geography* vol. 35.

Star, S.L. and Griesemer, J. (1989), 'Institutional Ecology, "Translations" and Coherence: Amateurs and Professionals in Berkeley's Museum of Vertebrate Zoology', *Social Studies of Science* 19, 1907–39.

Stehr, N. and Storch, H. von (eds). (2000), *Eduard Brückner – The Sources and Consequences of Climate Change and Climate Variability in Historical Times*, Dordrecht: Kluwer.

Svansson, A. (2006), *Otto Pettersson: Oceanografen, kemisten, uppfinnaren*, Göteborg: Tre böcker.

Weart, S. (2003), *The Discovery of Global Warming*, Cambridge, MA: Harvard University Press.

5
Genealogies of the Ecological Moment: Planning, Complexity and the Emergence of 'the Environment' as Politics in West Germany, 1949–1982

Holger Nehring

Environmental issues have gained a growing prominence in West European politics over the recent years: their scope is considered to be global, and our awareness of these issues transcends national boundaries, for example when the mass media discuss the global impact of climate change. Yet thinking along those lines is relatively novel. While traces of it can be detected in more remote periods of history, earlier political and media discussions rarely focused on 'the environment', but rather on nature conservation in specific issue areas. In political debates, the term 'environment' was used widely only from the mid- to late 1960s onwards and came centre stage during the 1970s and early 1980s. This short period of time might be called the ecological moment.

This chapter focuses on the Federal Republic of Germany and seeks to trace the meaning of the ecological moment for West German politics and, in particular, for the way in which the West German government sought to govern West German society. The key issue that contemporaries, and now historians, have to grapple with is complexity. This chapter seeks to show that we can only meaningfully write the history of the environment during and after the ecological moment if we come to terms with this complexity.

Historians, in particular those of more contemporary time periods, have found it difficult to integrate 'the environment' into their stories. While environmental history has seen a boom over the most recent years, most textbooks of post-1945 European history do not include systematic accounts of the 'environment' as a political issue. The key reason for this is that historians have not sought to grapple with societal complexity, and instead have continued to work with holistic versions of society, such as the one suggested by Max

Weber around a hundred years ago that assumes a triad of economy, society and culture to which 'environment' could be added on.[1]

Environmental issues usually remain confined to issues of policy making or social movement activity.[2] Two main explanations for the emergence of the environment as a political issue in both West Germany and other countries have dominated historical debates over the last few years. Proponents of the first of these explanatory models regard environmental politics as a post-fascist phenomenon. The seeds of ecological thinking were sown, so this argument goes, in the anti-modernist and naturalist policies of the fascist and National Socialist movements. The ecologists of the 1970s and 1980s often merely adapted these anti-modernist ideas to a different context. In politics, traces of such arguments can be found in the conservative labellings of environmental movements as 'eco-fascist'.[3]

The second main explanation, by contrast, regards environmentalism as genuinely democratic and as a democratising force. This model highlights the new forms of protest championed by environmental movements and emphasises the global (rather than narrowly nation-centred) concerns of those who are concerned for 'the environment'. Its main actors are those, mainly younger, members of the middle classes whose views are no longer materialist, but are orientated towards so-called 'post-material' values, such as compassion and human rights.[4] These interpretations are essentially scientific versions of the environmentalists' self-descriptions as enlightened warners of the near end of the world.[5]

Both these models are based on rather problematic assumptions about a dualism between man-made culture on the one hand and nature on the other, an optic that is itself a historical result of societal debates.[6] More importantly, they present rather simplistic versions of the highly complex and differentiated web of politics in modern societies.

This chapter offers a suggestion for ways in which we can integrate 'the environment' into contemporary history. Its argument relies on the sociological insights from a complex systems theory championed by Niklas Luhmann and combines these with insights from the history of governmentality in the Western world.[7] Luhmann regards communication as the fundamental

---

[1] On the limitations of this triad cf. Ziemann (2003).
[2] Most recently for West Germany: Rödder (2003); Wirsching (2006).
[3] Staudenmaier and Biehl (1996); for more subtle re-interpretations cf. Mauch (2004); Radkau and Uekötter (2003); Uekötter (2006); Brüggemeier et al. (2005); Oberkrome (2004); Sievert (2000).
[4] Cf. most recently Inglehart (1997).
[5] For a critique cf. Uekötter (2004).
[6] Cf. Cronon (1995); Latour (1999); and the overview by McNeill (2003). On the historicity of this claim cf. Daston and Vidal (2004).
[7] Luhmann (1989); Burchell, Gordon and Miller (1991).

operation within social systems. Without communications, environmental factors, that is factors that lie outside these systems, cannot be known.[8] Environmental problems are, therefore, without exception, the result of social definitions of problems. Access of social actors to nature is always socially mediated. This was particularly true for the period from the mid-1960s onwards, when the mass media transformed the nature of experiences by medialising them.[9]

Such an approach allows us to bridge the gaps between histories of environmental knowledge, the cultural history of landscape and the history of environmental thought. The first of these approaches focuses on discourses about the environment; environment is thus merely a linguistic category and is usually subordinated to other discourses, such as the one about the nation, or the role of science and technology.[10] The latter two approaches focus on the environmental hard facts themselves and examine how they have influenced the course of human history. Human history thus becomes a dependent variable of the quasi-automatic workings of natural forces.[11]

By uncovering the ways in which 'ecological communication' was linked to the changing nature of government and politics in the Federal Republic of Germany after the Second World War, this chapter seeks to contribute to the conceptual history of the present. This chapter begins by uncovering changing images of nature in West German society as an indicator for the growing consciousness of an ecological moment. The second section follows the genealogies of planning nature during the 1950s and 1960s; while the third section examines the crises of planning during the 1970s and early 1980s. Analysing these periods will enable us to catch a glimpse at one of the most important epochal shifts in nineteenth- and twentieth-century history, which terms such as 'reflexive modernity' and 'post-modernity' do more to obscure rather than reveal.[12]

Changing images of 'nature'

The term 'environment' was not an invention of the 1970s and 1980s. Its history goes back to the late nineteenth century, most notably to the German scientist Ernst Häckel. Even before the 1970s, certain authors, both in Germany and around the world, used the term to highlight the ecological dangers

---

[8] Cf. Luhmann (1989), ch. 6; for a related, yet not entirely convincing suggestion cf. Engels (2006a); and Brüggemeier and Engels (2005).

[9] Luhmann (2000).

[10] Cf. as a brilliant example of this genre: Zimmer (1998).

[11] Classic: Braudel (1991). For the USA cf. Worster (1979). For an attempt at mediation cf. Nehring (2006).

[12] Cf. Beck, Giddens, and Lash (1994).

connected with technological progress.[13] What was novel in the 1970s and 1980s was, however, that the term was much more widespread and conveyed through the mass media. Moreover, it was no longer tied to anti-modernist and anti-technological discourses. Perhaps most importantly, 'environment' had, over the course of the 1970s and early 1980s, come to lie at the heart of the West German government's definition of politics and society. Before, politics had focused primarily on specific issue areas, such as the protection of nature and smoke abatement.[14]

The proliferation of the term 'environment' in relation to natural factors could only take place in the wake of changes of the ways in which the relationship between human beings and nature was perceived. From 1945 onwards, nature turned from a subject with almost human attributes that sought to impede cultural progress and civilisation to an object of human efforts of planning.[15] We can see this development by examining mass media perceptions of 'natural catastrophes' in the Federal Republic. During the floods in Bavaria in 1954 and the spring flood in Hamburg in early 1962, nature appeared to destroy the protected space of civilisation. Many commentators, including governmental spokesmen, referred to the actions of 'higher beings' that sought to teach post-National Socialist and increasingly affluent Germans a lesson.[16] Many demanded the establishment of clear boundaries between man and nature, for example through the building of dykes.[17]

But such images already indicated a subtle shifts in attitudes, since they implied that nature could, in theory, be controlled and planned. This became especially obvious during the Hamburg flood in 1962, when the then mayor Helmut Schmidt likened his management of the flood with the planning and waging of war.[18] Like in other European countries during this time period, technologies and science played a growing role in West Germans' perceptions of nature, not only amongst expert discourses as in the nineteenth- and earlier twentieth century, but in everyday-life discussions amongst the general public.[19]

---

[13] Huxley (1994), 86; Manstein (1961). More generally cf. Worster (1994). For France cf. Neboit-Guilbot and Davy (1996) and Charvolin (2003).

[14] Cf. Uekötter (2003).

[15] This follows Engels (2003).

[16] *Bulletin des Presse- und Informationsamts der Bundesregierung*, 15 July 1954; *Frankfurter Allgemeine Zeitung*, 14 July 1954; *Hamburger Echo*, 19 Feb. 1962; *Hamburger Abendblatt*, 9 Feb. 1962.

[17] *Frankfurter Allgemeine Zeitung*, 27 Feb. 1962.

[18] Cf. the Hamburg mayor Helmut Schmidt, quoted in *Stenographische Berichte über die Sitzungen der Bürgerschaft zu Hamburg, 1962*, 4th meeting, 21 Feb. 1962, pp. 99 and 101.

[19] *Handelsblatt*, 20 Feb. 1962; *Hamburger Abendblatt*, 30 Mar. 1962; Karl Lehmann, 'Lassen sich Deichbrüche vermeiden?', *Die Bautechnik*, 40 (1963), pp. 160–1; 'Sonderausschuß

By the late 1970s, the shift towards an objectified nature had moved full circle. During the so-called 'snow catastrophe' in the winter 1978/9, images of a nature without specific content dominated mass-media discourses. Commentators pointed out that the amounts of snow and the wind speeds were not particularly above average. Instead, they blamed deficiencies of planning for the problems. The maintenance of infrastructure, especially electricity provision, came centre stage.[20] Nature was now no longer an alien opponent of human civilisation. It had become inextricably intertwined with human actions. The very ability to formulate the scale of an environmental issue through its geographical and territorial location is a social product.[21]

The focus, from the late 1960s onwards, on the global scale of the 'environmental' problem has not just been the result of a correspondence between actual problems and scholarly efforts that recognise them as such. In these perceptions, it was virtually impossible to disentangle the social and the natural: in the age of the mass media, experiences with nature were increasingly medialised experiences.[22] For West Germany from the 1960s onwards, David Blackbourn's question 'But what about *real* geographies?' no longer makes sense, as real geographies in a social-historical sense could, at least from the late 1960s onwards, only be those that were communicated about.[23]

It was only in this context that connections between human actions and the catastrophes under the heading of 'environment' could be made. For as long as nature had been connected with a moral (if not Christian) message of a lack of power of human beings, it was impossible that at least some of these catastrophes could have been caused by civilisational progress itself. The calls for a return to nature that many extra-parliamentary actors issued during this time period were therefore not merely directed against technological modernity.[24] As commentators and politicians emphasised the importance of man-made problems, they very much argued from the vantage point of technological modernity.[25] While the references to man-made problems were themselves not novel, they had never played such an important role in public debates, nor had they been based on this particular assumptions about technological and scientific progress.

Hochwasserkatatrophe', in *Stenographische Berichte der Bürgerschaft zu Hamburg*, 1962; for parallels in France cf. Pritchard (1997); Bess (1995); Brookes et al. (1976).
[20] Sethe (1980); *Frankfurter Allgemeine Zeitung*, 31 Jan. 1979; *Stuttgarter Nachrichten*, 5 Jan. 1979; *Die Zeit*, 23 Feb. 1979.
[21] Cf. the French classic by Lefebvre (MA, 1991), 28–31.
[22] Cf. for the USA White (1999).
[23] Blackbourn (1999), 16–17; Blackbourn (2006).
[24] For examples cf. Wenke and Zilleßen (1978).
[25] *Der Spiegel*, 29 May 1978; *Frankfurter Allgemeine Zeitung*, 11 Feb. 1980.

It was on the basis of this objectivation of nature that the term 'environment' gained widespread political importance in the Federal Republic of the late 1960s. In a speech that outlined the programme of his Social-Democratic/Liberal coalition government, Chancellor Willy Brandt still envisaged a programme of 'natural protection' and demanded to direct more attention towards the protection of animals and nature reserves. When he referred to 'environment', he meant the risks posed by technological progress, such as those risks posed by a 'technicised and automatised environment'.[26]

Influenced by mass-media observations of similar discussions in Western countries, most notably in the United States and in Scandinavia, 'environment' soon came to refer to the dangers posed by the adverse effects of economic and technological growth and came to be endowed with a central importance for society and social policies.[27] The dangers of the systemic environment for the survival of West German society was increasingly phrased in terms of an ecological crisis, beginning with alarmist messages about the lack of air for breathing and culminating in the 1972 report of the Club of Rome about the finite nature of economic growth and the risks posed by further growth and human development for societal survival.[28] Transnational events, such as the ones organised by European institutions over the course of 1970 under the heading of the 'European year of the environment', played a key role in the proliferation of this kind of ecological communication. They peaked with Earth Day in April 1970 which was, according to one commentator, 'the largest environmental demonstration in history'.[29]

Can nature be planned?

Ecological communication was, however, not only a function of discourses about the changing relationship between human beings and nature. It was inextricably linked to the question of government of the complex and differentiated West German society and thus to an area that West German governments regarded as central: planning for the future in order to provide citizens with social and material security. Ecological communication blossomed first within the realms of the West German state-level (*Länder*) and federal governments:

---

[26] 'Regierungserklärung vor dem Deutschen Bundestag am 28.10.1969', *Bulletin des Presse-und Informationsamts der Bundesregierung*, no. 132 (1969), p. 1.

[27] Brandt (1970); For a later point cf. Hans-Dietrich Genscher (1972).

[28] *Frankfurter Allgemeine Zeitung*, 26 Sep. 1970; *Hamburger Abendblatt*, 23 Feb. 1972; *Süddeutsche Zeitung*, 8/9 Aug. 1970; *Stuttgarter Zeitung*, 20 Mar. 1971; *Jahrbuch der öffentlichen Meinung 1968–1973* (1974), 433–4; Meadows et al. (1972), 139; *Die Zeit*, 17 Mar. 1972; *Frankfurter Rundschau*, 17 Mar. 1972.

[29] McCormick (1989), 47.

social actors continued to be concerned with issue-specific areas of nature protection and conversation well into the early 1970s.[30]

Planning had been a key category for West German governments since the mid-1960s. It aimed at the rational ordering of society and policies on the basis of relevant knowledge, as one contemporary defined it. The dream, first formulated in the late nineteenth century and fully developed by the late 1950s, was to create a technologically driven management of modern society that would establish some sort of social equilibrium and thus avoid the negative consequences of modernity. Planning was directed towards the future. It was not directed at a final goal, but was understood pragmatically,[31] especially as new system-theoretical understandings of statehood gained prominence amongst West German policy-makers,[32] and it was increasingly influenced by discussions within the growing field of future studies.[33] Planning thus came to be directly connected to the generation of political legitimacy. This was not merely a top-down process, as James C. Scott has argued. Rather, it involved constant communication and recalibration between government and citizens.[34]

Planning had existed in earlier period of history as well. But the novelty of developments in the 1960s was characterised by five factors: the desire and ability to increase the information and steering capabilities of politics; the extension of future planning from the medium into the long term; the increasing awareness of interdependencies between different issue areas; a desire to draw on scientific expertise in order to handle these developments; and, not least, very optimistic assessments about the opportunities offered by planning.[35] Over the course of the 1950s and 1960s, planning had thus become almost synonymous with social reform. Such social planning had allowed West German politicians to gain legitimacy and address a twofold challenge. State intervention had been widely discredited in the wake of the National Socialist dictatorship. And it continued to be problematic, as the other German state, the German Democratic Republic, was founded on the desire to build a planned economy. Crucially, however, policies of planning, thus defined, neglected the importance of natural structures for human development.

An awareness for these structures could only emerge in light of the growing objectivation of nature in West German public discourses (outlined in the

---

[30] Engels (2006b).
[31] Lübbe (1966); *Die Zeit*, 10 Dec. 1971, 48.
[32] Wagner (1990), 376.
[33] Cf. Schmidt-Gernig (2003).
[34] Scott (1999); on conceptualising political legitimacy dynamically cf. Conway and Romijn (2004).
[35] Hockerts (2003), 249.

section 'Changing images of "nature"') and under changing political, social and economic conditions. Ecological communication thus rephrased the question of political, social and economic planning at a time when the predictions of the planners of the 1960s had become obsolescent, a fact that was symbolised by the end of the economic boom that had characterised the roughly 30 years after the end of the Second World War, the concomitant slow-down of economic growth and the two oil crises of the early and late 1970s. It soon became obvious to the planners that they had fallen to a 'short dream of ever-lasting prosperity' that was no longer based on reality.[36]

The West German government reacted to these challenges not in response to societal pressures, but within the framework of planning: the environment became yet another area that was added to the already existing areas of planning, and environmental issues were first attached to those portfolios that had taken a leading role in planning efforts of the late 1960s, such as economic affairs, the interior ministry and the chancellor's office. The reversal from the subjectivation of nature that controlled human beings to one that regarded 'the environment' as something that could be managed and planned meant that West Germans now transferred their desires for progress and planning the future onto the environment. As technology appeared to give them an ever-increasing power over the material world, they gradually came to see themselves as caretakers of their country's territory.

This process occurred first on the state (*Länder*) level, especially in Bavaria and North-Rhine Westphalia.[37] But it soon found an expression in federal policies of the Social-Democratic/Liberal coalition government as well, first with the so-called 'Immediate Programme' (*Sofortprogramm*) of 1970, then with the more comprehensive 'Environmental Programme' (1971)[38] and with the first attempts at environmental legislation during 1971 and 1972.[39]

The scope of these environmental issues was holistic and transcended national boundaries, and the emergence of environmental planning within the West German government was itself the result of communication across borders. This communication primarily took the form of observation and was facilitated by discussions in international organisations, both in Europe and beyond. The example of the United States played an especially important role.[40]

---

[36] Lutz (1984) and Ruck (2000); Lindlar (1997); Kaelble (1992).
[37] Müller (1986), 45; On Bavaria cf. Bergmeier (2002). On North-Rhine Westphalia cf. the important study by Hünemörder (2004), esp. ch. III.
[38] *Deutscher Bundestag, Drucksache* VI, 2710, 6.
[39] Cf. Ditt (2005).
[40] Bungarten (1978); Huntley (1976); on perceptions in West Germany: Letter Wilhelm Hennis to Horst Ehmke, 15 Dec. 1969: *Bundesarchiv Koblenz* [henceforth quoted as BAK], B136/5308. For predecessors cf. Wöbse (2003); Ghebali (2002).

In the sense that these communications transcended the boundaries of the political system, they were themselves ecological, in that they highlighted the connectedness between worldwide debates and national political discourses.[41]

The UNESCO conference that gathered 230 delegates from 63 member states and 90 delegates from international organisations in Paris in 1968 played a key role in the emergence of ecological communications in West Germany. It broadened the terms of reference amongst West German experts and scientists beyond a focus on sectoral problems, such as nature conservation and smoke abatement, and highlighted the relationship between mankind and nature as a sub-problem of the relationship between mankind and its global environment.[42]

Discussions within the Western military alliance NATO were particularly important in the West German context. The institutionalisation of environmental issues within the West German government can be directly linked to the emergence of environmental issues as security issues within NATO discussions, a policy that was driven by Daniel P. Moynihan, an adviser of US President Richard Nixon and that culminated in the creation of a NATO 'Committee on the Challenges of Modern Society' in which the West German chancellor's office was involved.[43] West German politicians, most notably the Social Democrat Erhard Eppler, discussed similar ideas at the United Nations environmental conference that took place in Stockholm in 1972.[44] The conference further increased the proliferation of 'environment' in West German political debates.[45]

These developments and transnational observations – particularly of the creation of a federal environmental agency in the United States – boosted the ongoing attempts within the West German government to develop formalised interministerial links that could expand planning to 'environmental issues'.[46] The first stages of the institutionalisation of the environment, however, did not result in the creation of a special department within the chancellor's office,

---

[41] On the general background cf. Hünemörder (2003).

[42] UNESCO (1970), 4, 262–3, 271.

[43] Internal memos Federal Ministry of the Interior, 1 July 1969 and 10 July 1969; 'Minutes of the inter-ministerial meeting in the Foreign Office regarding environmental problems', 23 June 1969; Joseph Glazer, 'An Urbanologist in the White House', n.d.: BAK, B142/5009; Ausschuß zur Verbesserung der Umweltbedingungen A 431, 11/69 ff., in *Handbuch der NATO* (1969), p. 23; report by the Foreign Office to the other Ministries, 18 Sep. 1969: BAK, B142/5009; Memo by Ulrich Sahm, 1 Dec. 1969: BAK, B136/5308.

[44] Eppler (1971). On the West German role at the conference cf. Hünemörder (2004), ch. VIII.

[45] 'Wir fordern: Bürgerrecht auf Wahrheit', 7 May 1972: *Hauptstaatsarchiv Düsseldorf*, NW 267–12.

[46] Cf., for example the suggestions by the nuclear physicist Carl Friedrich von Weizsäcker to Horst Ehmke, 18 Nov. 1968: BAK, B136/5308; Horst Ehmke to Ertl, 17 Dec. 1969: BAK, B116/22950.

but it came to fall into the portfolio of the Ministry of the Interior under Hans-Dietrich Genscher of the liberal FDP (Free Democratic Party), who had gained the areas of water and air protection from the Ministry of Health and who is often credited with the introduction of the word '*Umweltschutz*' as a translation from the American 'environmental protection' into West German public discourse.[47] A federal governmental Department for the Environment (also charged with 'nuclear reactor security') was only created by the Christian-Democratic/Liberal coalition government in 1986 as a direct consequence of the meltdown of the nuclear reactor in Chernobyl in the Ukraine.

The Secretary of State responsible for environmental questions in the Federal Ministry of Interior, Peter Menke-Glückert, had studied history and political science in the United States in the 1950s and followed contemporary discussions there, especially in the fields of future research and environmental protection.[48] Accordingly, he promoted a systems-based ecological approach that went beyond nature protection, but intended to go beyond traditional boundaries between portfolios and integrate economic and social policies with scientific results and technologies of government.[49]

Yet in the early days of ecological communication, very much in accordance with the objectivation of nature, West German policy-makers and the public more generally continued to be concerned with the planning against environmental dangers by improving the available infra-structure. While the origins of the word 'infrastructure' lie in the railway engineering in nineteenth-century France, it entered post-Second World War West European politics through the use for NATO support operations from the early 1950s onwards.[50] Yet it soon became a general technological category that promised to link and connect different systems and that, over the course of the 1960s, became almost synonymous with state provisions in economy and society. In public perceptions, infrastructure became a key factor for economic growth and thus was the endowed with providing the conditions upon which planning had to be built.[51] The meaning of the term broadened more and more, so that, by the end of the 1960s, 'structure' was the key part of the word. Organic notions of society, still widespread especially amongst conservatives during the 1950s,[52] were now replaced by cybernetic ones that stressed connections, but still emphasised

---

[47] Derlien (1978), 77.
[48] Menke-Glückert (1969); Menke-Glückert (1973).
[49] Menke-Glückert, 17 Dec. 1970, BAK, B106/29403; Menke-Glückert (n.d.).
[50] Communiqué of the NATO Council, *European Review* (1951), 2. Cf. on this context Laak (1999), especially 296–8.
[51] Jochimsen (1996), 100. For the context of economic development cf. Rostow (1960); Bonesteel (1953); Kapferer (1955).
[52] Cf. Laak (2003).

structures. It came to mean a holistic understanding of the state's material and immaterial efforts.[53] The period found its intellectual expression in the growing popularity of structuralist theories of society, culture and language.[54]

Just as the seeds of modern conservationism were planted in the European encounter with the natural world in the colonies,[55] so the developments in the 1970s and 1980s still expressed the desires for planting the human garden: the mastery of nature, now conceptualised as global environment of human activities, has often been linked to the conquest of others and was now endowed with the task of ordering one society, when the existence of such an holistic approach towards the social had itself come under attack.[56]

Political legitimacy came increasingly under strain. Planning, as defined by the government, was primarily about efficiency, but not always about participation. Especially the growing awareness of transnational environmental problems put political legitimacy, based on territorial planning within a national context, under stress. The emergence of social movements concerned with environmental questions on the local and transnational levels and the demands for more participation were thus the direct corollary of the expansion of planning towards ecological issues. By the mid-1970s, there were too many financial and organisational demands on the state, while the resistance against big infrastructure projects, most importantly those against nuclear energy, thematised the importance of civic participation in decisions of infrastructural planning and in ecological communications more generally.[57] Yet such demands could only gain political salience when the infrastructure itself was seen as deficient or not working, most famously with the decree on car-free Sundays and the empty West German motorways in West Germany during the 1973 oil crisis.[58]

Crises of planning and ecological moment

The West German government responded by stressing the ecological nature of social problems and their organic connections. Notably, however, such analyses still took the form of thinking in terms of planning and organising space.[59] Yet still, the problems of planning that the government encountered and a rapidly growing scientific awareness of the interconnectedness of the world

---

[53] Genscher (1970).
[54] Schiwy (1969), 16 and 228.
[55] Grove (1995); Bassin (1999); Sunderland (2004); Ford (2004).
[56] Blackbourn (2006), 348; Glasheim (2006).
[57] Schäfers (1974).
[58] Cf. Hohensee (1996).
[59] 'Bundesraumordnungsbericht', *Bundestags-Drucksachen* VI/1340, 4 Nov. 1970, 39–40.

heralded a fundamental departure in the nature of politics and society in West Germany. While some assumptions of planning were maintained, the issue was now discussed on a different level. Three elements were novel. There existed a growing societal awareness of the complexity of the West German social system. Moreover, environmental issues now obtained a global scope, as illustrated by the ubiquitous metaphor of the 'spaceship earth'. Finally, the expectations of environmental problems were expanded even further into the future. As a consequence, everyday life in West German society became increasingly permeated by technological knowledge and vocabularies.[60]

While, with some exceptions, no politically influential fundamental critique of technology had existed in West German politics and society in the 1950s and 1960s, from the late 1960s onwards, risks of a novel scale and scope thus appeared on the horizon.[61] They threatened the destruction of everyone in case of accidents, and there the consequences of the disasters could not be reversed.[62]

This development could be seen especially clearly in the field of nuclear energy. During the 1950s and 1960s, nuclear energy and progress had been synonymous, even amongst those who opposed nuclear weapons.[63] While such an energetic view of the world had seen technological innovations as indicators of social progress, economic growth and progress were now no longer linked, and there were increasingly widespread fears about the irreversibility and unpredictable consequences of current policies.[64] A key event for the emergence of such interpretations were the protests against the building of a nuclear reactor in the south-West German town of Wyhl in 1975 that united both local campaigners and environmentalists from across the country. The protests against the building of nuclear waste sites at Brokdorf on the lower Elbe in 1976 and at Gorleben (near Hanover) that continue to the present day generated similar debates in the local communities and the national mass media.[65]

The observation of these protests led the government towards a gradual adaptation of governmental ideas of planning. Discussions about the 'environment' rather than 'nature' signalled a paradigm shift that regarded natural protection no longer as a cultural concern or as an idea linked to the management and planning of space. Rather, it connected such views with ideas that highlighted

---

[60] Beck (1981); with examples from this time period: Radkau (1987); for agriculture cf. Uekötter 2006. More generally: Raphael (1996).

[61] Traube (1978), 242 and 246.

[62] Beck (1992).

[63] Cf. Nehring (2004).

[64] Eppler (1975); Gaul (1974). For an overview of the contemporary debate cf. Stange (1993).

[65] Cf. Rucht (1980).

human beings' dependency in the biosphere and the perception of dangers for the survival of humankind.[66]

Such perceptions frequently led to analyses that stressed the acceleration of historical time that found its expression in discussions about 'environmental crisis' and alarmist warnings of the end of the world.[67] Notably, the images of such crises conveyed in the mass media focused on man-made environmental catastrophes that brought the dangers from far away into one's own society, most notably with the discussions about the leak of poison in the Italian town of Seveso (1976), the explosion of a tank full of pesticides in the Indian city of Bhopal in 1984 and the meltdown of a nuclear reactor in the Ukrainian city of Chernobyl in 1986. The triumphs of engineering appeared to be replaced by chains of catastrophes.[68] The fact that most of these catastrophes happened in countries that the West German public regarded as 'under-developed', yet rapidly industrialising, only reinforced the message of connections between 'environment' and socio-economic development and highlights the connections between ecological communication and the growing awareness for North–South imbalances during the late 1970s and 1980s.[69]

Attempts at planning and organising society, the economy and politics no longer seemed to be adequate. Instead, contemporary analyses came to be dominated by assessments of societal complexity and a 'dangerous modernity' (Ulrich Beck). Governmental policy makers glanced back melancholically at the planning euphoria of the 1960s.[70] There was a growing realisation that the planners of the 1960s had inscribed a human story of progress on a landscape and geography that could not support it.[71]

On the one hand, communicating about the environment expressed this realisation. On the other hand, however, it also revealed a desire to establish some kind of stability, a stability that only the hard facts of nature and natural processes appeared to be able to give. In short, it revealed the desire to control and to colonise the material world at a time when political and economic ideas appeared to run up against the complexities of modern society. It was a grasp for territory at a time when territoriality began to matter less and less as a decision space and identity space.[72] The politics of the environment during and after the ecological moment was thus a form of political power at a time when the

---

[66] Radkau (1989), 50.

[67] Cf. Hohensee and Uekötter (2004).

[68] Brüggemeier (1998).

[69] Cf., most notably, Brandt (2006).

[70] Mielke (1999); Metzler (2002).

[71] This argument follows Cronon (1992); who is inspired by Todorov (1984).

[72] On these ambiguities cf. federal programme of the German Green Party (1981), printed in: Wilharm (1989), 226–30. On this context cf. Maier (2000).

traditional form of political power encapsulated in the Enlightenment phrase of 'gouverner est prevoir' ('governing means foreseeing') no longer appeared to be grounded in reality.

  This paradox can also be gleaned through a close analysis of the term 'environment' that only emerged in political debates in West Germany from the late 1960s onwards and that contained an odd balance between local and global. 'The environment' is perhaps one of the vaguest historical terms. While its meaning in everyday language is fairly obvious, its historical uses might encompass everything that lies outside the political, social and economic systems, although today we have come to connect it with the protection of nature, wildlife, the biosphere as well as with the preservation of the quality of air and water, if not life in general. Its scope is therefore potentially global, in terms of both its geographical scale and the scope of issues addressed by it. Although global terms such 'environment' or the German *Umwelt* emerged in the 1880s as part of discussions about the nature modernity and 'the modern' that gave rise to thinking along ecological lines, the term 'environment' only gained *political* salience in most West European countries over the course of the 1970s and early 1980s. It became firmly linked to what contemporaries envisaged as a global society, as a world without borders. Yet what is striking if we look at the history of environmental politics is the term's connection to national, if not local, and territorial factors.

  The proliferation of the term 'environment' in West German politics from the early 1970s onwards meant that all political and social relationships now came to be communicated with regard to the unity of the political and social systems and their difference from their environments: these environments were not necessarily connoted as natural, they could also refer to everything that lay outside the immediate frame of reference, such as material conditions or economic growth. At the same time, there was a growing awareness of ecological connections, connections between the different areas. Political communication about the environment thus came to discuss the unity of difference between system and environment, as Niklas Luhmann has called it. The ancient Greeks had defined what we now call 'the environment' as an all-embracing body (*sôma periéchon*) (if not as a visible cosmos) and had imagined our world as a small body contained in and protected by a larger body. By the early 1970s social systems (such as politics, the economy and religions) now defined their boundaries themselves, most importantly by differentiating themselves from their environments.[73]

  In such complex systems, it is difficult to attribute causal relationships to specific interactions between environment and system. For example, the greenhouse effect can no longer be attributed to one clearly definable and particular

---

[73] Luhmann (1989), ch. II.

human action. Instead, in complex and differentiated societies such as the one of the Federal Republic in the 1970s and 1980s, the attribution of causalities mainly took the form of an observation of the ways in which systems observed themselves. A good example for this was the various parliamentary and governmental commissions or the more recent Intergovernmental Panel on Climate Change that sought to bring together the explanations of different scientists. Ecological communication thus added a whole new dimension of complexity to the already complex West German society.

This had important implications for political legitimacy. 'Trust' became an increasingly popular part of the political vocabulary of the hypercomplex society of the Federal Republic.[74] Both advocates and opponents of nuclear energy as well as politicians realised that ecological communications had pushed the relationship between governments and citizens had come off kilter.[75] While the Social-Democratic/Liberal coalition governments under Chancellor Willy Brandt and Helmut Schmidt and the Christian-Democratic/Liberal coalition government led by Helmut Kohl sought to make planning more efficient and thus more legitimate, its opponents in the environmental movements saw an all-powerful 'atomic state'.[76] Amongst many on the West German political left, technocracy had now become a code for the fixation on the future without regard for the consequences,[77] an assessment that spread to free-market liberals during 1980s.[78]

The protesters felt increasingly colonised by their governments and constrained by the natural environment the government planned for them, just like the natural environment used to be an important tool of social control for the colonial mother countries.[79] It is, therefore, no coincidence that their forms of protest and their advocacy of grass-roots democracy emerged from an engagement with anti-colonial protests and that many activists denied the systemic structuralism of the governmental planners by highlighting what came to be called 'post-structuralist' ideas.[80]

Unlike in the 1950s, scientists could now no longer generate trust in technology and economic development through their public appearance alone.[81] They now took on an increasingly political role by appearing as symbolic supporters

---

[74] Wagner (1994).
[75] Erhard Eppler in Wyhl, 26 June 1975: *Archiv der sozialen Demokratie*, Bonn, 1/EEAC 000153, pp. 24–6.
[76] Jungk (1977).
[77] Koch and Senghaas (1970).
[78] Habermann (1984).
[79] Sunseri (2003); Neumann (1998).
[80] For predecessors cf. Brantlinger (1996); and the links between post-colonial Indian politics, concepts of modernity and the nation cf. Zachariah (2004), 139–203 and Zachariah (2005).
[81] Weisker (2003).

of the one or the other side in the battle for legitimacy. But they realised themselves that they could only manage access to scientific information, rather than help create trust.[82] The different risk assessments by different experts, widely distributed through the mass media, led to a direct politicisation of science and a further scientification of politics.[83]

Conclusions

While geography and politics had become disentangled in West German public discourses after 1945, not least in response to the National Socialist policies of *Lebensraum* ('living space'), the ecological moment re-connected these two disciplines, yet now conceptualised space globally as 'the environment'.[84] As in France, therefore, the ecological moment in West Germany has not led to a reappraisal of the theme of bureaucratic planning.[85] Rather, the novelty of the ecological moment lay in the fact that planning continued, yet under different auspices. In the name of the return to nature and the recreation of the environment, the abstract power of governmental agencies has been increased. And new technological products are now marketed through their green credentials; this allowed social actors to maintain their beliefs in progress and thus bridge the horizon between experiences and the apocalyptical expectations. But it also means that any serious historical writing after the ecological moment cannot ignore the 'common assumption that human experience has been exempt from natural constraints'.[86]

'Environment' emerged as a political (rather than scientific) issue at a juncture in contemporary German and West European history that can be clearly defined: the end of the economic boom that had characterised the 1950s and 1960s and the beginning of concerns about the sustainability of economic growth, most prominently with the eventual breakdown of the Bretton Woods monetary system, the two oil crises which led to soaring oil prices and worries about energy security and the report of the Club of Rome on the finite reservoir of natural resources.[87] This historical juncture can be termed the 'ecological moment'[88]: communications within the political, social, economic and cultural systems became increasingly concerned about the factors outside these systems

---

[82] Müller (1977).

[83] Weingart (1983).

[84] On this insight cf. Ford (2001).

[85] Cf. Bess (1995); Bess (2003).

[86] Worster (1990), 1088. For recent evidence on the popularity of environmental history cf. Ford (2007).

[87] On this complex cf. Doering-Manteuffel (2007).

[88] On the character of this juncture as a 'turning point' cf. Engels (2008).

that threatened to hinder system operations. Not least, in the wake of the eco-logical moment, global developments became inextricable parts of local and national politics. The issue was no longer the preservation of 'nature' within the political system – it was now the challenges that the environment outside the system posed to politics that came centre stage. The challenge was ecologi-cal in that the connectedness of politics and the economy to natural processes across the world appeared to be revealed.

If we regard contemporary history as a pre-history of our current problem constellations, as suggested by the German historian Hans Günter Hockerts[89], then historians – and environmental historians in particular – can no longer write the history of societies as if these societies possessed a focus and a centre. Instead, they should place complexity and diversity at the centre of their anal-yses.[90] Ecological communication only added another and qualitatively new layer to this complexity, as it brought the unity of difference between the sys-tems and their environments and societal self-observations about this into play. Ecological communication thus became an essential part of social integration in the various sub-systems. In a sense, imagining the environment as the mate-rial and natural world – even if this takes the form of the biosphere – endowed social relations with some form of material and physical structure: it helped keep the illusion of a societal focal point intact. Environment thus served as the site at which the complexity of risk societies was negotiated.

This has had major repercussions for political legitimacy. But the reasons for the impact on political legitimacy did not lie, as James C. Scott argues, in the simplification of reality and the consequent failure by governments to understand it. Nor did it lie in the 'colonisation of lifeworlds' by governmen-tal bureaucracies as the German philosopher Jürgen Habermas has termed the phenomenon.[91] These explanations contain a false dichotomisation between power and powerless, and they still assume that modern complex and differ-entiated societies have central focal points. At the end, they merely replicate contemporary positions towards the problem.

The problem of political legitimacy generated by the ecological moment lay deeper: it lay in the bifurcation of different systems of progress and think-ing about time and its implications for politics. To be sure, the complexity of society meant that different historical times affected different areas of soci-ety, and the notion of environment was but a shadow of the seeming unity of society. Yet, as part of ecological communication, there now emerged a fundamentally novel thinking about progress and of the future. Since the emergence of the modern understanding of 'progress' in the late eighteenth

---

[89] Hockerts (1993), 124.
[90] Cf. Ziemann (2004).
[91] Scott (1999); Habermas (1981), 447–593.

century, the term referred to the difference between past and future: progress can only happen if people want it and plan it. Future offered the horizon for these planning activities. By the 1970s, however, experience of this world and the horizon of expectations for the future were even further divided than ever. The traditional understanding of progress could no longer bridge present and future – this gap could now only be managed.[92] The proliferation of a rhetoric and semantics of 'crisis' serves as evidence for this phenomenon: since it was in 'crises' that time appeared to be accelerated and the gap between present and future bridged.[93] In Sweden, by contrast, this management of the future did not take the form of crisis communication. Rather, there was an appeal to Swedish models of solidarity and equality, a Swedish 'model' that was supposed to maintain beliefs in planning when the structural conditions no longer held.[94]

This signalled a fundamental departure for modern politics whose significance cannot be over-estimated. Over the course of the 1970s and early 1980s, a new system of thinking about historical time and progress emerged, the system of ecology, that now stood next to 'system of progress' that had characterised previous historical periods since the French Revolution. This system had regarded the future as plannable. It implied that there would either be progression and further development or regression in terms of social and economic backwardness would follow. This was dependant on technological progress that meant intervening in nature. This solved some problems, but also created new dangers due its consequences. By the 1970s, however, the highly complex and differentiated society of the Federal Republic had begun to reflect on these dangers. This was, on the one hand, itself the consequence of planning and expert advice; on the other hand, it reflected the aporias of the current system of progress that assumed that nature could not only be used, but also be changed. Yet by the 1970s and 1980s, apocalyptic visions had become part of everyday life, going hand in hand with the realisation that governing nature might also mean destroying it.

This aspect is emphasised by the system of ecology. This system of ecology argues with a different pattern of time and looks beyond planning phases and election cycles. The key question of politics now became how to mediate between the two different prognoses offered by the two systems of thinking.[95]

---

[92] Cf. Koselleck (2003).

[93] Cf. Koselleck (1958).

[94] Andersson (2006).

[95] This argument follows Reinhart Koselleck, 'Allgemeine und Sonderinteressen der Bürger in der umweltpolitischen Auseinandersetzung', in idem, *Begriffsgeschichten* (Frankfurt/Main, 2006), 516–26.

In our current world, the system of ecology has gained almost universal acceptance. Yet even within that system, it is still not clear how to mediate politically between the enormously complex scientific prognoses about environmental damages that are now even more widely available in our medialised society. This remains the key legacy of the ecological moment and the key challenge of politics in the current world.

References

Andersson, J. (2006), 'Choosing Futures: Alva Myrdal and the Construction of Swedish Future Studies, 1967–1972', *International Review of Social History*, 51, 277–95.

Archiv der sozialen Demokratie, Bonn, 1/EEAC 000153, pp. 24–6.

Bassin, M. (1999), *Imperial Visions: Nationalist Imagination and Geographical Expansion in the Russian Far East, 1840–1865*, Cambridge.

Beck, U. (ed.) (1981), *Soziologie und Praxis*, Göttingen.

Beck, U. (1992), *Risk Society: Towards a New Modernity*, Cambridge.

Beck, U., Giddens, A. and Lash, S. (1994), *Reflexive Modernization. Politics, Tradition and Aesthetics in the Modern Social Order*, Cambridge.

Bergmeier, M. (2002), *Umweltgeschichte der Boomjahre 1949–1973: das Beispiel Bayern*, Münster.

Bess, M.D. (1995), 'Ecology and Artifice: Shifting Perceptions of Nature and High Technology in Postwar France', *Technology and Culture*, 36, 830–62.

Bess, M. (2003), *The Light-Green Society. Ecology and Technological Modernity in France, 1960–2000*, Chicago.

Blackbourn, D. (1999), 'A sense of place: New Directions in German History' (1998 Annual Lecture), German Historical Institute, London.

Blackbourn, D. (2006), *The Conquest of Nature: Water, Landscape, and the Making of Modern Germany*, London.

Bonesteel, P.H. (1953), 'NATO and the Underdeveloped Areas', *Annals of the American Academy of Political and Social Science*, no. 288, July, 67–73.

Brandt, W. (1970), 'Gesellschaftspolitische Bedeutung eines wirksamen Umweltschutzes', *Bulletin des Presse und Informationsamtes der Bundesregierung*, no. 167, 1 Dec.

Brandt, W. (2006), *Über Europa hinaus. Dritte Welt und Sozialistische Internationale*, ed. Rother, B. and Schmidt, W., Bonn.

Brantlinger, P. (1996), 'A Postindustrial Prelude to Postcolonialism: John Ruskin, Wiliam Morris, and Gandhism', *Critical Inquiry*, 22, no. 3, 466–85.

Braudel, F. (1991), *The Identity of France, vol. 1: History and Environment*, New York.

Brookes, S.K., Jordan, A.G., Kimber, R.H., and Richardson, J.J. (1976), 'The Growth of the Environment as a Political Issue in Britain', *British Journal of Political Science*, 6, no. 2, 245–55.

Brüggemeier, F.-J. (1998), *Tschernobyl, 26. April 1986, Die ökologische Herausforderung*, Munich.

Brüggemeier, F.-J. and Engels, J.-I. (eds) (2005), *Natur- und Umweltschutz in Deutschland nach 1945. Konzepte, Konflikte, Kompetenzen*, Frankfurt/Main.

Brüggemeier, F.J., Cioc, M., Zeller, T. (eds) (2005), *How Green Were the Nazis?* Athens, OH.

Bulletin des Presse- und Informationsamts der Bundesregierung, 15 July 1954; no. 132, 1969.

Bulletin des Presse- und Informationsamts der Bundesregierung, Burchell, G., Gordon, C. and Miller, P. (eds) (1991), *The Foucault Effect: Studies in Governmentality*, Chicago.

Bundesarchiv Koblenz, B136/5308; B142/5009; B116/22950; B106/29403.

'Bundesraumordungsbericht', *Bundestags-Drucksachen* VI/1340, 4 Nov. 1970.

Bungarten, H. (1978), *Unmweltpolitik in Westeuropa. EG, internationale Organisationen und nationale Umweltpolitiken*, Bonn.

Charvolin, F. (2003), *L'invention de l'environnement en France: Chroniques anthropologiques d'une institutionalisation*, Paris.

Conway, M. and Romijn, P. (2004), 'Introduction', *Contemporary European History*, 13, no. 4, 377–88.

Cronon, W. (1992), 'A Place for Stories: Nature, History, and Narrative', *Journal of American History*, 78, no. 4, 1347–76.

Cronon, W. (1995), 'The Trouble with Wilderness; or, Getting Back to the Wrong Nature', in Cronon, W. (ed.), *Uncommon Ground: Rethinking the Human Place in Nature*, New York, 69–90.

Daston, L. and Vidal, F. (eds) (2004), *The Moral Authority of Nature*, Chicago.

Derlien, H.-U. (1978), 'Ursachen und Erfolg von Strukturreformen im Bereich der Bundesregierung unter besonderer Berücksichtigung der wissenschaftlichen Beratung', In Böhret, C. (ed.), *Verwaltungsreformen und politische Wissenschaft*, Baden-Baden, 67–87.

Deutscher Bundestag, Drucksache VI, 2710.

Ditt, K. (2005), 'Die Anfänge der Umweltpolitik in der Bundesrepublik Deutschland während der 1960er und frühen 1970er Jahre', In Frese, M., Paulus, J. and Teppe, K. (eds), *Demokratisierung und gesellschaftlicher Aufbruch. Die sechziger Jahre als Wendezeit der Bundesrepublik*, 2nd edition, 305–47.

Doering-Manteuffel, A. (2007), 'Nach dem Boom. Brüche und Koninuitäten der Industriemoderne seit 1970', *Vierteljahrshefte für Zeitgeschichte*, 55, no. 4, 559–581.

Engels, J.-I. (2003),'Vom Subjekt zum Objekt. Naturbild und Naturkastrophen in der Geschichte der Bundesrepublik Deutschland', In Groh, D., Kempe, M. and Maulshagen, F. (eds), *Naturkatastrophen. Beiträge zu ihrer Deutung, Wahrnehmung und Darstellung in Text und Bild von der Antike bis ins 20. Jahrhundert*, Tübingen, 119–42.

Engels, J.-I. (2006a), *Naturpolitik in der Bundesrepublik. Ideenwelt und politische Verhaltensstile in Naturschutz und Umweltbewegung 1950–1980*, Paderborn.

Engels, J.-I. (2006b), 'Umweltgeschichte als Zeitgeschichte', *Aus Politik und Zeitgeschichte*, no. 13, 32–8.

Engels, J.-I. (2008), 'The ecological turn', in Radkau, J., and Uekötter, F. (eds), *The Turning Points of Environmental History*, Lanham.

Eppler, E. (1971), *Wenig Zeit für die Dritte Welt*, Stuttgart.

Eppler, E. (1975), *Ende oder Wende. Von der Machbarkeit des Notwendigen*, Stuttgart.

European Review (London), no. 12 (Oct. 1951).

Ford, C. (2001), 'Landscape and Environment in French Historical and Geographical Thought: New Directions', *French Historical Studies*, 24, no. 1, 125–34.

Ford, C. (2004), 'Nature, Culture and Conservation in France and Her Colonies 1840–1940', *Past and Present*, no. 183, 173–98.

Ford, C. (2007), 'Nature's Fortunes: New Directions in the Writing of European Environmental History', *Journal of Modern History*, 79, no.1, 112–33.

Frankfurter Allgemeine Zeitung, 14 July 1954, 27 Feb. 1962, 26 Sep. 1970, 31 Jan. 1979, 11 Feb. 1980.

Frankfurter Rundschau, 17 Mar. 1972.

Gaul, E. (1974), *Atomenergie oder Ein Weg aus der Krise?* Reinbek.

Genscher, H.D. (1970), 'Infrastruktur als öffentliche Aufgabe', lecture, 8 June 1970, Bonn.

Genscher, H.-D. (1972), 'Umweltschutz und Umweltpolitik als weltweite Aufgabe', *Bulletin des Presse- und Informationsamtes der Bundesregierung*, no. 86, 13 June.

Ghebali, V.Y. (2002), 'Before UNESCO and WHO', *Contemporary European History*, 11, no. 4, 659–63.

Glasheim, E. (2006), 'Ethnic Cleansing, Communism, and Environmental Devastation in Czechoslovakia's Borderlands, 1945–1989', *Journal of Modern History*, 78, 1, 65–92.

Grove, R.H. (1995), *Green Imperialism: Colonial Expansion, Tropical Island Edens and the Origins of Environmentalism, 1600–1860*, Cambridge.

Habermann, G. (1984), *Der Wohlfahrtsstaat*, Frankfurt/ Main: Propylaen.

Habermas, J. (1981), *Theorie des kommunikativen Handelns*, vol. 2, Frankfurt/Main.

Hamburger Abendblatt, 9 Feb. 1962, 30 Mar. 1962, 23 Feb. 1972.

Hamburger Echo, 19 Feb. 1962.

Handbuch der NATO, Frankfurt/Main, 1969.

Handelsblatt, 20 Feb. 1962.

Hauptstaatsarchiv Düsseldorf, NW 267–12.

Hockerts, H.G. (1993), 'Zeitgeschichte in Deutschland. Begriff, Methoden, Themenfelder', *Historisches Jahrbuch*, 113, 98–127.

Hockerts, H.G. (2003), 'Einführung', In Frese, M. Paulus, J. and Teppe, K. (eds), *Demokratisierung und gesellschaftlicher Aufbruch. Die sechziger Jahre als Wendezeit der Bundesrepublik*, Paderborn, 249–57.

Hohensee, J. (1996), *Der erste Ölpreisschock 1973/74. Die politischen und gesellschaftlichen Auswirkungen der arabischen Erdölpolitik auf die BRD und Westeuropa*, Suttgart.

Hohensee, J. and Uekötter, F. (eds) (2004), *Wird Kassandra heiser? Die Geschichte falscher Ökoalarme*, Stuttgart.

Hünemörder, K.F. (2003), 'Vom Expertennetzwerk zur Umweltpolitik. Frühe Umweltkonferenzen und die Ausweitung der öffentlichen Aufmerksamkeit für Umweltfragen in Europa (1959–1972)', *Archiv für Sozialgeschichte*, 43, 275–96.

Hünemörder, K.F. (2004), *Die Frühgeschichte der globalen Umweltkrise und die Formierung der deutschen Umweltpolitik (1950–1973)*, Stuttgart.

Huntley, J.R. (1976), *Das atlantische Bündnis und die Umweltkrise*, Brussels.

Huxley, A. (1994), 'Die zweifache Krise', In Huxley, A. (ed.), *Streifzüge. Ansichten der Natur und Reisebilder*, Koppenfels, W. von, Munich, 86–95.

Inglehart, R. (1997), *Modernization and Postmodernization: Cultural, Economic, and Political Change in 43 Societies*, Princeton.

Jahrbuch der öffentlichen Meinung 1968–1973 (1974), ed. Noelle, E. and Neumann, E.P., Allensbach.

Jochimsen, R. (1966), *Theorie der Infrastruktur. Grundlagen der marktwirtschaftlichen Entwicklung*, Berlin.

Jungk, R. (1977), *Der Atom-Staat*, Munich.

Kaelble, H. (ed.) (1992), *Der Boom 1948–1973. Gesellschaftliche und wirtschaftliche Folgen in der Bundesrepublikn Deutschland und in Europa*, Opladen.

Kapferer, C. (1955), 'Die unterentwickelten Gebiete: eine internationale Verpflichtung', *Wirtschaftsdienst*, 35, no. 1, 1–2.

Koch, C. and Senghaas, D. (eds) (1970), *Texte zur Technokratiediskussion*, Frankfurt/Main.

Koselleck, R. (1958), *Kritik und Krise - Eine Studie zur Pathogenese der bürgerlichen Welt*, Frankfurt/Main.

Koselleck, R. (2003), 'Sozialgeschichte und historische Zeiten' in his *Zeitschichten. Studien zur Historik*, Frankfurt/Main, 317–35.

Koselleck, R. (2006), 'Allgemeine und Sonderinteressen der Bürger in der umweltpolitischen Auseinandersetzung', In Koselleck, R. (ed.), *Begriffsgeschichten*, Frankfurt/Main, 516–26.

Laak, D. van (1999), 'Der Begriff "Infrastruktur" und was er vor seiner Erfindung besagte', *Archiv für Begriffsgeschichte*, 41, 280–99.

Laak, D. van (2003), 'From the Conservative Revolution to Technocratic Conservatism', In Müller, J.W. (ed.), *German Ideologies since 1945: Studies in the Political Thought and Culture of the Bonn Republic*, New York, 147–60.

Latour, B. (1999), *Politiques de la nature*, Paris.

Lefebvre, H. (1991), *The Production of Space*, Cambridge, MA.

Lehmann, K. (1963), 'Lassen sich Deichbrüche vermeiden?', *Die Bautechnik*, 40, 160–1.

Lindlar, L. (1997), *Das mißverstandene Wirtschaftswunder. Westdeutschland und die westeuropäische Nachkriegsprosperität*, Tübingen.

Lübbe, H. (1966), 'Herrschaft und Planung. Die veränderte Rolle der Zukunft in der Gegenwart', In Rombach, H. (ed.), *Die Frage nach dem Menschen*, Freiburg/Munich, 188–211.

Luhmann, N. (1989), *Ecological Communication*, Chicago.

Luhmann, N. (2000), *The Reality of the Mass Media*, Cambridge.

Lutz, B. (1984), *Der kurze Traum immerwährender Prosperität*, Frankfurt/Main.

Maier, C.S. (2000), 'Consigning the Twentieth Century to History: Alternative Narratives for the Modern Era', *American Historical Review*, 105, 807–31.

Manstein, B. (1961), *Im Würgegriff des Fortschritts*, Frankfurt/Main.

Mauch, C. (ed.) (2004), *Nature in German History*, New York.

McCormick, J. (1989), *The Global Environmental Movement. Reclaiming Paradise*, London.

McNeill, J.R. (2003), 'Observations on the Nature and Culture of Environmental History', *History and Theory*, 42, no. 4, 5–43.

Meadows, D.L., Meadows, D.H. and Zahn, E. (1972), *Die Grenzen des Wachstums. Bericht des Clubs of Rome zur Lage der Menschheit*, Stuttgart.

Menke-Glückert, P. (1969), *Friedensstrategien. Wissenschaftliche Techniken beeinfluessen die Politik*, Reinbek.

Menke-Glückert, P. (1973), 'Anoforderungen an die Wissenschaft', In Ernst, W. and Thoss, R. (eds), *Planung zum Schutz der Umwelt*, Münster, 12–28.

Menke-Glückert, P. (n.d.), 'Zur systemtheoretischen Analyse politischer Bürokratien', In *Systemtheorie*, Berlin, 141–9.

Metzler, G. (2002), 'Am Ende aller Krisen? Politisches Denken und Handeln in der Bundesrepublik der sechziger Jahre', *Historische Zeitschrift*, 275, 57–103.

Mielke, G. (1999), 'Sozialwissenschaftliche Beratung in den Staatskanzleien. Ein Werkstattbericht', *Forschungsjournal Neue Soziale Bewegungen*, 12, no. 3, 40–8.

Müller, E. (1986), *Innenwelt der Umweltpolitik: Sozial-liberale Umweltpolitik: Ohnmacht durch Organisation?* Opladen.

Müller, W.D. (1977), 'Kernenergie und Öffentlichkeit', *Atomwirtschaft*, 22, 19–20.

Neboit-Guilbot, R. and Davy, L. (1996), *Les Français dans leur Environnement*, Paris.

Nehring, H. (2004), 'Cold War, Apocalypse and Peaceful Atoms. Interpretations of Nuclear Energy in the British and West German Anti-Nuclear Weapons Movements, 1955–1964', *Historical Social Research*, 3, 150–70.

Nehring, H. (2006), 'Politics and the "Environment" in Twentieth-Century Germany', *Minerva*, 44, no. 3, 338–54.

Neumann, R.P. (1998), *Imposing Wilderness: Struggles over Livelihood and Nature Preservation in Africa*, Berkeley.

Oberkrome, W. (2004), *'Deutsche Heimat'. Nationale Konzeption und regionale Praxis von Naturschutz, Landschaftsgestaltung und Kulturpolitik in Westfalen-Lippe und Thüringen (1900–1960)*, Paderborn.

Pritchard, S.B. (1997), 'Reconstructing the Rhône: The Cultural Politics of Nature and Nation in Contemporary France, 1945–1997', *French Historical Studies*, 27, no. 4, 765–99.

Radkau, J. (1987), 'Die Kernkraft-Kontroverse im Spiegel der Literatur', In Hermann, A. and Schumacher, R. (eds), *Das Ende des Atomzeitalters?* Munich, 307–34.

Radkau, J. (1989), *Technik in Deutschland. Vom 18. Jahrhundert bis zur Gegenwart*, Frankfurt/Main.

Radkau, J. and Uekötter, F. (eds) (2003), *Naturschutz und Nationalsozialismus*, Frankfurt/Main.

Raphael, L. (1996), 'Die Verwissenschaftlichung des Sozialen als methodische und konzeptionelle Herausforderung für eine Sozialgeschichte des 20. Jahrhunderts', *Geschichte und Gesellschaft*, 22, 165–93.

Rödder, A. (2003), *Die Bundesrepublik Deutschland 1969–1990*, Munich.

Rostow, W.W. (1960), *The Stages of Economic Growth. A Non-Communist Manifesto*, Cambridge.

Rucht, D. (1980), *Von Wyhl nach Gorleben. Bürger gegen Atomprogramm und nukleare Entsorgung*, Munich.

Ruck, M. (2000), 'Ein kurzer Sommer der konkreten Utopie – Zur westdeutschen Planungsgeschichte der langen 60er Jahre', in Schildt, A., Siegfried, D. and Lammers, K.C. (eds), *Dynamische Zeiten. Die 60er Jahre in den beiden deutschen Gesellschaften*, Hamburg, 363–401.

Schäfers, B. (1974), 'Zur Genesis und zum Stellenwert von Partizipationsforderungen im Infrastrukturbereich', *Raumforschung und Raumordnung*, 32, no. 1, 1–6.

Schiwy, G. (1969), *Der französische Strukturalismus*, Reinbek.

Schmidt-Gernig, A. (2003), 'Forecasting the Future. Future Studies as International Networks of Social Analysis in the 1960s and 1970s in Western Europe and the United States', In Gienow-Hecht, J. and Schumacher, F. (eds), *Culture and International History*, New York and Oxford, 157–72.

Scott, J.C. (1999), *Seeing Like a State. How Certain Schemes to Improve the Human Condition have Failed*, New Haven.

Sethe, H. (1980), *Der große Schnee. Der Katastrophenwinter 1978/9 In Schleswig-Holstein*, Husum.

Sievert, J. (2000), *The Origins of Nature Conservation in Italy*, New York.

Der Spiegel, 29 May 1978.

Stange, S. (1993), 'Die Auseinandersetzung um die Atomenergie im Urteil der Zeitschrift "Der Spiegel"', In Hohensee, J. and Salewski, M. (eds), *Energie – Politik – Geschichte*, Stuttgart, 127–52.

Staudenmaier, P. and Biehl, J. (1996), *Ecofascism: Lessons from the German Experience*, Oakland, CA.

Stenographische Berichte über die Sitzungen der Bürgerschaft zu Hamburg, 1962, 4th meeting, 21 Feb. 1962, 99 and 101.

Stuttgarter Nachrichten, 5 Jan. 1979.

Stuttgarter Zeitung, 20 Mar. 1971.

Süddeutsche Zeitung, 8/9 Aug. 1970.

Sunderland, W. (2004), *Taming the Wild Field: Colonization and Empire in the Russian Steppe*, Ithaca, NY.

Sunseri, T. (2003), 'Reinterpreting a Colonial Rebellion: Forestry and Social Control in German East Africa, 1874–1915', *Environmental History*, 8, no. 3, 430–51.

Todorov, T. (1984), *The Conquest of America*, New York.

Traube, K. (1978), *Müssen wir umschalten? Von den politischen Grenzen der Technik*, Reinbek.

Uekötter, F. (2003), *Von der Rauchplage zur ökologischen Revolution. Eine Geschichte der Luftverschmutzung in Deutschland und den USA 1880–1970*, Essen.

Uekötter, F. (2004), 'Wie neu sind die Neuen Sozialen Bewegungen? Revisionistische Bemerkungen vor dem Hintergrund der umwelthistorischen Forschung', *Mitteilungsblatt des Instituts für soziale Bewegungen*, no. 31, 109–31.

Uekötter, F. (2006), 'Did They Know What They Were Doing? An Argument for a Knowledge-Based Approach to the Environmental History of German Agriculture', In Zelko, F. (ed.), *From* Heimat *to* Umwelt. *New Perspectives on German Environmental History*, Bulletin of the German Historical Institute, Washington, DC, suppl. 3, Washington, DC, 145–66.

Uekötter, F. (2006), *The Green and the Brown. A History of Conservation in Nazi Germany*, Cambridge.

UNESCO (ed.) (1970), *Use and Conservation of the Biosphere. Proceedings of the Intergovernmental Conference of Experts on the Scientific Basis for Rational use and Conservation of Resources of the Biosphere, Paris, 4–13 Sept. 1968*, Paris.

Wagner, G. (1994), 'Vertrauen in die Technik', *Zeitschrift für Soziologie*, 23, 145–57.

Wagner, P. (1990), *Sozialwissenschaften und Staat. Frankreich, Italien und Deutschland 1870–1980*, Frankfurt/Main.

Weingart, P. (1983), 'Verwissenschaftlichung der Gesellschaft – Politisierung der Wissenschaft', *Zeitschrift für Soziologie*, 12, 225–41.

Weisker, A. (2003), 'Expertenvertrauen gegen Zukunftsangst. Zur Risikowahrnehmung der Kernenergie', In Ute Frevert (ed.), *Vertrauen. Historische Annäherungen*, Göttingen, 394–421.

Wenke, K.E. and Zilleßen, H. (eds) (1978), *Neuer Lebensstil – verzichten oder verändern?*, Opladen.

White, R. (1999), 'The Nationalization of Nature', *Journal of American History*, 86, no. 3, 976–86.

Wilharm, I. (ed.) (1989), *Deutsche Geschichte 1962–1983. Dokumente in zwei Bänden*, vol. 1, Frankfurt/Main.

Wirsching, A. (2006), *Abschied vom Provisorium. Die Geschichte der Bundesrepublik Deutschland 1982–1989/90*, Munich.

Wöbse, A.-K. (2003), 'Der Schutz der Natur im Völkerbund – Anfänge einer Weltumweltpolitik', *Archiv für Sozialgeschichte*, 43, 177–90.

Worster, D. (1979), *Dust Bowl. The Southern Plains in the 1930s*, New York: Oxford University Press.

Worster, D. (1990), 'Transformation of the Earth: Toward an Agroecological Perspective in History', *Journal of American History*, 76, no. 4, 1087–106.

Worster, D. (1994), *Nature's Economy. A History of Ecological Ideas*, New York.

Zachariah, B. (2004), *Nehru*, London, 139–203.

Zachariah, B. (2005), *Developing India. An Intellectual and Social History, c. 1930–1950*, Delhi.

Die Zeit, 10 Dec. 1971, 17 Mar. 1972, 23 Feb. 1979.

Ziemann, B. (2003), 'Sozialgeschichte, Geschlechtergeschichte, Gesellschaftsgeschichte', In Richard van Dülmen (ed.), *Das Fischer Lexikon Geschichte*, Frankfurt/Main, 84–105.

Ziemann, B. (2004), 'Vom Bierstreik zur Komplexitätsreduktion. Die Gesellschaftsgeschichte der Bundesrepublik in den 60er Jahren', *Mittelweg 36*, 13, 45–52.

Zimmer, O. (1998), 'In Search of Natural Identity: Alpine Landscape and the Reconstruction of the Swiss Nation', *Comparative Studies in Society and History*, 40, 637–65.

Part II

History and the Environmental Sciences

6
The Environmental History of Mountain Regions

Robert A. Dodgshon

There is a case for arguing that mountain areas have been neglected in general studies of how the European countryside has evolved, at least when seen in relation to how much of the countryside actually comprises mountain areas. Fernand Braudel's point that 'Mountains come first' may have been heeded in his own study of the Mediterranean during the sixteenth century but it is not advice that has been freely taken up by others.[1] In fact, if we look at standard pan-European economic and social histories, there are comparatively few references to mountain areas and certainly no attempt to ask whether farming communities in such areas had a different experience compared to those in lowland areas so that we rarely find mountain landscapes separated out for treatment in their own terms. Slicher van Bath's pioneering study of the agricultural history of Europe (1963), for instance, had no explicit reference to mountain areas as farming areas. In trying to explain this perceived neglect, Braudel thought that the historian tended 'to linger over the plains, which is the setting for the leading actors of the day, and does not seem eager to approach the high mountains nearby'.[2] Arguably, it is not as simple as this.

To start with, we cannot attribute it to any basic lack of research. Even more so now than when Braudel penned his *Mediterranean in the Age of Philip*, there exist many fine studies exploring the history of particular mountain communities, localities or themes, in addition to the pioneering studies of the *alpwirtschaft*.[3] More persuasive as an explanation is the fact that standard social and economic histories of the countryside have tended to emphasise areas of change, innovation and growth, an emphasis that inevitably marginalises the story of most mountain areas no matter how much has been written about

---

[1] Braudel (1972), 25.
[2] Slicher van Bath (1963); Braudel (1972), 29.
[3] Netting (1973); Netting (1981); Loup (1965); Viazzo (1989); McNeill (1992); Carrier (1932); Frödin (1940–1).

them. Admittedly, some recent approaches offer the prospect of a more accommodating approach since they have placed the question of how society has interacted with particular habitats closer to the core of their analysis. By this, I do not mean that more is being done on long-established themes like the clearance of the wood, draining of the fen or reclamation of the heath. Rather is my point that some recent histories of the countryside have shifted the theme of how society has interacted with particular habitats from being a sub-theme in history of the countryside to being the very framework around which its wider history is organised, a change of emphasis that is well illustrated by recent studies in Britain.[4] Revealingly though, none of these studies actually have a separate category dealing with mountain or upland habitats. At best, mountain habitats are bolted on to discussions that are primarily about lowland habitats. Amongst these recent studies, only Smout's study of Scotland and northern England (2000) treats upland and mountain habitats in a substantive way, but that said, it is a study that seeks to be a study of environmental history *per se*.

I want to suggest that this understatement of mountain areas in general studies of how the countryside has developed may have a more fundamental cause, one that arises from a lack of any real discussion over how mountain communities might differ from their lowland counterparts. Too easily, such areas are treated in the same terms as lowland areas and, as a consequence, as having less to contribute to general histories of the countryside, whether such histories stress economic growth or habitat change. In response, some might claim that mountain communities have employed too diverse a range of adaptations, and have experienced too diverse a history, to be boxed neatly together so as to generate broad generalisations about their character. However, asking how they differed in detail from each other is not the same as asking whether there were aspects of their cultural ecology that distinguished them from that of their lowland counterparts. Taking this point further, I want to argue that if we are to produce an effective history of how mountain communities interacted with their environment, then it needs to be written more in its own terms rather than as an impoverished version of lowland histories. It goes without saying that all rural communities have to engage with their ecology or environmental setting but traditional mountain communities faced a particular mix of problems and a particular degree of challenge that deserves separate treatment. Indeed, such is their mutuality, it is difficult to write a history of such communities that is not also an environmental history. For this reason, establishing a clearer understanding of their particular problems and challenges not only provides a firmer basis for understanding how mountain

---

[4] Rackham (1986); Thirsk (2000); Williamson (2003).

communities survived, but also, an agenda for the environmental history of such areas.

Altogether, I want to stress four particular aspects. First, we need to understand the inner frontier that runs through the resource space of mountain and upland communities. Second, we need to elaborate on the basic conflicts that existed over their traditional systems of resource use and the different ways in which these conflicts could be resolved. Third, we need to clarify how most traditional mountain communities pursued a strategy of engaged risk in all aspects of their farming owing to the risk-laden nature of the environments in which they farmed. Fourth and finally, we need to identify and define how farming impacts produced a distinct cultural input to the habitat diversity of mountain habitats, one that is more subtle and more elusive when set beside the cultural impacts on lowland habitats.

Mountain communities: their inner frontier

Across Europe as a whole, farming has to contend with different types of frontier. In lowland areas, there is a northern frontier to crop growth that shifts back and forth on a longitudinal axis, as conditions flux year-on-year, or shifts more fundamentally with climate change. Of course, within the overall limits to plant growth set by the required amount of accumulated temperatures, different plants had different tolerances so that there are different frontiers for different crops and plants.[5] Again, in lowland areas, these individual frontiers are always linear in form though, in practice, they are best seen as represented by fuzzy rather than a sharply defined line simply because the point at which farmers abandon cropping is usually based on a social or economic calculation involving the frequency of crop failure or low returns. Yet however defined, what matters about these low-ground frontiers is that they affect only a comparatively small sample of farms out of the total that exist, or those that lie along them and are affected by their annual and long-term fluxes of position. They are, therefore, selective frontiers in terms of the limits to output or farm viability. Mountain areas add a vertical plane to these frontiers, but do so with this effect. It means that most mountain or upland farms had these frontiers to crop and plant growth running through their usable resource space, effectively adding an inner frontier to *all* such farms.[6] Of course, it partly does so because many such farms are arranged so as to run from valley bottoms up to the surrounding watershed, enabling each farm unit to share in the different types of land available, a configuration of resource allocation that – in some areas – can be extended back to the first settled colonisation of upland areas.

---

[5] For example, Mead (1953), 134–6.
[6] Parry (1981), 319–36.

When trying to understand how these internal frontiers affected mountains farms, we should also keep in mind the fact that the limitations on plant growth were not just about the limitations on cultivated crops. Grazings also had frontiers. Ecologically, mountain pastures embrace a wide range of conditions, owing to their variations of height, aspect and climate. If we draw the comparison with arable, grazings involve a much more diverse range of plants and conditions than the relatively simple agro-ecologies of arable even allowing for arable weeds. This diversity of plants and conditions inevitably means that the output of grazing was shaped by a much more varied frontier than was the case with arable. It is not simply a case of defining a sub-alpine limit to grazing, where no or little grazing was possible, but of defining the seasonal limits to pasture growth at different heights and in different situations (e.g. south-facing/north-facing slopes or *adret/ubac*). In an alpine setting, the better pastures, the so-called *mayen*, would yield grazings by May and would still be capable of being grazed late in September and October.[7] Pastures on the high alp meanwhile, have a much shorter growing season but can be extremely productive, or strategically valuable, during it. In some cases, particular pastures served as shelter. This was the case even in Atlantic areas of Europe, such as the Scottish Highlands, a point well made by court testimony in 1740 that detailed areas routinely used by herdsmen when they were caught out by snow during July and August whilst their stock grazed shieling grounds around Loch Ericht.[8] In effect, the mountain farmer had to cope with frontiers to growth that varied significantly between the different parts of their high ground. For Alpine farmers especially, these differences had to be tracked closely, with stock being moved onto the pastures a matter of weeks behind the retreat of snow and the return of pasture growth in late spring and early summer and then kept just ahead of the cessation of growth or advance of snow in late summer and autumn. In one sense, these were frontiers that farmers actively distanced themselves from, yet there is also a sense in which – because of the seasonal movements involved – they were inner frontiers that farmers routinely crossed back and forth. The range and variability of mountain climates, prime features of their character, meant that the engagement of farmers with them was fraught with miscalculation and risk. Yet this was something over which they had no choice if they were to maximise the output of their alp. If they wanted to take advantage of the output from their highest grazings, in particular, then they had to graze stock during the relatively short window of the year when it was possible and to engage with the risks involved.

---

[7] For example, Carrier (1932), 330–40; Weinberg (1975), 16, 21.
[8] National Archives of Scotland, hereafter NAS, Macpherson of Cluny Papers, GD80/384/12–26.

Mountain resource conflicts

Mountain areas can also be characterised through a fundamental conflict that lay at the heart of their agro-ecology. In lowland areas, the problem of land use for a traditional village was largely a problem of deployment: which land to use for which land use: arable, pasture, meadow or wood. This is because, except where factors like poor drainage impose an over-riding control, or soils were impoverished, most parts were potentially usable for each land use. The real challenge for such communities was maintaining the balance between the different land uses. This was especially critical as regards that between arable and pasture, since it determined the necessary flow of nutrients from the latter to the former. Mountain areas faced a different type of choice. Only a limited amount of their total resource set was physically or climatically suited to being homefield or infield, that is, for arable or productive meadow. Eighteenth-century data for the canton of Bern suggests that arable in southern Alpine areas occupied less than 20% of available land in some parts but fell to less than 10% in the interior or higher valleys.[9] For comparison, data for the Scottish Highlands and Islands shows that – at the peak of cultivation for the region ca.1800 – arable (infield plus outfield) formed only ca.9% all available land.[10] In these circumstances, expansion beyond such limits depended on how effective communities were in exploiting their non-homefield resources. They could expand their supplies of hay and winter feed by converting higher pastures into meadows, expand homefield arable by swapping high-ground meadow for homefield meadow, make more use of available grazings by pushing stock onto the higher or more remote pastures, harvest resources from non-homefield land (i.e. peat, turf, etc.) that could be used as manurial supplements and use it to gather extra sources of food, like mushrooms, nuts, berries and herbs.

I want to argue that these physical restraints meant that mountain communities, and especially alpine communities, had to cope with a distinct opportunity space (Figure 6.1). For such communities, the resource space available to them was rarely divided evenly between low and high ground, or graduated in a way that saw a simple decline in opportunities as one moved higher. In many cases, they had access to more usable land and more resource opportunity as they moved up off the lower valley ground. Thus, to give an example, the land available to communities in the Val d'Anniviers all lies above 527m. They have access to 11.2 sq. km below 1000m, that is, roughly the height at which their traditional homefields began. Between 1000m and 2200m, the range over which most of their homefields, meadows and alp-usable pasture were laid out, they had access to a further 83.21 sq. km, whilst above that,

---

[9] Pfister (1995), 158–60; Pfister and Egli (1998), 106–7; Viazzo (1989), 19.
[10] Dodgshon (1993), 679.

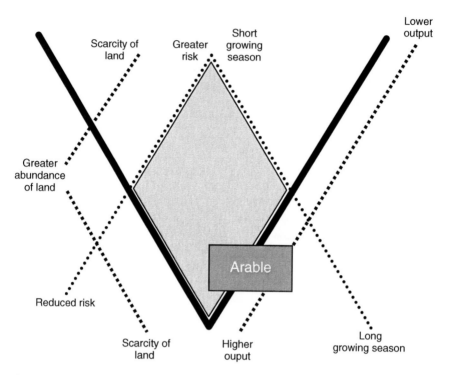

Figure 6.1 Factors shaping the opportunity space of mountain communities.

they had access to a further 149.77 sq. km, an amount that exceeded all that below.[11] Of course, whilst the supply of usable land may initially increase as one moves up out of the valley bottoms, ultimately, the growing seasons for cultivated crops and eventually pasture become shorter and the risks associated with their use greater, raising the frequency of crop failure and reducing the output of pastures. Compounding these problems is the greater effort involved in exploiting the higher ground, since it raises the threshold at which what is invested matches what is gained.[12] Bringing these various points together enables us to see how the opportunity space of many mountain resource systems has something of a diamond rather than simple 'V' shape to it (Figure 6.1), with expanding opportunities as one moves upslope out of the valley bottoms, but contracting opportunities and increasing risks as one reaches the higher ground.

Seeing their opportunity space in this way helps make two points. The first arises from the fact that the land most suited to arable or meadow is highly

[11] Based on data published in Loup (1965), 32–3; see also Viazzo (1989), 19.
[12] Cole (1972), 163.

prescribed. Up to a point, the needs of both sectors – arable and stock – could be accommodated without conflict. If expansion was sustained though, a point was reached at which the needs of the two sectors conflicted.[13] The sort of land needed to expand arable was also the very land on which you could develop your best meadows to provide the hay for wintering stock indoors. To expand arable was to sacrifice such meadows. Conversely, to expand meadow was to sacrifice arable. Communities had to find ways of displacing this conflict or to accept that the one was expanded at the expense of the other. In Alpine areas, the standard solution to the dilemma was to displace the supply of hay by establishing high-ground meadows, even though it required an increased level of effort on land that yielded less.[14] However, this was not the only strategy. We can see this by comparing sample data drawn from the Scottish Highlands and the Swiss Alps. At first sight, they appear as similar systems (Figure 6.2), with roughly the same hectarage of meadow per holding. Seen in terms of the proportion of land within each settlement that was under meadow though, real differences emerge, with a far higher percentage under meadow in Swiss alpine systems than in traditional Scottish Highland systems (Figure 6.3). Arguably, they are the differences that one might expect between a system that emphasised hay output compared with one that emphasised arable. These differences though, were not fixed. I have argued elsewhere that townships in the Scottish

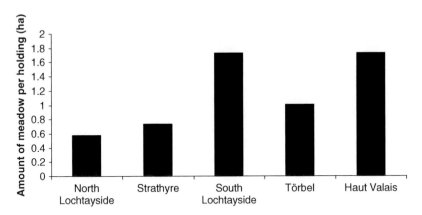

Figure 6.2   Average hectarage of meadow per holding in the Scottish Highlands and Swiss Alps.
Source: Based on data from McArthur, M.M. *Survey of Lochtayside 1769*, Scottish History Society, 3rd series, xxvii, Edinburgh, 1936; Wills, V. (ed.), *Statistics of the Annexed Estates, 1755–1756*, Scottish Record Office, Edinburgh, 1973, 1–3; Netting, 1981, 16–17; Loup, 1965, Appendix II.

---

[13] Cole (1972), 164–5; Cole and Wolf (1974), 123–4.
[14] Cole (1972), 163.

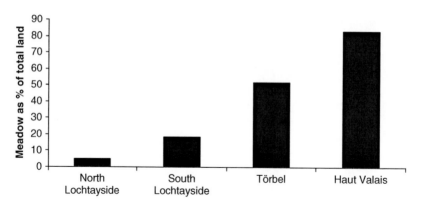

Figure 6.3 Meadow as percentage of total land within settlements.
Source: Based on McArthur, M.M. *Survey of Lochtayside 1769*, Scottish History Society, 3rd series, xxvii, Edinburgh, 1936; Netting, 1981, 16–17; Loup, 1965, Appendix II.

Highlands probably saw a significant expansion of arable in the century or so prior to the mid-eighteenth century, an expansion that was probably at the expense of their better pasture and meadow, a shift in land use that must have restricted their stock carrying capacity.[15] When we actually look at what was left as meadow by the end of this expansion, such as in townships along the northside of Loch Tay, it comes across as the most marginal bits and pieces left in between arable fields, once the latter had been pushed to their limits.[16] By contrast, we know that some central Swiss cantons probably saw an expansion of meadow over the same period and a corresponding expansion of stock numbers.[17] Such choices clearly had different consequences in terms of their environmental impacts, not least the abundance of species-rich meadow in the latter and their relative scarcity in the former.

The way in which the opportunity space of mountain communities was structured leads to a further point. Whether settlements began as valley-side sites, with a capacity to expand downwards as well as upwards, or as sites on the lower valley ground, any sustained expansion would have forced them upwards, to make more intensive use of their higher ground. As an open frontier of unoccupied and under-used land, their higher ground would always have encouraged expansion from communities under pressure. In time, what restrained such expansion was not the full exploitation of high ground, but the increasing risks of farming at altitude. In other words, when under pressure from any growth in numbers, mountain communities had a tendency to press up against their inner frontiers of risk.

---

[15] Dodgshon (1998a), 196–7, 223–7.
[16] For example, NAS, RHP 973 and 974; see also, Dodgshon (2004c), 32–4.
[17] Viazzo (1989), 184.

Mountain communities and engaged risk

In any comparison of risk between upland and lowland areas, it can be argued that the former faced far higher levels of risk, because of their climatic regimes and the greater variability of these regimes, the way in which their lower crop yields lowered the threshold at which crises kicked in, and because of the way in which any sort of sustained expansion led communities to ultimately interact with their inner frontiers and increased risk.[18] All traditional farming communities faced risks but, compared to lowland communities, mountain communities faced more extreme and more endemic levels of climatic risk. These endemic climatic risks required coping strategies that dealt with them as a matter of routine. What we find is that a great deal of their cultural practice and schemes of resource exploitation were engaged with different forms of risk aversion and risk buffering.

The point can be illustrated by looking at the various forms of risk faced by traditional farming communities in the Scottish Highlands prior to 1800 and the buffering strategies used to combat them. Taking arable first, past discussions has tended to see this as the most risk-laden sector of mountain areas. Its risks are more easily assessed because we can mark discrete limits for particular crops on a map and calculate the probability of crop failure for given heights. The changing limits to growth though, are only part of the problem. As Parry argued in his study of Lammermuir, storminess and short runs of adverse weather (i.e. wets summers) could be as much of a problem as any onset of colder conditions.[19] This is certainly my own reading of what may have happened in the Scottish Highlands during the Little Ice Age. When we look at settlement during the worst phase of the Little Ice Age, the so-called Maunder Minimum, 1645–1715, we find evidence for the abandonment of land but it is largely for the *temporary* abandonment of land, lasting no more than a few years at most. Furthermore, whilst we can find examples for the interior of the Highlands, for exposed areas like Lochaber, Rannoch and Glenorchy, the more abundant surviving evidence is for exposed, western areas, such as Tiree, Mull, Ardnamurchan and Kintyre, including low lying as well as high-ground townships.[20] The most likely explanation for this pattern of temporary abandonment is that it was brought about by relatively short phases of severe storminess and wetness. In fact, work on the Little Ice Age in Scotland by both Lamb and more recently by Dawson *et al.* has stressed the role played by increased storminess.[21]

We need to ask how communities could buffer themselves against increased storminess if it could affect low ground no less than high ground, since

---

[18] Dodgshon (1993), 688–92; Hubert (1991), 34.
[19] Parry (1978); Parry (1981), 319–36.
[20] Dodgshon (2005), 328–33.
[21] Lamb (1988), 104–31; Dawson et al. (2003), 381–92.

retreating from high-ground sites would not have removed the risk. The answer lies in something obvious. Most Highland townships had a simple cropping of their core arable: one part under bere or barley and two parts under oats. At first sight, such a simple scheme hardly seems to offer any protection or buffering against storminess. In fact, there are two reasons why it does. First, we know from the very detailed yield data that is available for islands like Tiree that because bere or barley yielded approximately 50% more than oats, their differences in acreage were effectively equalised when we add their differences in output. Communities ended with the same yield of bere in hand as they did of oats. Second, bere had a significantly shorter growing season compared to oats. Oats could be in the ground for 4–5 months or more before being harvested, whereas bere could take as little as 10 weeks from seed to harvest, though more usually it needed 3–3½ months. These differences were not based on a common point of sowing or a common point of harvest. Oats was always sown earlier and harvested later than bere. If we relate this to patterns of storminess, the later sowing and earlier harvesting for bere meant that it missed out on a significant part of the equinoctial storms, reducing risks significantly.[22] Seen in a wider context, it was this ability to produce a crop over a shorter growing season that explains why we find barley being cropped further north in Europe than oats and why, in a country like Iceland, oats disappeared from cropping regimes during the early phases of the Little Ice Age, but not barley.

Discussions of how communities coped with the Little Ice Age tend to deal almost exclusively with how it affected arable, yet mountain communities were equally dependent on stock. Stock management faced equal risks. Except where no herbage existed whatsoever through bare rock or scree, or where snow lay all year round, most mountain grazings were open to use at some point in the year. The emphasis here is on 'at some point of the year'. The fact that the annual cycle of plant growth commences later and later as one moves higher and ceases first in those places where it was last to start means that to maximise available grazing requires an elaborate strategy of movement. One is not, as with most lowland systems, faced with a resource space is which all parts of the farm have a growing season of roughly the same length so when one chooses to exploit grazing is largely a matter of farming choice. In mountain systems, there are grazings that have to be exploited at a particular time or not at all. The choice is uncompromising in that it takes labour away from the best land and redirects it to what, in terms of *total* overall energy yield, may be the least productive land and it does do so at the height of the growing season simply because there is no other window of opportunity for such land. It is because of this conflict over labour allocation that many mountain communities adjusted

---

[22] Dodgshon (2004a), 8–10.

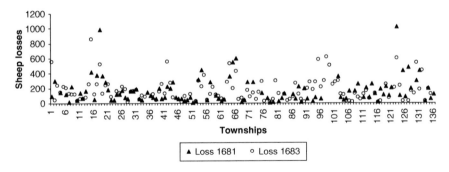

Figure 6.4 Rent abatements on sample of Buccleuch Farms in the southern uplands, 1683.
Source: Based on data from NAS, GD 2214/243/7.

their social structure in their adaptation to resource use, with only part of the community moving with herds and flocks to the summer grazings and why it was beneficial to operate a system of common grazings.

The main risk for stock loss was through the unpredictability of weather, especially in spring and autumn, and the loss experienced over winter brought about by poor levels of winter feed. Detailed data for the Buccleuch estate in the southern uplands of Scotland enables us to assess the scale of loss experienced during some of the worst winter storms on record for Scotland, those of the early 1680s. A sample of the data is shown in Figure 6.4. The figure of 300% rent rebate has been highlighted in the graph because it represents an amount equivalent to an entire year's output. In most cases, the loss was made up of stock loss, both sheep and cattle. One early estimate put the frequency with which this sort of severe storm afflicted stock levels as being about once every 20 years.[23] In addition, we need to factor in the regular annual loss that came from the persistently poor levels of winter feed reported for the upland areas of Scotland, a loss that can be tied in with the limited acreage managed as meadows.

Human impacts and mountain habitats

Over the past decade, ecological perspectives across all European mountain areas have shifted from seeing such habitats as natural to seeing them as semi-natural and as owing part of their character to some form of farming impact.[24] Describing them as semi-natural habitats though, is only the start. We need

[23] Scott (1886), 125.
[24] For example, Thompson, Hester and Usher (1995); Duclos and Mallen (1998), 96–8; Chemini and Rizzoli (2003), 3–4; Tasser, Prock and Mulser (1999), 235–46; Tasser and Trappeiner (2002).

more understanding of how human impacts have worked to change their natural character. Part of the problem is that some of the big-event changes that have been such a feature of the debate over human impacts in lowland areas (i.e. draining the fen, reclamation of the heath) have not had the same role in the environmental history of mountain areas, though phases of forest clearance have patently been a factor in some mountain areas.[25] Instead, understanding how human impacts have produced semi-natural habitats in mountain areas is more about understanding how everyday routines of husbandry and land management have produced shifts or modifications in the character of habitats.

All writers on mountain or upland ecology stress the diversity and species richness of their habitats. There were obvious reasons for this diversity and species richness: differences in height, slope, aspects, insolation, wetness, etc, all combine to produce a mosaic of base conditions and, therefore, of habitats. These factors alone ensured that mountain habitats are species rich. Admittedly, thanks to overgrazing, heavy programmes of pasture improvement or the abandonment of meadowing, recent changes have reduced some grazings to species-poor pastures, molinia deserts, etc. However, if we concentrate on the impact of traditional farming systems, there is a case for arguing that their impact in creating and managing meadows and on open-range pastures may have worked to accentuate diversity. Of course, meadowing and low-intensity grazing could have had a similar impact in lowland areas but the scale of meadowing and low-intensity grazing in mountain areas meant that their impact was spread more extensively.

Traditional farming affected the diversity of meadow and open-range pasture in mountain areas in different ways. First, it did so through the straightforward movement of stock and resources across the system, movements that could involve substantial distances and, as a consequence, seed exchanges across a range of different habitats. Second, it could do so through the nature of grazing practices, including variations in the types of different animals involved, different levels of stocking and variations in the pattern of herding and grazings. Third, it had a notable impact through the practice of meadowing or hay cutting. Fourth, we need to take account of the extent to which the large-scale harvesting and movement of resources like peat, turf, heather and ferns impacted on habitats. Fifth and finally, farming could change habitats through routine processes of land management, such as heather and grass burning. As already noted, these upland impacts – once woodland had been cleared – were more about modifications rather than transformations. Whilst some of these modifications undoubtedly led to species reduction, there is a case for arguing that their gross impact was to enhance the species richness of mountain

---

[25] For example, Smout (2000), 37–63.

habitats. Further, because many impacts arose out of routines of husbandry that were light and uneven in their effects, many involved modifications that were both patchy and temporary, so that one had a repeated cycle of impact and recovery played out on a patchwork basis. If we had to sum up such impacts, it would be that they not only accentuated the mosaic of mountain and upland habitats but, along with natural fires, they turned it into what H. Walter described as made up of 'innumerable variants of different degradation and regeneration stages'.[26]

I want to illustrate some of these points by reference to the Scottish Highlands and the Alps. In the past few years, I have tried to research various forms of human impact in the Scottish Highlands. Because they arise out of mundane routines of husbandry and land management, such impacts are not easily documented. However, enough is available to draw out some of the variables involved as regards the impacts of stock grazing and the management of open-range pasture and to indicate how these variables might have changed over time. The impact of stock grazing is a key aspect of the region's recent environmental history owing to the debate that surrounds how the clearances and the spread of specialised sheep farming may have impacted on grazings, with some arguing that it led to a deterioration of pastures by the mid-nineteenth century. Any answer to this question though, must ask how stock grazing impacted on such pastures prior to the clearances, as well as after them. Altogether, we can arrange the potential impacts of grazing under four headings: those related to stock balance, levels of stocking, grazing strategies and the role played by stock in the cross-habitat movement of seed.

Stock balance affects habitats through the grazing habits of different stock in terms of where, what and how they graze. Prior to the clearances and spread of sheep, highland townships pursued a mixed stocking strategy, with cattle, horses, sheep and goats present in most townships. Whilst we find numerous contemporary references to the damage that stock, particularly goats, could cause to woodland, stripping bark and suppressing tree re-growth, the overall impact of mixed stocking across open habitats was probably to sustain or even enhance species diversity.[27] Animals like cattle, sheep and goats each graze selectively but what they prefer to eat is different, with cattle preferring grass, sheep preferring forbs and goats, shrubs and bushes. At the very least, the consensus is that mix grazing works to maintain species diversity if only because it spreads their respective effects, preventing particular plants from becoming dominant. With the clearances for sheep, a process that started in the mid-eighteenth century, farms became stocked with sheep. In practice, the initial

---

[26] cited in Hubert (1991), 32.
[27] For example, NAS, Breadalbane Muniments, GD112/16/10/2/20; Armadale Castle, Lord Macdonald Papers, GD221/3695/2)

transformation was not as sharp as some have assumed, with many farms maintaining a fold of cattle and, because of the continued cropping of grain, working horses.[28] It was only with the renewed surge in sheep numbers in the 1830s–40s that one had the significant spread of sheep-only farms. This had the effect of concentrating their ecological impact, so it is not surprising that some link the spread of sheep to mid-nineteenth century reports of pasture deterioration, a deterioration marked by the spread of species-poor pasture.[29]

The problem with this interpretation is that it is not possible from the data available to disentangle the contribution of mono-grazing from that made by any increase in stocking levels. Establishing the stocking level of pre-clearance townships using soums as a form of standard livestock unit (SLU) and using the same calculation of SLUs to establish post-clearance stocking levels is straightforward. Unfortunately, obtaining pre- as well as post-clearance stocking data is only possible for a relatively small number of townships and their number is reduced still further when we rule out those units that underwent indeterminate changes in acreage through the post-clearance rationalisation of farms and grazing areas. However, where we can compare pre- and post-clearance stocking, it suggests that many farms saw a doubling of stocking levels, at the very least.[30] The increase doubles again if we allow for the widespread changes in animal size that occurred with the shift to larger breeds.[31] Of course, to say that stocking levels may, at the very least, have risen fourfold is only part of the problem. It was an increase loaded onto sheep, so that when we consider its impact on those grazings favoured by sheep – herb rich pastures – the scale of impact would have been enhanced.

Adding to the role of the clearances as a process of ecological change were the changes that accompanied them as regards strategies of grazing. Under traditional, pre-clearance systems, stock actually spent only 3–4 months grazing open-range pasture, whilst milk cattle spent even less time on it. The fact that part of this time was taken up by milking and the making of cheese and butter probably meant that stock tended to graze more closely in proximity to the main shielings, a fact that would help explain the evident nutrient loading around such sites. Further, reports suggest that all animals grazed together, without any attempt to separate out types or ages of animal. By contrast, when we reconstruct grazing strategies after the clearances, we find that stock tended to be on the hill ground for significantly longer time. Coupled with the adoption of new systems of shepherding and hirsels,[32] it meant that the grazing

[28] Dodgshon (2004b), 74.

[29] Innes (1983), 141–9; Smout (2000), 125–31; Dodgshon (2004b), 67–78.

[30] Dodgshon (1998b); Dodgshon (2004b), 67–78.

[31] Gibson (1988), 162–71.

[32] For example, National Library of Scotland, Sutherland Papers, 313/993.

of hill pasture became more prolonged and more systematically managed across all available grazings.[33] When combined, changes in stock composition, grazing levels and grazing strategies probably meant that open-range pasture shifted from being lightly grazed prior to the clearances to being heavily grazed afterwards.

The management of upland heaths and heather moor adds a further dimension to the impact of traditional farming on open-range pasture. The regular management of heather moor through cycles of burning has long been seen as fundamental to its character and the reason why it can be classed as a cultural landscape.[34] Yet though ecologists have established the relationship between heather moor and burning, there is a persistent belief that the practice of burning – and therefore the cultural management of heather moors – was introduced only during the nineteenth century. In fact, moor burning extends back to at least 1400, being documented through the acts passed by parliament and local barony courts.[35] These acts were directed at when and where burning could take place. What stands out about their infringement is the fact that those who carried out the burning, and who were being fined for burning at the wrong time or in the wrong place, were ordinary farmers. It is also clear from entries that burning was targeted as much at hill grass as heather.[36] As a management practice, it was probably intended to encourage younger more herbaceous growth for stock, but – as well as helping to suppress tree regeneration – it would have had the effect of encouraging a switch from grass to heather owing to the fact that heather was more fire resistant, a switch that finds support in palaeobotanical studies.[37] In other words, heather or, more accurately, moor burning was a long-standing form of land management, one that would have greatly contributed to the patchiness of habitats and the extent to which they were subjected to change and recovery. It was only in the nineteenth century, with the more active management of heather moor for game birds, that we find burning being used, first and foremost, to manage game moors and shifting from being practised by farmers to being practised by grouse moor managers. Previously, the interests of game birds had only been considered in regard to the dates when burning could take place.

Traditional farming had comparable impacts in Alpine areas. These impacts were enhanced by the fact that Alpine environments contain a rich mosaic of habitats, owing to their wide local variations of height, slope, aspect and climate, combined with a wide range of soil depth, moisture and fertility. As

---

[33] Dodgshon and Olsson (2006), 24–5, 29.
[34] Thompson, Hester and Usher (1995).
[35] Dodgshon and Olsson (2006), 27–8.
[36] Dodgshon and Olsson (2006), 27.
[37] Stevenson and Thompson (1993), 70–6; Stevenson and Birks (1995), 224–39.

environments in which the risks to farming were generally high, many communities pursued a conscious strategy of spreading their subsistence across as wide a range of habitats as possible, creating what Netting called 'a quilt of potential land use'.[38] Nor was it a case of spreading a simple structuring of land uses across the various habitats, with homefield arable and best meadow on the better sites around the winter settlement, *mayen* pasture for spring and autumn just above the homefields, extra meadow created from clearings set amidst the forest land that lies above the homefields and, finally, high-ground pastures on the high alp that lies beyond the forest line. In practice, a much finer-scale adaptation of resource exploitation was pursued, with broad scale differences of land use being elaborated by local differences over when crops might be harvested, meadows mown and grazings used. In a settlement like Törbel in Valais, arable was ranged from 1100m to just over 1600m, with the rye crop on the lower fields being harvested as early as mid-June but the last of the grain not being harvested until August, though as Cole and Wolf noted, the rotational systems used meant that the maximum height of cultivation was usually set by the least tolerant crop.[39] Beyond home fields, the lower, better land was always used for meadows to supply winter feed and the higher, less productive land for summer grazings, an allocation of use that underlined just how critical it was to have sufficient hay to keep stock over winter,[40] not least because stock had to be kept indoors from November to May (housing for ca.250 days being common). Because of their height range, the mowing of meadows – like the harvesting of grain – was spread across a number of months, especially where they were productive enough, or were encouraged through irrigation, to produce a second and even a third crop in late summer.[41] As regards land used solely for grazing, different grazings were grazed at different times and in different ways. The better, more accessible grazings on the alp were used for milk cattle. As Orland has recently pointed out, what was used for grazing milk cattle on the high ground may have been less productive compared to meadows, but they added something distinct to the cheese and butter produced from it.[42] Non-milk cattle were put on the next best grazing, with sheep and goats on the highest and least productive pasture. Likewise, the better, lower grazings were grazed for longer, often with animals being tethered or enclosed so as to graze land more closely. Animals put to the highest pastures, those stretching up to 2500m, were moved between distinct areas of graze in a systematic way, spending barely a few days on some pastures but much longer on others.[43] In

---

[38] Netting (1981), 21; Weinberg (1975), 104.
[39] Netting (1981), 32; Cole and Wolf (1974), 123; Cole (1972), 166.
[40] Cole and Wolf (1974), 125.
[41] Netting (1972), 138; Cole and Wolf (1974), 174; Reynard (2002), 343–61.
[42] Orland (2004), 333.
[43] Netting (1981), 12 and 65; Orland (2004), 336.

this way, even pastures of the high alp that flourished briefly in summer were incorporated into a community's system of husbandry. Overall, this 'verticality' to husbandry, as Viazzo termed it,[44] the structuring and shaping of seasonal activities according to height, spread the labour demands of husbandry out across the summer, making it easier for communities to cope with labour intensive tasks like hay making and to deal with the fragmented and dispersed nature of their opportunity space. It was not just a case of 'doing different things at different altitudes' as McNeill put it, but of doing them at different times.[45] It had the effect not just of spreading the impact of traditional farming across the mosaic of different habitats but of affecting different parts in different ways at different times. Mountain areas may have started out with a highly irregular array of different habitats, but the impact of traditional farming probably accentuated this variegation.

In addition, traditional farming impacts served to increase species diversity. Recent research has confirmed that there were at least three ways in which they could have done so: through the mowing of meadows, low-intensity grazing and stock movements. Mown meadows and nutrient-poor pastures that were lightly grazed were once extensive features of alpine farming systems. Both were associated with species-rich habitats that usually contained a high percentage of forbs, habitats that included some of the richest biodiversity in Europe. Though there were meadows that were simply mown and nutrient-poor pastures that were simply grazed, the distinction between the two was not everywhere so clear-cut. Many meadows were grazed after a crop of hay had been taken and after time had been allowed for regrowth. On some of the highest meadows in the central Alps, the so-called *mähder* lying between 1800 and 2300m, a hay crop was taken only every alternate year and the pasture grazed during the intervening summer, making it difficult to draw a hard and fast line between what was meadow and what was grazing.[46] Such variations, coupled with the varying dates at which a hay crop was taken from meadows and the fact that meadows, if grazed as well as mown, were likely to be grazed by cattle whilst nutrient-poor grazings that were wholly grazed were more likely to be grazed by sheep and goats, meant that the management inputs to these habitats varied in a way that is likely to have accentuated differences in terms of species composition. However, whilst there is ample work demonstrating that mowing meadows at the right time and the low-intensity grazing of subalpine pastures both maintained species diversity, the precise mechanisms by which they *increased* species diversity remains unclear, as this would presuppose a local seed source for new plants. It is in this context that recent work by Poschlod and WallisDeVries becomes significant because they have highlighted

---

[44] Viazzo (1989), 20.
[45] McNeill (1992), 105.
[46] Fischer and Wipf (2002), 1–11.

the potential role played by stock movements, including transhumance movements, in the transfer of seeds between widely separated sites.[47] Their work particularly focussed on calcareous grasslands along the edge of the Swabian Alps and the role of sheep as a means of seed transfer, but the mechanisms involved probably have wider application.

Recent ecological discussion of mountain habitats has stressed the extent to which the species richness of meadows and low-intensity grazed pastures probably required centuries of stable management for their formation and that their formation may have played on the low dynamics or slow rate of change experienced by such habitats. Yet it would be wrong to overstate their unchanging nature in the centuries prior to the recent contraction of such habitats following land abandonment, the disappearance of regular mowing and meadow management and, for those meadows and pastures that have remained in use, their improvement through programmes of heavy fertilisation. From when documentary sources first become available in the medieval period, a range of ongoing changes and adjustments in land use and management can be identified. The latter part of the medieval period saw the widespread expansion of arable and homefields, an expansion that was sustained in some areas by an expansion of meadows and haymaking beyond the homefields so as to provide manure for arable.[48] In the drier parts of the Swiss Alps, the need for more hay was addressed by the widespread construction of irrigation canals or bisses so as to supercharge the output of meadows, a development under way by 1200 whilst other areas saw the manuring of meadows in an attempt to boost output.[49] Where we have sufficient data, it is also clear that stock numbers were not static. Population growth and demand, the manurial needs of an expanded arable and climate change could all work together to flux stock numbers between states of over and under stocking on the higher ground. This has been particularly well shown by Crook *et al.* in their work on stocking levels between the sixteenth and nineteenth centuries in the Petit du Lac catchment near Annecy.[50] Significant changes also came about through shifts in the nature of stocking, as with the shift out of goats and sheep and the greater concentration on cattle that accompanied the growth of dairying in the central cantons parts of Switzerland from the sixteenth century onwards, a switch that also saw a widespread switch of arable into meadows as communities tried to increase the winter support for stock.[51] Taken together, such changes served to vary human impacts on mountain habitats over time and to accentuate the degree

---

[47] Poschlod and WallisDeVries (2002), 361–76.
[48] Cole and Wolf (1974), 148; Cole (1972), 160.
[49] Netting (1981), 43–4.
[50] Crook et al. (2004), 261–2 and 273.
[51] Viazzo (1989), 184; Orland (2004), 337; Hausmann et al. (2002), 287.

to which such habitats were continually in a state of response or adjustment rather than stabilised around fixed management strategies.

Conclusion

I have tried to argue in the foregoing essay that one cannot have a history of farming communities in mountain areas that is not also – in someway – an environmental history. Each mirrors the other. Mountain environments presented local communities with a diverse portfolio of opportunities. Survival meant mapping as much of this diversity into their strategies of resource exploitation as possible, mapping it not just in a geographical sense but via an elaborate week-by-week patterning of tasks and routines once growth recovered in spring. As they pushed upwards though, communities would increasingly have faced sites that offered less but demanded more. We can only grasp their persistence with these declining rates of return by appreciating the way resources and opportunities were pieced together strategically, sometimes via a complex account of substitutions, trade-offs and the seizure of short windows of opportunity. Some high pastures may only have been available for a short-lived burst of growth but, if exploited, they could take pressure of a lower pasture which could then be used as meadow which, in turn, could have better secured their winter feed or enabled a critical switch of land use within their homefield area. The relatively limited area of arable meant that the impact of most traditional mountain communities was mostly spread out across meadow, pasture and woodland and was spread lightly. Much recent research has been directed at the habitats created and maintained by this extensive but light impact on mountains habitats and how recent land abandonment and husbandry has lead to their loss. Too easily, the assumption is made that such habitats developed out of systems of management that had had been stable in the long term. For a variety of reasons, from documented changes in husbandry to the continual flux in stock numbers brought about by the vagaries of mountain climate, we cannot expect human impacts to have been so stable. Documenting such changes not only contributes a vital theme to the environmental history of mountain areas but also can play a significant role in helping to understand how the habitats produced under such conditions might be sustained.

References

Braudel, F. (1972). *The Mediterranean and the Mediterranean World in the Age of Philip II*, vol. 1, London.

Carrier, E.H. (1932). *Water and Grass. A Study of the Pastoral Economy of Southern Europe*, London.

Cernusca, A., Tappeiner, U. and Bayfield, N. (1999). *Land-Use Changes in European Mountain Ecosystems. Ecomont – Concepts and Results*, Berlin.

Chemini, C. and Rizzoli, A. (2003). 'Land use change and biodiversity Conservation in the Alps', *Journal of Mountain Ecology*, 7 (Suppl.), 1–7.

Cole, J.W. (1972). 'Cultural adaptation in the eastern Alps', *Anthropological Quarterly*, 45, 158–76.

Cole, J.W. and Wolf, E.R. (1974). *The Hidden Frontier: Ecology and Ethnicity in an Alpine Valley*, New York.

Crook, D.S., Siddle, D.J., Dearing, J.A. and Thompson, R. (2004). 'Human impact on the environment in the Annecy Petit Lac Catchment, Haute-Savoie: A documentary approach', *Environment and History*, 10, 247–84.

Dawson, A.G., Elliot, L., Mayewski, P., Locket, P., Noone, S., Hickley, K., Holt, T., Wadhams, P. and Foster, I. (2003). 'Late-Holocene North Atlantic climate "seesaws", storminess changes and Greenland ice sheet (GISP2) palaeoclimates', *The Holocene*, 13, 381–92.

Dodgshon, R.A. (1993). 'Strategies of farming in the western highlands and islands of Scotland prior to crofting and the clearances', *Economic History, Rev.*, xlvi, 679–701.

Dodgshon, R.A. (1998a). *From Chiefs to Landlords. Social and Economic Change in the Western Highlands and Islands, c.1493–1820*, Edinburgh.

Dodgshon, R.A. (1998b). 'Livestock production in the Scottish Highlands before and after the clearances', *Rural History*, 9, 32–3.

Dodgshon, R.A. (2004a). 'Coping with risk: Subsistence crises in the Scottish Highlands and islands, 1600–1800', *Rural History*, 15, 1–25.

Dodgshon, R.A. (2004b). 'The Scottish Highlands before and after the clearances: An ecological perspective', In Whyte, I.D. and Winchester, A. (eds), *Society, Landscape and Environment in Upland Britain*, Society for Landscapes Studies Monograph, 67–78.

Dodgshon, R.A. (2004c). 'Researching Britain's remote spaces: Themes in the history of upland landscapes', In Baker, A.R.H. (ed.), *Home and Colonial*, HGRG Research Series, 29–38.

Dodgshon, R.A. (2005). 'The Little Ice Age in the Scottish Highlands: The documentary evidence', *Scottish Geographical Journal*, 12, 321–37.

Dodgshon, R.A. and Olsson, G.A. (2006). 'Heather moorland in the Scottish highlands: the history of a cultural landscape, 1600–1880', *Journal of Historical Geography*, 32, 21–37.

Duclos, J.-C. and Mallen, M. (1998). 'Transhumance et biodiversité: du passé au présent', *Revue de Géographie Alpine*, 89–101.

Fischer, M. and Wipf, S. (2002). 'Effect of low-intensity grazing on the species-rich vegetation of traditionally mown subalpine meadows', *Biological Conservation*, 104, 1–11.

Frödin, J. (1940–1). *Zentraleuropas Alpwirtschaft*, 2 vols. Oslo.

Gibson, A.J.S. (1988). 'The size and weight of cattle and sheep in early modern Scotland', *Agricultural History Review*, 36, 162–71.

Hausmann, S., Lotter, A.F., van Leeuwen, J.F.N., Ohlendorf, C., Lemcke, G., Grönlund, E. and Sturm, M. (2002). 'Interactions of climate and land use documented in the varved sediments of Seebergsee in the Swiss Alps', *The Holocene*, 12, 279–89.

Hubert, B. (1991). 'Changing land use in Provence (France): Multiple use as a management tool', *Options Méditerranéennes*, Seminar series no 15, 31–52.

Innes, J.L. (1983). 'Landuse changes in the Scottish Highlands in the 19th century: The role of pasture degeneration', *Scottish Geographical Magazine*, 79, 141–9.

Lamb, H.H. (1988). 'The Little Ice Age period and the great storms within it', In Tooley, M.J. and Sheail, G.M. (eds), *The Climatic Scene*, London, 104–31.

Loup, J. (1965). *Pasteurs et Agriculteurs Valaisans*, Grenoble.

McNeill, J.R. (1992). *The Mountains of the Mediterranean World An Environmental History*. Cambridge.

Mead, W.R. (1953). *Farming in Finland*, London.

Netting, R.M. (1973). 'Of men and meadows: Strategies of alpine land use', *Anthropological Quarterly*, 45, 132–44.

Netting, R.M. (1981). *Balancing on an Alp Ecological Change and Continuity in a Swiss Mountain Community*, Cambridge.

Orland, B. (2004). 'Alpine milk: Dairy farming as a pre-modern strategy of land use', *Environment and History*, 10, 327–64.

Parry, M.L. (1978). *Climatic Change, Agriculture and Settlement*. Folkestone.

Parry, M.L. (1981). 'Climatic change and the agricultural frontier: A research strategy', In Wigley, T.M.L., Ingram, M.J. and Farmer, G. (eds), *Climate and History*, Cambridge, 319–36.

Pfister, C. (1995). *Im Strom der Modernisierung. Bevölkerung, Wirtschaft und Umwelt im Kanton Bern 1700–1914*, Bern.

Pfister, C. and Egli, H.-R. (1998). *Historisch-Statisticher Atlas des Kantons Bern 1750—1995*, Bern.

Poschlod, P. and WallisDeVries, M.F. (2002). 'The historical and socioeconomic perspective of calcareous grasslands-lessons from the distant and recent past', *Biological Conservation*, 104, 361–76.

Rackham, O. (1986). *The History of the Countryside*, London.

Reynard, E. (2002). 'Hill irrigation in Valais (Swiss Alps): Recent evolution of common-property corporations', In Pradhan, P. and Gautam, U. (eds), *Farmer Managed Irrigation Systems in the Changed Context*, Kathmandu, 343–61.

Scott, C.S. (1886). 'Wintering hill sheep', *Transactions of the Highland and Agricultural Society*, 4th series, xvii, 124–48.

Slicher van Bath, B.H. (1963), *The Agrarian History of Europe A.D. 500–1850*, London.

Smout, C. (2000). *Nature Contested. Environmental History in Scotland and Northern England since 1600*, Edinburgh.

Stevenson, A.C. and Thompson, D.B.A. (1993). 'Long-term changes in the extent of heather moorland in upland Britain and Ireland: Palaeoecological evidence for the importance of grazings', *The Holocene*, 3, 70–6.

Stevenson, A.C. and Birks, J.J.B. (1995). 'Heaths and moorland: Long-term ecological changes and interactions with climate and people', In Thompson, D.B.A., Hester, A.J. and Usher, M.B. (eds), *Heaths and Moorland: Cultural Landscapes*, Edinburgh, 224–39.

Tasser, E., Prock, S. and Mulser, J. (1999). 'The impact of land-use on vegetation along the Eastern Alpine transect', In Cernusca, A., Tappeiner, U. and Bayfield, N. (eds), *Land-Use Changes in European Mountain Ecosystems*, 235–46.

Tasser, E. and Tappeiner, U. (2002). 'Impact of land use changes on mountain vegetation', *Applied Vegetation Science*, 5, 173–84.

Thirsk, J. (ed.) (2000). *Rural England. An Illustrated History of the Landscape*, Oxford.

Thompson, D.B.A., Hester, A.J. and Usher, M.B. (eds) (1995). *Heaths and Moorland: Cultural Landscapes*, Edinburgh.

Viazzo, P.P. (1989). *Upland Communities Environment, Population and Social Structure in the Alps Since the Sixteenth Century*, Cambridge.

Weinberg, D. (1975). *Peasant Wisdom. Cultural Adaptation in a Swiss Village*, Berkeley.

Williamson, T. (2003). *Shaping Medieval Landscapes. Settlement, Society and Environment*, Macclesfield.

7

Interdisciplinary Conversations: The Collective Model

A. Hamilton, F. Watson, A.L. Davies, and N. Hanley

Introduction

Environmental history prides itself on being explicitly interdisciplinary. This is usually considered to mean the fostering of dialogue 'between humanistic scholarship, environmental science, and other disciplines'.[1] This position has emerged largely because of the subject's obvious relationship with environmental and biological/ecological sciences, which have long dominated investigations into the natural world. Once documentary historians began to 'discover' and discuss the relationship between society and the environment, it was natural that they should incorporate the findings and some of the methods of ecology in particular. One need only think of the titles of two important environmental history texts – *Ecological Imperialism* by Alfred Crosby and *The Ecological Indian* by Shepard Krech – to note the explicit link, particularly in the United States. Crosby is a historian and Krech an anthropologist. Some have moved the other way: Jared Diamond, an evolutionary biologist and physiologist, has earned international acclaim for his syntheses of human history, placing biology at the explanatory heart of key historical questions; biologist Daniel Botkin believes that his subject cannot be divorced from human cultural activity, past and present. All have attempted to integrate more traditional historical narratives with scientific findings, achieving impressive results.[2]

However, this chapter will argue strongly that environmental history's interdisciplinarity should not rest on a general principle of a single-disciplinary practitioner becoming bi- or even multi-lingual, by seeking to understand and incorporate elements of the natural sciences into traditional history, or vice-versa. Perhaps, in works of synthesis, using the research of a wide range of other scholars to draw new conclusions or create provocative new theories like

---

[1] See, for example, http://www.aseh.net/about-aseh and http://eseh.org/.
[2] Crosby (2004); Krech (2000).

those mentioned above, it matters less, as Botkin says, since the authors are often acting as expert witnesses rather than as primary researchers themselves.[3] However, when tackling the issue of interdisciplinary research (IDR) of a *primary* nature, the act of directly analysing past human activity or ecological processes, we argue that disciplinary integrity – the maintenance of demonstrably excellent disciplinary standards – is fundamental. While some remarkable individuals are able to maintain sufficient disciplinary integrity in more than one discipline to undertake primary IDR by themselves, as exemplified by the botanist-turned-historian Oliver Rackham,[4] generally speaking, it is our view that the best way to maintain interdisciplinary rigour is through collaboration. We argue this for two reasons.

Firstly, serious questions remain about the academic rigour of attempts by non-scientists to truly understand evolving debates and methodologies within ecology and environmental sciences. This is equally true for scientists attempting to place environmental change within complex social, economic, political and cultural contexts. If any particular analysis cannot be defended before its disciplinary practitioners, rather than other non-specialists, then this will weaken environmental history's value within the academy. Secondly, it is perhaps time to assert explicitly that environmental history should, where appropriate, encompass much more than a simple relationship between history and ecology. An environmental historian from whatever disciplinary background should – where appropriate – seek to work with the practitioners of other disciplines relevant to the research question(s) in hand. If environmental history is a question-driven research endeavour, rather than led by select 'preferred' disciplines, it ceases to be defined by which disciplines and/or methods are 'in' or 'out', 'core' or 'peripheral' to its concerns. Instead, it can remain firmly focussed on the key issues and questions in human–environment relations and how best to address them. Environmental *and* human responses can then both be addressed within any environmental history. An environmental history which does not encompass appropriate aspects of the complexity of human responses runs the risk of allowing environmental determinism to become the explanatory force. On the other hand, an environmental history which does not explore the equally complex influences of environmental/ecological processes reduces nature to a merely passive element, the canvas upon which anthropogenic activity is drawn but which possesses no power of its own. As our case study, discussed later in this chapter, shows, questions about the relationship between upland Scottish farmers and their environment over time seemed to us to be best addressed (given that money was certainly

---

[3] Botkin (2007).
[4] See, for example Rackham (1989) and Rackham (2000).

an object) by a team consisting of a palynologist (pollen scientist), historians, ecologists and environmental economists. If money had been no object, then it might also have been beneficial to have included experts on literature, place names, art history and anthropology, but this was not strictly necessary within the parameters of our specific objectives and hypotheses.

As an example of the practical development and implementation of this theoretical approach, we will describe our experiences of an IDR project which examined human impacts on upland biodiversity in Scotland over the last 400 years. First, we will outline some issues to be aware of in all IDR projects. We then describe the difficulties and incongruities that emerged in trying to bring these disparate disciplines together into a coherent working relationship, and the insights gained in solving them. We illustrate these through some of our results, which, above all show how such interdisciplinary work produces significantly more as a whole than any single discipline could hope for individually. In describing our approach to IDR, it is not our aim to provide an in-depth review of past or existing practices, but rather to offer for discussion a different and methodologically stronger alternative.

Interdisciplinary issues

The increasing number of disciplines addressing questions under the broad banner of environmental history undoubtedly reflects a growing awareness of the complexities involved in understanding human–environment systems and the relevance of understanding such systems, as well as, perhaps, the array of 'tools' needed to further that understanding. However, in practice, the prevailing tendency seems to be that primary research is done in an individual author's own discipline, and then combined with a synthesis (usually cherry-picked from secondary sources) of particular arguments from a complementary discipline. From a methodological point of view, this is far from satisfactory. As noted above, it is questionable whether individuals can, in today's climate of increasing specialisation and rapid advances, become sufficiently conversant in another discipline to understand the current state of critical thinking within it to enable them to know what is truly applicable to their own research. It is also the case that scientific studies particularly are usually designed to answer a specific set of questions and are not necessarily applicable in other contexts. Finally, there is an increased tendency to look for evidence within the complementary discipline that supports a pre-determined argument generated in the author's core discipline, an approach that is clearly problematic. We argue that collaboration can overcome these fundamental methodological issues. Obviously collaboration has methodological issues of its own, which we will address throughout the course of this chapter.

The first requirement of a research project that may cross-disciplinary boundaries is that someone has an initial 'idea' encompassing environment and people that has the potential to expand through the consideration of different disciplinary perspectives. Such ideas often start with just one person or perhaps a small group of people who recognise the limitations of the starting discipline, and thus seek input from wider sources. When considering the ways in which such initial research ideas may be expanded, refined and ultimately addressed, several options may be available:

1. To expand the questions and methods, but rely on the individual(s) concerned gaining some expertise in other subjects or using existing secondary sources;
2. To break up the project into several discrete sub-projects, nominally tied together in the funding application, but in reality operating independently until some attempt in the later stages to tie together findings;
3. To bring in the necessary disciplines early on to develop common research questions, and build a project which is methodologically justifiable and robust for each discipline, and in which constant communication between disciplines is integral.

Adopting option 1 would result in an essentially single-discipline focus project, which may be capable of producing good environmental history, but could not claim to be interdisciplinary in terms of primary research. For example, much good work has been done on understanding the effects of grazing in the Scottish uplands from both a historical and ecological perspective.[5] However, as these studies stand, they cannot merely be added together to provide a fuller picture. That would have required detailed discussion before the research was undertaken to ensure that they were able, where appropriate, to compare like with like in geographical and temporal terms. The other options require recognition that there is a role for researchers from different disciplines to be involved in the same study, and can be equated with different levels of interdisciplinary work. Although in option 2 it is possible for different disciplines to independently address the same question, and is a form of IDR, unless an effort has been made to address issues of objectives, scale and compatibility, the results cannot be effectively brought together. Option 3, where the different disciplines involved make an ongoing contribution from the start, provides a stronger approach to truly IDR.

The need to develop a project in an appropriate manner is clear, but what does interdisciplinary work actually involve? Any IDR project will work somewhere along what is effectively a continuum of 'degrees of integration'. Several authors have recognised different levels or stages in IDR, as shown in Table 7.1. The collaboration achieved at stage one in the model shown here, roughly the

---

[5] For historical and ecological perspectives, see Dodgshon and Olsson (2006) and Stevenson and Thompson (1993), respectively.

Table 7.1   Interdisciplinary developmental model.[6]

Dimensions	Stage one	Stage two	Stage three
Discipline orientation	Dominant	Parallel	Integrative
Knowledge engagement	Expert	Co-ordinated	Collaborative
Work orientation	Individual	Group	Team
Leadership	Top down	Facilitative, inclusive	Web-like or servant

STAGE DESCRIPTIONS

Stage one: Single-discipline orientated – information exchange but no integration. Disciplines and individuals considered to be competing.

Stage two: Work still single-discipline focused, but within overall co-ordination. Individuals have more understanding of other disciplines. Competition is replaced by coexistence.

Stage three: Shared understanding and decision-making occurring in an adaptive team, with increased communication at all levels. Individuals listen and reflect, and are motivated by learning as much as task completion. Coexistence is replaced by integration.

result of proceeding with option 2 above, is perhaps reflected in many supposedly integrated studies. Although the separation of disciplines means that disciplinary rigour is maintained, the lack of ongoing communication and co-ordination, and perhaps the presence of 'competition' (individuals defending their discipline or results), means that a full synthesis of all available results is unlikely, and what is happening in reality may be a loose gathering of related projects, each headed by an individual discipline. At the other end of the spectrum, stage three represents a situation where there is a breakdown of disciplinary boundaries, with a completely collaborative work ethic and methodology. Whilst this seems ideal in theory for an IDR project, it remains important to consciously maintain individual disciplinary rigour, for reasons discussed already.

In contrast, stage two or somewhere between stages two and three represents a situation that we feel strikes a balance between maintaining disciplinary rigour and integration of disciplines. Although there is overall discipline and individual co-ordination and co-operation (facilitated by effective understanding and communication), individual disciplines and their roles are still well defined whilst working towards common and defined goals – we refer to this as the collective model of IDR. This does require that individuals have some understanding of other disciplines, which can be developed by

---

[6] Adapted from Amey and Brown (2005).

fostering a collaborative working environment. This is necessary for the initial appreciation of the possibilities of IDR research, and to allow development of effective methods where individuals assess and accept the contributions which can be made by other disciplines. Such understanding is also needed to appreciate the results produced by other disciplines, as well as the manner in which these results are interpreted.

Having noted the need for understanding other disciplines, it should be stressed that individuals must also possess a considered and honest understanding of their own field. Whilst this may seem obvious, individuals must be able to explain their disciplines to other non-experts, and must be aware enough of the strengths and weaknesses of their discipline to be able to embed it within an IDR framework. This requires a candid assessment of what each discipline can and cannot achieve, including the caveats or conditions attached to different methodologies and, most importantly, to results and the ensuing interpretations or conclusions. There is no place within IDR for individuals or disciplines claiming to be able to achieve more than what is actually possible within the confines of any particular project.

In practice – a research framework

In this and the following section we illustrate the methodology and findings of our research, in which we implemented the collective model of IDR. We are not proposing that the precise methodology described here is applicable in all situations, nor that the method was implemented ideally throughout our research. Rather we present our work as an example of real-life implementation, pointing out the necessary compromises, and the strengths and weaknesses in terms of achieving our aims. We then discuss integration problems and general recommendations, and identify issues of which all collective IDR practitioners should be aware.

Identifying questions and disciplines

The initial impetus for the project in Scotland developed from a palaeoecological[7] interest in the relationship between upland farmers and their landscape and natural resources, and the potential that primary historical evidence offered in terms of understanding how they dealt with the agricultural potential and limitations of that environment. Upland landscapes – in Scotland and elsewhere – tend to be presented as 'marginal' for both settlement and agriculture, certainly in comparison with lowland arable areas, and our interest

---

[7] 'Palaeo' means old; so palaeoecology is the study of past ecological communities, including past resource management and land-uses.

derived in part from a concern about the accuracy of such a label.[8] In particular, it was hoped that historical sources would provide the socio-economic context within which management decisions were made in rural communities and therefore help explain the reasons underlying changes in the palaeoecological record, and improve (or disprove) existing ecological inferences concerning the relationship between past grazing regimes and current upland vegetation diversity. Associated with this was an interest in the implications of past ecological change for understanding current landscapes; a secure understanding of the role which people in the past have played in shaping the present, largely uninhabited, heaths and mountains may inform the development of sound management and conservation strategies for the future. For that reason, we included an ecological survey of current conditions in order to allow us to assess current ecological conditions in relation to the historical evidence (both documentary and palaeoecological).

The theme of the work thus developed into an investigation of the relationships between people and their environment (particularly past plant diversity), encompassing land management practices as well as the environmental, social and economic factors which shaped them. For instance, how did farmers respond to changing pressures within the constraints of the environment and society in which they lived? How did their responses and changing methods of land management alter the quality and diversity of the natural resources on which they relied? How, in turn, did they respond to shortages or changes in the quality of key resources, such as pastures and woods? The chronological span of the project was limited to the period after 1600 due to the fact that the surviving historical record in rural Scotland for the pre-1600 period was comparatively poor. In spite of these limitations, however, this 400-year period encompasses some of the most drastic periods of change in agriculture and technology, society and economy in Scotland.[9] By contrast, their environmental impacts have been poorly researched to date. These transformations lie at the heart of the intensification and industrialisation of economies throughout the Old and New Worlds.[10]

As a result of initial discussions, it was deemed useful to attempt the creation of a statistical model, based on techniques used by environmental economists, to assess which of a range of potential causal factors had the most effect on the palaeoecological measures of plant diversity on a national scale over time and across sites. Environmental economists are not the only ones to use such 'panel data' models, but they have adapted this technique to look at particular socio-environmental issues such as drivers of deforestation, or changes in

---

[8] For example, Parry (1978) and Barber (1997) for environmentally deterministic views of upland communities.
[9] See Hunter (1976); Devine (1994); Dodgshon (1998) and Richards (2000).
[10] For example, Simmons (2001); Dovers (1994); Dovers et al. (2002).

pollution over time. The intention was to adapt the method to look at changes in biodiversity over time. This would provide a useful analytical tool in view of the wide and varied range of data which we envisaged the project providing, and a valuable adjunct to more conventional interpretation methods. The environmental economists were thus not concerned with data gathering, but only with data analysis. However, their requirements had to be taken into account by those engaged in the data gathering.

Our first aim was to develop a research design which allowed each discipline to contribute as fully as possible to the common themes. In question-driven research, subjects of common interest need to be addressed by each discipline through identical research questions to ensure that data collection and subsequent integration are targeted towards the same ends. This led us to formulate several hypotheses to ensure that the project-level strategic objectives were clear to each member of the project. These hypotheses formed the overall research focus for all disciplines. In our experience, it is at the level of data gathering, which was still conducted within the methodological parameters of each discipline, that each researcher may need to reassess conventional methods in their field to ensure that the type and resolution of the data generated or collected are appropriate to the aims of the overall project (discussed later). Collaborative IDR thus has to operate at two levels: joint development of objectives for the overall project, and discipline-level decision making as to how data are gathered and analysed in a manner that best satisfies the needs of the project as a whole while maintaining the standards for that discipline.

Communication is especially important from the beginning in this approach in order to help foster closer integration from the initial planning to final dissemination. The extra time required to achieve a level of common communication between disciplines should not be underestimated.[11] In addition, we appointed an external advisory board to evaluate our approach independently, from both the individual disciplinary stance, and from an integrated perspective. The project hypotheses provided a means for the board to assess our approach and progress.

Finally, in an ideal world project discussion, development and implementation would be engaged upon purely from an intellectual point-of-view, identifying the best way forward. In the real world, we must also take into consideration which grant-funding body or bodies are best suited for such potential research, not least because their criteria may not be familiar to all the collaborators.[12]

---

[11] For further discussion of issues in interdisciplinary work, see the end of this chapter and associated references.

[12] See Davies and Watson (2005) for a discussion of the infrastructural challenges facing European interdisciplinary research.

Developing the methodology

Site selection

Site selection should be based, theoretically at least, on the criteria devised to answer a particular set of research questions. In practice, however, single-disciplinary research tends to focus on sites that will yield the best results for that particular discipline's techniques, usually based on the quality of known evidence. This is the archaeological or ecological equivalent of an historian selecting case studies based on the apparent richness of an archive and is entirely reasonable. In interdisciplinary work, it is inevitable that compromises need to be made in order to allow each disciplinary practitioner to perform to the best of his/her ability without seriously compromising the effectiveness of their colleagues. To that end, the key data-gathering disciplines need to under-take a preliminary investigation of a number of potential sites that would fulfil the criteria for the research questions in order to ascertain whether they are likely to yield results across those disciplines. This approach does not guarantee that unforeseen problems may not arise once data gathering has begun, as ini-tial surveys are by their nature rapid and hence not exhaustive. However, that would be true in single-disciplinary research too.

In our case, the needs of palaeoecology and documentary history were given priority, since the state of the current ecology could be assessed at any site. The sites selected for the project had to fulfil two primary disciplinary-level requirements by providing:

1. archives with sufficient depth and quality of historical documentation to understand the changing management practices in each site, which could then be set into a wider socio-economic context.
2. waterlogged sediments which could be analysed at an appropriate spatial and temporal resolution, since this determines the degree of comparability with historical, archaeological and ecological sources. As indicated below, it was essential that the sites selected were sensitive to vegetation change on the permanent fields around a farm and on detached areas of grazing, rather than at a coarser, landscape or regional scale.[13]

To summarise, we therefore required undisturbed peat sequences which pre-serve a palynological (pollen-based) record of change in the fields at sites discussed and described in sufficient detail in the archival sources.

In order to investigate human and ecological responses under different con-ditions, the study sites were located in three areas across Scotland in order to incorporate a range of gradients: environmental (e.g. climatic, ecological) and socio-economic (e.g. proximity to markets). This allows us to contribute to

---

[13] Davies and Tipping (2004).

current debate about the future of the uplands across a wide biogeographical spectrum.[14]

In order to go back the full 400 years necessary for the project, it was essential to use the estate (an extensive area of landholding) as the main unit to be studied, since their owners' archives (potentially) provide the depth and range of material necessary for us to answer our research questions. In that respect, the project was necessarily biased towards human, rather than ecological, circumstances. Based on the prior experience of the historians involved in the project and the existing historical literature, our selected sites were farms and shielings (summer pastures, at a variable distance from the farm) as these were likely to reflect a range of agricultural activities (pastoral and arable) and provide evidence for how past communities balanced potentially competing resource uses. This approach would allow us to tie the documentary evidence with geographical areas investigated by the palaeoecologist and ecologists, so providing a common geographical focus for integrating our results. In addition, two farm-shieling pairs were selected on each estate to allow us to assess whether the changes in agricultural practices and the ecological responses observed at the farm or the shieling level were characteristic or atypical of each particular environment and estate. The original aim was thus to generate 12 individual sub-sites (2 separate farm-shieling pairs [4 sites] within 3 estates). It was accepted that over a 400-year study period, the function of each site was unlikely to remain constant in all cases. This would allow us to determine how estate management plans were put into practice at the farm level and to establish how their environmental impacts (especially on plant diversity) varied over space and time. Based on these criteria, the geographical/estate areas initially selected were: the Sutherland Estate (northern Scotland), the Breadalbane Estate (Central Highlands) and the Buccleuch Estate (Southern Uplands) (Figure 7.1). Although some of our solutions may be unique to the location of our research sites in rural upland Scotland, others should be broadly applicable to inter-disciplinary projects in any country.

Data acquisition

Following the joint site selection procedure, data acquisition was carried out independently by the three data-gathering disciplines. Team meetings were, however, held at regular intervals to monitor the success of this process and deal with any methodological/practical problems as they arose. Aspects of the methodology are described below, but more detail can be found in other articles about this project.[15] National or regional data sources were also collated, particularly for the post-c.1850 period, as the statistical model brought home the lack of archival data for this period.

---

[14] Brown et al. (1993); Usher and Balharry (1996).
[15] Davies and Watson (2007); Hamilton and Davies (2007); Hanley et al. (2008).

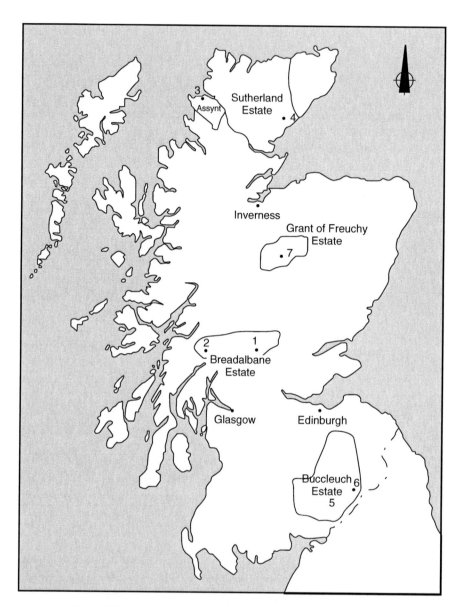

Figure 7.1   Map of project study sites across the Scottish uplands.

The documentary evidence

As is usually the case, a number of problems were encountered with the state of the historical records during the course of data acquisition, including the destruction of material within an archive and the inaccessibility of some material still held in private hands. However, the archives of the Breadalbane and Buccleuch estates are extremely extensive, entailing, as usual, the selection of material because of time constraints. If the historians had been tackling the project on their own, they would not have had to, for example, ensure that they had consistent temporal coverage (see below). However, the fact that some non-site specific estate information was not collected, was a source of some later regret.

As some of the chosen sites were amalgamated with other townships to form larger farm units during the period c.1750–1850, it was decided to also transcribe earlier records relating to the constituent parts of these enlarged farms to gain as complete a picture as possible of all the different historical variables that pertained to each site and to understand whether the chosen sites were managed in comparable or anomalous fashion to neighbouring farms/townships. However, deciding more generally and systematically on what exactly was 'relevant', given the considerable interrelationships between the individual farm, the estate and wider socio-economic factors (including Scottish parliamentary legislation) became extremely challenging owing to the complexity of the project and the strict timescale (3 years) involved. It was difficult to deduce whether specific legislation was adhered to at our site-specific level, although material such as court records can indicate which regulations (local or national) were ignored (or, by implication, being followed) at particular points in time. When combined with evidence from the range of study sites, this approach was designed to allow the social, economic and agricultural history of each site to be placed within their local, estate and national contexts. That this did not always happen was a reflection of the ambitiousness of the project and, it must also be admitted, the fact that both the historians involved were really specialists in the much earlier (medieval) period and so did not bring separate specialisms to bear on the project.

The pollen evidence

As with the documentary material, peat in which pollen records are preserved can also be subject to disturbance and loss, due to peat-cutting and artificial drainage, for example. In addition, naturally good drainage can simply result in an absence of the waterlogged conditions necessary for the formation of peat deposits and this of the pollen archive which they preserve. This necessitated the rejection of some sites due to a lack of suitable sediments and others on the grounds of insufficient historical evidence. We therefore added another estate to maintain the range or sites and quantity of data for the statistical model: including the Grant of Freuchy Estate in NE Scotland also allowed us to explore environmental and socio-economic relations in an additional biogeographical zone.

As indicated above, the pollen sites were chosen to reflect local or field-level land-use and vegetation change. This fine resolution approach provides a detailed understanding comparable with community-scale ecological surveys, but this information can be difficult to extrapolate to wider spatial scales, such as the estate level. While larger pollen-collecting sites (e.g. lakes or large peat bogs) provide a regional picture, such records amalgamate pollen from a potentially wide range of plant communities and land-uses, providing a somewhat blurred view of past landscapes. Ideally, a combination of both local and wider scale views would be desirable,[16] providing an ecological-scale perspective as well as a landscape-scale overview. However, pollen analysis, like the use of written archives, is a time-consuming process and this could not be accomplished within the constraints of the project. In the Breadalbane study area, few other published pollen diagrams are currently available for comparison with our data,[17] while in Sutherland, most published pollen sequences simply had too few samples and limited dating evidence for the historical period to provide more than a basic landscape context.[18] We were more fortunate in the Buccleuch study area, as pollen diagrams from elsewhere in the Southern Uplands provided a useful comparison for extrapolating to broader scales.[19] A simplified summary of the palynological results is shown in Figure 7.2, to illustrate the kind of results available for integration with historical evidence. The pollen data were statistically analysed to provide a proxy measure for changes in plant diversity over time.[20]

Integration issues

As already mentioned, issues of scale are vitally important to integrating these two datasets. Given that pollen records cannot be precisely dated to a selected year most easily reconciled with documentary evidence, each pollen sequence was analysed at c.20 year intervals, as far as possible, to capture change in vegetation cover on a timescale of human generations.[21] This spatial and temporal approach ensured that the work met the needs of joint historical-palaeoenvironmental research and also contributed to issues within palynology. However, there is no doubt that the relatively constant availability of pollen data through time, albeit at a different temporal resolution to written records, contrasted with the intrinsic variability of the material provided by the

---

[16] Lindbladh et al. (2000).
[17] See Davies and Watson (2007).
[18] For example, Froyd (2001).
[19] Tipping (2000); Davies and Dixon (2007).
[20] Birks and Line (1992).
[21] In addition, since year-to-year pollen production fluctuates depending on growing conditions (including weather and management), annual pollen records can be complex to interpret and are not necessarily ideal.

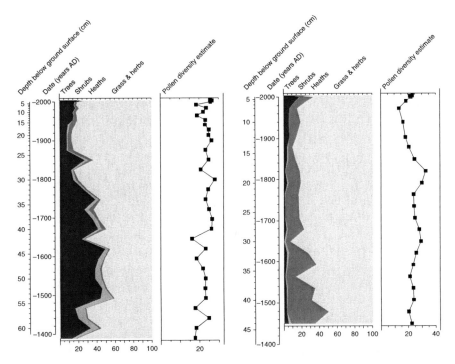

Figure 7.2 Pollen results from Leadour farm (left) and shieling (right), summarising patterns of vegetation change and estimated pollen diversity around each site over the last c.600 years. Trees were more abundant by the farm, while open grass and heather dominated the hillsides around the shieling. The increase in tree pollen at the farm around 1460–1620 reflects the birchwood of Leadour which was destroyed by cutting between 1614 and 1620. Note the prolonged decline in pollen diversity at the shieling from c.1820, which suggests that sheep farming had a detrimental impact on the richness of the pastures, in contrast with shorter-lived fluctuations at the farm.

historical sources. Equally, the documentary material related to the farm-unit scale at best, often failing to differentiate between different management units, such as infield and summer pasture grounds; by contrast, the pollen records chosen here are most reliable in describing these as separate systems, but it was more difficult to extrapolate from them to the landscape-level, as indicated above. Nevertheless, this left a significant degree of overlap between the two datasets to provide new insights into upland grazing regimes.

The ecological evidence

Vegetation surveys were carried out to establish the modern ecological context, conservation value and condition of each site. Conservation value was assessed by comparing each site with national designation criteria, and measures of diversity and heterogeneity were calculated. Nationally used Site Condition Monitoring methods were modified to allow target habitats at each site to be designated in 'good' or 'poor' condition. From an ecological and scientific viewpoint, these sites were selected primarily to meet the historical and palynological needs and are not a random or representative sample of upland sites in Scotland. Consequently, care must be taken with any ecological inferences that are drawn, in particular any generalisations or extrapolations made from the results about the uplands as a whole.

The statistical model

The intention here was to use the pollen-derived estimates of plant diversity across space and time, and assemble a dataset of livestock density, land-use change, intrinsic environmental conditions, technological improvements and changes in property rights to act as the variables most likely to have produced changes in diversity. Panel regression analysis was then used to explore relationships between the diversity estimate and these environmental, economic and social drivers.[22]

This was a novel and exploratory application of this analysis technique, and problems of course exist. The first is simply that of missing information, most importantly perhaps on the density of grazing animals on our sites over time. We also note the loss of information which results from the need to transform varied written sources (information rich, structurally poor) and complex pollen data to quantitative variables (information poor, structurally rich) for use in the regression model. For example, once research got underway it became glaringly obvious that the vitally important element (so far as being a potential cause of change in diversity was concerned) of livestock density (i.e. stocking numbers in relation to a definable area) could not be sufficiently recovered from the documentary evidence. It was therefore decided to use animal prices as a proxy for livestock density, having accounted for both supply – and demand – side drivers of local prices.[23] However, it is essential that the statistical model is viewed alongside the more detailed and complex datasets.

Results

As our research progressed, we were confident that we had developed an appropriate methodology to ensure that our research questions could be effectively addressed by all disciplines, as appropriate, and that the data generated was

---

[22] Hanley et al. (2008).

[23] It is not possible here to discuss how it was shown that livestock prices could be used as a proxy for stocking density but for a detailed discussion of this important point, see Hanley et al. (2008).

as compatible as possible. However, we quickly realised that, in order to avoid circular or flawed arguments (caused particularly by looking in one set of evidence for trends and/or incidents observed in another), we needed to maintain disciplinary integrity for the initial stages of that process. Integration therefore only took place once key trends and/or incidents had been identified through independent analysis of each dataset.

To present even selected elements of these individual analyses in this chapter would not necessarily be helpful for illustrating the relevance of our approach to IDR in environmental history. The following discussion therefore deals with a factor of national and continuing significance in upland management and conservation – namely grazing – and highlights some of the conclusions drawn once the datasets had been placed alongside each other. This includes the ecological evidence and the value of the basic results derived from the statistical model, which covered all sites across time. In particular, we draw attention to issues that would have been ignored, impartially or incorrectly understood if they had only been studied by one discipline. In order to show how much more can be understood by putting together the evidence from palaeoecology and history particularly, we have chosen to focus mainly on one site, Leadour on Loch Tayside in the eastern part of the Breadalbane Estate (Figure 7.1), drawing in relevant evidence from the study site at Corries, also owned by the Earl/Duke of Breadalbane, to show that even within the same estate, environmental histories can vary, not least because of diverse environmental conditions and the variability of decision making under similar socio-economic circumstances. The pollen results for the farm and shieling areas at Leadour are summarised in Figure 7.2, which shows the changing incidence of the main pollen types (indicative of the dominant vegetation).

Although grazing and cultivation were established around these study sites during prehistory, our narrative commences just before 1600, when written sources become detailed enough for us to begin understanding farming and tenurial practice in detail. Around 1400 a wood became established at the farm township of Leadour, possibly on the upper slopes around what is now the head dyke. At the same time, the last few trees at the shieling, some 2 km further up the valley, died out, probably due to sustained grazing. Unfortunately the surviving documents do not indicate either how or why this birch-dominated wood became established at the farm, though we might surmise, from the lack of evidence, that it had not, for example, been planted as a result of estate policy. Since trees were owned by the landowner, and used by the tenant, we might presume too that the tenant was also not responsible, since he had no incentive to engage in the expensive job of planting something that would not belong to him. A similar situation occurred at Corries shieling, further west in the Breadalbane Estate, where the written existence for local woodland growth is somewhat ambiguous from the 16th century onwards. If we were reliant on written sources only, then we would be on very thin ice indeed when commenting on the when and where of woodlands, let alone their species composition.[24]

However, when we are able to compare the evidence for wood-grazing dynamics in written and pollen records, this highlights the complex array of factors which governed the balance between natural resources. This is particularly pertinent when considering estate regulations relating to both grazing and wood use and providing evidence of the priorities of landowners' attitudes (and the corresponding responses of the tenants) towards animals and trees. The Leadour birchwood reached its maximum extent around 1560–1630. The pollen record indicates the continuation of grazing by cattle and sheep, and cultivation of oats and bere (barley) throughout this time, suggesting that the wood owes its longevity to active management, and this concurs with the documentary evidence for estate policy on afforestation and the protection of trees. The trees were cleared around 1630 and there is little doubt as to the cause: court books record an exceptional 24 entries relating to the destruction of the birchwood between 1614 and 1620. Unfortunately it is not clear exactly why the tenants should have undertaken such a collective act of vandalism; other evidence suggests that the laird of Glenorchy kept very firm control over the management of his estate.[25]

Fortunately, the documentary evidence now provides insight into why the value of woods relative to livestock may have fallen during the mid-17th century. Stocking information through the late 16th to early 18th centuries is confined to bow books (which record the livestock and seed corn provided by the estate) and soums (stipulating the maximum number of livestock units allowed to graze each year on a defined area of land). Consequently, total livestock numbers are not available, but changes in stock types and markets are important factors in understanding ecological and economic responses at several of our study sites. Leadour appears to have been stocked with cattle in 1655, in contrast to the neighbouring farm, which continued to stock sheep, and cattle appear to have remained the main stock type until around 1700 or possibly even until the 1760s.

Of the bow cattle provided in 1660, some were 'English' suggesting that the estate was experimenting with different breeds. The market opportunities provided by the need to feed garrisons of Cromwellian troops during the civil war and, in the longer term, by the growing cattle trade with England, may have stimulated this emphasis on grazing, and on cattle in particular, at the expense of the trees at Leadour. If so, this dramatic move into cattle paid off: in 1707 the

---

[24] Davies and Watson (2007).

[25] See Watson (1997), pp. 108–111.

tenant, Ewen Ban McAphie, was able to wadset (mortgage) Leadour from the landowner (now the Earl of Breadalbane) for the sum of £1000 (c.£83 sterling). The wadset continued throughout much of the 18th century.[26]

These changes in stocking are paralleled by the pollen evidence for vegetation changes dated to the mid-17th century. Around 1650, a change in pastoral management is inferred on the farm, as a different range of herbs assume dominance in the pastures, perhaps in response to the grazing preferences of cattle, compared to a mix of cattle and sheep.[27] Peak levels of pastoral and arable activity are also recorded around the shieling at c.1680, causing vegetation cover to become increasingly species-poor and more homogeneous from c.1650–1760, in contrast with rising diversity at the farm from c.1650–1700. In other words, the same change in management regime appears to have had both positive and negative ecological impacts, depending on the location, and presumably the management regime of the grazings below and beyond the head dyke. It can be hypothesised that closer regulation of stock and soil management techniques such as manuring and the recycling of materials from the hearth, buildings and byre to the infield prevented the depletion of fertility and diversity around the farm compared with naturally more acidic conditions and less intensive soil maintenance which produced greater sensitivity or a lower threshold for change at the shieling.

From c.1680 to 1720 similar incentives may have led to increased grazing in the wood at Corries shieling.[28] Certainly the McIntyre tenants were also able to wadset their lands from Breadalbane, a testament to their abilities, like the McAphies at Leadour, to generate hard cash through cattle.[29] These comparable patterns at Leadour and Corries may be contrasted with the oakwoods around nearby Loch Awe (belonging to the Campbells of Argyll), which appear to have been more systematically exploited from around 1700, moving away from a system of communal use.[30] This may reflect the different values placed on various woodland types, from precious oakwoods of commercial value to the landowner to the 'common woods' that existed near many farms and needed by the tenants as shelter for livestock and for timber. Fragmentary information on livestock and woodland resources can thus be more productive when combined with both a broader view of changing socio-economic conditions and pollen evidence for the changing pattern of resources.

The description of Leadour and its shielings in the 1769 survey of Loch Tay compares well with the palynological data. The author mentions that the

---

[26] National Archives of Scotland, GD112/2/102/1–10.
[27] The term 'herbs' refers to herbaceous (i.e. not woody) plants, not to the culinary variety.
[28] Davies and Watson (2007).
[29] NAS GD112/2/51 15–16.
[30] Sansum (2004), p. 332.

meadows are poor, in common with other farms on the south side of Loch Tay, perhaps consisting of the sedges and grasses on the numerous damp flushes in the outfield.[31] The documentary record also provides increasing evidence for growing sheep stocks and the reorganisation of land holdings. These changes had immediate impacts on the landscape: from c.1760–1790, the fields at Leadour show evidence of heavier grazing, forming closely cropped grassland. The pollen record suggests that cultivation on the farm may have intensified after c.1790 and particularly after c.1870. Certainly the estate stipulated in 1797 that tenants had to grow peas, potatoes, turnips and flax in addition to barley and oats. Some of these crops, including turnips and potato, are undetectable through pollen analysis. Several legume pollen types occur at c.1810 but possible sources include vegetables (beans, peas), wild species or clover in sown grassland. Historical evidence is important for establishing cultivation patterns since the replacement of grain by crops which produce no pollen or insufficiently distinctive pollen could lead to erroneous conclusions regarding the diversity of crop cultivation.

However, contrary to expectations, the shielings at Leadour and also Corries show no intensification at this time; in fact, the woods at Corries shieling appear to have come under more intensive management, including tree planting.[32] This suggests that the first 'improvements' may have been aimed at permanent farmland, rather than the wider hill ground. However, the shieling pastures at Leadour subsequently underwent a strong decline in diversity from c.1820, as a result of efforts to increase productivity. By c.1860 many species had been lost locally, converted into increasingly homogeneous grassland, with less heather and a reduced range of herbs. Many of these were common plants, which are normally tolerant of disturbance, indicating the scale of grazing pressures as well as suggesting a reduction in the quality and fertility of the pastures due to overgrazing. This combined picture, while far from complete, provides new insights into the landscape effects of decision-making processes during a period of radical changes in the rural landscape, the ecological effects of which have previously been little researched.

The brief discussion above has, hopefully, provided a flavour of the kind of synthesis that can result from combining pollen and documentary evidence particularly. The ecological surveys conducted for all our sites helped to set the historical patterns into a current ecological context, which indicated that all our sites, generally speaking, currently comprise typical upland vegetation communities, although shaped by specific dominant land-uses. This has enabled us to establish that, although our sites may have reached a degree of homogeneity

---

[31] McArthur (1936).
[32] Davies and Watson (2007).

by the present, their past (as deduced through both documents and pollen) has been one of greater diversity, in every sense.

There is no doubt that analysing in detail all of our sites was a complex and time-consuming process. However, the statistical model has enabled us to bring together the data from across all of our sites.[33] The main conclusions that emerged were that agricultural prices (and therefore stocking densities) exerted a significant influence on biodiversity over the period 1600–2000, as did the extent to which sites were farmed year-round. However, no significant effects were found for climatic variables, or for extreme civil events or technology change.[34] This emphasises the long-term primacy of economic and market drivers in upland agricultural and landscape history, a conclusion that could not be as strongly supported based on results from any of the disciplines involved alone.

Interpretation and integration issues

Data interpretation is seldom unequivocal and, where different disciplines can contribute data of appropriate quality, interdisciplinary work has the potential to provide sources that can test and inform the interpretation of other datasets. The potential rewards of this can result in a reassessment of commonly accepted theories or provide new insights into established ideas. This can provide innovative and more nuanced research agendas for understanding complex human–ecosystem interrelations, thus moving research forward across disciplines and providing a sound justification for undertaking IDR. Corroboration across disciplines is evident in our work, where the data and conclusions from the historical sources backed up those from the palynological, and vice versa. Of course, contradictions can also arise, and in dealing with these the relative strengths of the data, analytical methods, and ways of drawing conclusions and associated (un)certainties must be taken into account. If there is no clear reason to support one conclusion over another then it must be made clear that contradictory evidence (or interpretations) exist. The ability of the different data/disciplines to fill in gaps in the other dataset is a clear advantage.

However, there is an important caveat to the process of integration. We regard independent disciplinary analysis as essential to the IDR process since attempts at integration during the initial analytical stage could potentially, or perhaps even inevitably, result in undue preference being given to one specific interpretation, particularly whenever one type of data is more ambiguous or apparently

---

[33] See Hanley et al. (2008).

[34] A count of innovations per 20-year period was included to allow us to recognise key periods of change, especially during the Improvement period era, which nevertheless did not take place at exactly the same time across our sites.

'stronger' than another. This recommendation strongly contrasts with the unquestioning use of generalised archaeological or secondary historical evidence in some palaeoenvironmental and ecological work, where scientific rigour is not applied to data from other sources. Here, imprecise dating or low temporal resolution may invalidate correlations between palaeoenvironmental and historical data,[35] or inferences about the long-term impacts of historically documented management from present ecology may involve significant inference and extrapolation if direct evidence of past vegetation is absent.[36] Scientists are aware that research is carried out to meet specific aims, which may make the data unsuitable for answering other questions, and the same principle must be applied when evidence from other disciplines is used. This issue also underlines our initial contention that the collaborative, rather than single practitioner, approach is preferable in interdisciplinary research, since it is unlikely that one person can undertake the entirely separate and impartial analysis of two datasets without knowledge of the first influencing the second.

In many interdisciplinary projects, the results from each discipline are presented in separate chapters before one or two members of the team write the discussion. Inevitably, more weight is given to some areas than others, dependent on the clarity with which specialists have written their own work and the interests of the writer(s) responsible for the completion of the research. Assimilating the results from a range of specialists is time consuming but writing a joint narrative in which these specialists are equally involved, and their respective sources are given equal weight, is an even more complex task, but one we believe is worth the effort, as evinced by the popularity of synthetic works such as Diamond (2005). As part of that process, it is also important not to summarise or simplify the extent to which uncertainties turn into certainties, or that important nuances in the data or interpretation are lost.

In the first instance, the contributors must read and assimilate the reports of their colleagues and then reflect on the implications of these analyses on their own disciplinary interpretation. The next stage should begin a new phase of interdisciplinary discussion, resulting in the generation of an integrated timeline showing the points of agreement in events and interpretations across the relevant disciplines, the points of disagreement and the points of ambiguity. From this, particular themes will begin to emerge, to be moulded into a range of narratives, depending on which disciplines wish to take the lead in writing up particular aspects and which disciplinary audience they wish to target. In essence, these efforts allow the complex, complementary and sometimes contradictory histories of the same landscapes derived from different disciplines to come together in a coherent and methodologically convincing manner. Although there are some journals and presses interested in interdisciplinary

---

[35] For example, Baillie (1991); Dumayne et al. (1995); Tipping (2004).
[36] Tipping et al. (1999).

work *per se,* the writing-up process should also allow the results to be presented in a range of disciplinary publications, introducing the potential of collaboration to a wider audience. This also satisfies current demands, certainly in the UK, for the individual researcher to publish in as high a quality journal as possible within their own discipline, though this also implies an open-mindedness on the part of editors to include innovative work across boundaries as well as within them.

General guidelines and conclusions

The nature of IDR means that every project will, to some degree, be unique. Consequently, there can never be one correct overall methodological template to adopt or follow. However, the kinds of issues which we had to confront will be appropriate to most collaborative IDR projects and will require similar anticipation and responses to deal with them. We have summarised

Table 7.2   Issues arising in interdisciplinary research.

Issue	Comments
Select appropriate people	A flexible attitude and an openness to new ways of thinking and approaching problems may be more important than the level of discipline expertise/experience.
Early planning and participation	Involve all parties/disciplines as soon as possible in all aspects of planning the project. This can help identify areas of misunderstanding early.
Support structures/attitudes	Identify and use institutional support structures for interdisciplinary work. If lacking, can any new initiatives be started? Are colleagues/superiors supportive?
Time and funding	Studies will take longer, so build in more time via funding (to allow people/disciplines to familiarise, for extra discussion time, and for writing/publishing).
Aims, objectives and overall boundary	Aims should be explicit and understood by all. Although crossing boundaries is expected, an overall boundary should be defined to stop the project 'drifting' off the target.
Milestones and outputs	Specify early in project the milestones and outputs, as part of defining targets to aim for and to judge success.
Contribution of different disciplines	Identify where and how each discipline will contribute, and for different project parts identify whether one discipline is necessarily more dominant.
Required degree of integration	Integrate work to the required degree, but maintain discipline integrity as required.
Communication and language problems	Individuals must have time available to learn the necessary basics of other disciplines, including philosophy, language, methodology and conventions

Table 7.2   (Continued)

Issue	Comments
Group participation	Expect meetings to be a time consuming but vital part of developing understanding. Schedule more than usual, preferably regular (not just nearing project completion), and face-to-face.
Integration of data	Explore ways of integrating different data types, by addressing differences in scale, applicability and precision, and dealing with complementarities and contradictions.
Integration of conclusions	Different disciplines have different routes to, and certainties attached to, conclusions. Work out how to tie these different 'stories' together, especially if they have different certainty weightings.
Publishing protocols	Interdisciplinary results may be harder to publish (than single-discipline papers). Identify outputs and target journals early, and also who or which discipline takes the lead for each.
Leadership and organisation	The leader cannot be an expert in all fields, but takes a co-ordinating role in keeping the team together, communicating, integrated, and focused on the aims and objectives.

Further perspectives can be found in appropriate papers (e.g. Turner and Carpenter 1999; Frost and Jean 2003; Bruce et al. 2004; Amey and Brown 2005; Boulton et al. 2005; Tress et al. 2005).

further issues in Table 7.2, drawing on our own experience and that of other authors. We strongly recommend this table as a general 'checklist' to anyone undertaking an IDR project.

In essence, the key to conducting IDR, no matter who is collaborating with whom, is an open mind, an honest, reflective understanding of the mechanics of your own discipline, combined with a willingness to be appropriately flexible when interacting with others, and, above all, a desire to – in the appalling jargon of our times – think outside the diminishing boxes of our own knowledge bases. We would argue strongly that, in order to get the best out of IDR, it is necessary to avoid the 'post-study' model, whereby the data is gathered and interpreted by the different disciplines, and only then is there some attempt made to stitch the different stories together. IDR needs to be planned from the start, to make sure that the different parts of the project really will fit together, that the data gathered is as compatible as possible and designed to address the same aims and objectives, and thus advance our collective understanding.

We advocate that a collaborative approach should, on occasions, be considered as the best method of engaging in primary research of an interdisciplinary nature. In a subject like environmental history, where a considerable number of disciplines are already addressing very similar questions from their own perspectives, it makes a lot of sense, on occasions, to try to answer them together.

We argue that there are considerable methodological implications in *not* engaging in collaborative work, implications that can only be overcome by becoming sufficiently expert in another discipline so that one's analyses will stand up to scrutiny by peers within that discipline.

Collaborative IDR does, of course, engender methodological implications of its own. We hope that we have addressed many of them through both our general observations and the case study drawn from our own experience. Some of the most important issues include the compatibility of data, the repercussions for documentary history of the site-driven approach required by scientists and social scientists, and how to handle holes in the dataset and still generate a credible analysis and interpretation. These are all difficult and important questions but ones which can be overcome (though we must acknowledge when they cannot). The results, we sincerely believe, are more than a sum of their parts (though the parts also make interesting reading). Certainly, the more we do of such collaborative work, the more powerful will become the tools at our disposal. If we truly believe that the past can inform the present, then it is incumbent on all of us engaged in analyses of historical conditions, whether of the environment or the humans living within it, to produce the best and most complete work we can. There will always be a place for expert witnesses able to synthesise the work of others into useful new paradigms, but they can only do so if the primary research from which they draw their conclusions continues to be superlative in all its aspects. Working together ensures that past centuries of disciplinary refinement and methodological authority are maintained but disciplinary boundaries are no barrier to knowledge.

Acknowledgements

The authors wish to thank the Leverhulme Trust for funding this project. We would also like to acknowledge the role played by the Centre for Environmental History, funded by the Arts and Humanities Research Council at the University of Stirling, where this research took place and which acted as an essential environment in which such interdisciplinary collaboration could be instigated and seen through.

References

Amey, M.J. and Brown, D.F. (2005), 'Interdisciplinary collaboration and academic work: a case study of a university-community partnership', *New Directions for Teaching and Learning* 102, 23–35.

Baillie, M.G.L. (1991), 'Suck-in and smear: two related chronological problems for the 90s', *Journal of Theoretical Archaeology* 2, 12–16.

Barber, J. (ed.) (1997), *The Archaeological Investigation of a Prehistoric Landscape: Excavations on Arran 1978–81*, Edinburgh.

Birks, H.J.B. and Line, J.M. (1992), 'The use of rarefaction analysis for estimating palynological richness from quaternary pollen-analytical data', *The Holocene* 2, 1–10.

Botkin, D.B. (2007), 'Scientific opinion and the opinion of scientists'. In *People and Nature*, (http://www.danielbbotkin.com/archives/category/people-nature).

Boulton, A.J., Panizzon, D. and Prior, J. (2005), 'Explicit knowledge structures as a tool for overcoming obstacles to interdisciplinary research', *Conservation Biology* 19, 2026–2029.

Brown, A., Horsfield, D. and Thompson, D.B.A. (1993), 'A new biogeographical classification of the Scottish Uplands. 1. Descriptions of vegetation blocks and their spatial variation', *Journal of Ecology* 81, 207–230.

Bruce, A., Lyall, C., Tait, J. and Williams, R. (2004), 'Interdisciplinary integration in Europe: the case of the Fifth Framework programme', *Futures* 36, 457–470.

Crosby, A.W. (2004), *Ecological Imperialism: The Biological Expansion of Europe, 900–1000*, Cambridge.

Davies, A.L. and Dixon, P. (2007), 'Reading the pastoral landscape: palynological and historical evidence for the impacts of long-term grazing on Wether Hill, Cheviot foothills, Northumberland', *Landscape History*.

Davies, A.L. and Tipping, R. (2004), 'Sensing small-scale human activity in the palaeoecological record: fine spatial resolution pollen analyses from West Glen Affric, northern Scotland', *The Holocene* 14, 233–245.

Davies, A.L. and Watson, F. (2005), *Environmental History: Problems and Potential in the Integration of the Sciences and Humanities*. Scientific report, ESF Exploratory Workshop EW04-158, held November 2005 (http://www.esf.org/generic/2167/04158Report.pdf).

Davies, A.L. and Watson, F. (2007), 'Understanding the changing value of natural resources: an integrated palaeoecological-historical investigation into grazing-woodland interactions by Loch Awe, Western Highlands of Scotland', *Journal of Biogeography* 34, 1777–1791.

Devine, T.M. (1994), *Clanship to Crofter's War*, Manchester.

Diamond, J. (2005), *Collapse: How Societies Choose to Fail or Survive*, Oxford.

Dodgshon, R.A. (1998), *From Chiefs to Landlords: Social and Economic Change in the Western Highlands and Islands, c.1493–1820*, Edinburgh.

Dodgshon, R.A. and Olsson, G.A. (2006), 'Heather moorland in the Scottish Highlands: the history of a cultural landscape, 1600–1880', *Journal of Historical Geography* 32, 21–37.

Dovers, S., Edgecombe, R. and Guest, B. (eds) (2002), *South Africa's Environmental History: Cases and Comparisons*, Cape Town.

Dovers, S. (ed.) (1994), *Australian Environmental History: Essays and cases*, Oxford.

Dumayne, L., Stoneman, R., Barber, K. and Harkness, D. (1995), 'Problems associated with correlating calibrated radiocarbon-dated pollen diagrams with historical events', *The Holocene* 5, 118–123.

Frost, S.H. and Jean, P.M. (2003), 'Bridging the disciplines: interdisciplinary discourse and faculty scholarship', *The Journal of Higher Education*, 74(2), 119–149.

Froyd, C.A. (2001), *Holocene Pine (Pinus sylvestris L.) Forest Dynamics in the Scottish Highlands*. Unpublished PhD thesis, University of Cambridge.

Hamilton, A. and Davies, A.L. (2007), 'Written in the hills: an environmental history project in the Scottish uplands', *History Scotland*, 25–32.

Hanley, N., Davies, A., Angelopoulos, K., Hamilton, A., Ross, A., Tinch, D. and Watson, F. (2008), 'Economic determinants of biodiversity change over a 400 year period in the Scottish uplands', *Journal of Applied Ecology*, 45, 1557–1565.

Hunter, J. (1976), *The Making of the Crofting Community*, Edinburgh.

Krech, S. (2000), *The Ecological Indian: Myth and History*, New York.

Lindbladh, M., Bradshaw, R. and Holmqvist, B.H. (2000), 'Pattern and process in south Swedish forests during the last 3000 years, sensed at stand and regional scales', *Journal of Ecology* 88, 113–128.

McArthur, M.M. (1936), *Survey of Lochtayside 1769*, Edinburgh.

Parry, M.L. (1978), *Climatic Changes, Agriculture and Settlement*, Folkestone.

Rackham, O. (1989), *The Last Forest: The Story of Hatfield Forest*, London.

Rackham, O. (2000), *The History of the Countryside*, London.

Richards, E. (2000), *The Highland Clearances: People, Landlords and Rural Turmoil*, Edinburgh.

Sansum, P.A. (2004), 'Historical resource use and ecological change in semi-natural woodland: western oakwoods in Argyll, Scotland'. Unpublished PhD thesis. University of Stirling, Stirling.

Simmons, I.G. (2001), *An Environmental History of Great Britain: From 10,000 Years Ago to the Present*, Edinburgh.

Stevenson, A.C. and Thompson, D.B.A. (1993), 'Long-term changes in the extent of heather moorland in upland Britain and Ireland: palaeoecological evidence for the importance of grazing', *The Holocene* 3, 70–76.

Tipping, R., Buchanan, J., Davies, A. and Tisdall, E. (1999), 'Woodland biodiversity, palaeo-human ecology and some implications for conservation management', *Journal of Biogeography* 26, 33–43.

Tipping, R. (2000), 'Palaeoecological approaches to historical problems: a comparison of sheep-grazing intensities in the Cheviot Hills in the medieval and later periods'. In Atkinson, J., Banks, I. and MacGregor, G. (eds), *Townships to Farmsteads: Rural Settlement Studies in Scotland, England and Wales*, British Archaeological Reports (British Series), 130–143.

Tipping, R. (2004), 'Palaeoecology and political history; evaluating driving forces in historic landscape change in southern Scotland'. In Whyte, I.D. and Winchester, A.J.L. (eds), *Society, Landscape and Environment in Upland Britain*, Society for Landscape Studies supplementary series 2, 11–20.

Tress, B., Tress, G. and Fry, G. (2005), 'Ten steps to integrative research projects', In Tress, B., Tress, G., Fry, G. and Opdam, P. (eds), *From Landscape Research to Landscape Planning: Aspects of Integration, Education and Application*. Heidelberg, 241–257.

Turner, M.G. and Carpenter, S.R. (1999), 'Tips and traps in interdisciplinary research', *Ecosystems* 2, 275–276.

Usher, M.B. and Balharry, D. (1996), *Biogeographical zonation of Scotland*. Perth.

Watson, F. (1997), 'Rights and Responsibilities: Wood-management as seen through Baron Court Records', In Smout, T.C. (ed.), *Scottish Woodland History*, Edinburgh, 101–114.

8

New Science for Sustainability in an Ancient Land

Libby Robin

Global ideas, local contexts

In this chapter, I want to consider the history of the overlapping revolutions in science and society in Australia that have come together in the partnerships for 'sustainability'. Although this is a global phenomenon, its local manifestations are subject to the strong biophysical constraints of the Australian continent. Australian environmental conditions, I argue, make for distinctive nuances in the way the science of conservation biology has evolved there. My chapter is in a sense a 'worked example' of William M. Adams's argument that '[c]onservation is . . . geographically diverse, historically changing and contested'.[1] It is as Adams says, 'fundamentally a social phenomenon or social practice'. R.A. Kenchington, an Australian coastal zone scientific manager, put it succinctly: 'We do not manage the environment', only the human behaviours that affect its structure and processes.[2] It is not just the human behaviours, it is also the nature of the ecological crisis itself. Crisis demands more than 'just science', as Michael Soulé, a doyen of conservation biology, commented:

> Conservation biology differs from most other biological sciences . . . it is often a crisis discipline . . . In crisis disciplines, one must act before knowing all the facts; crisis disciplines are thus a mixture of science and art, and their pursuit requires intuition as well as information.[3]

The politics of Aboriginal Australia also places distinctive constraints on environmental management, particularly in the Northern Territory, where approximately 85% of the coastline and 44% of the total land mass is held under

---

[1] Adams, 'Separation, Proprietorship and Community in the History of Conservation', Chapter 2, p. 50.
[2] Kenchington (1994), 245–50.
[3] The term 'science of crisis' is discussed in Soulé (1985), 727.

Aboriginal title.[4] This in turn, affects the practice of conservation biology. My argument draws both on the history of science and environmental history, showing how an environment shapes the cultural practice of science as the scientists go in to battle for it. It is environmental history in that both the science and the society are needed to explain the environment, and the environment is itself a force shaping science in society. Sörlin and Warde recently challenged environmental historians to define more clearly the 'problem' of environmental history, specifically the theoretical linkages between environment, politics, society and culture: Science studies, particularly those that consider 'sustainability sciences', provide a natural place to articulate some of these connections.[5]

This chapter brings into dialogue the several distinct strands in environmental history: the broad history of environmental ideas (including science), the late twentieth-century and twenty-first century obsession with 'sustainability' (which includes development), and the context-dependent historical study of a particular science (conservation biology) in a particular place (Australia). Australia's ancient land carries a number of diverse but intersecting 'exceptionalisms', including a very long human history, a first world economy, very high extinction rates, megadiversity and high scientific literacy in the general population (including politicians). Increasing engagement with Aboriginal understandings of 'country' is also seen to be essential to developing paths to sustainability in Australia. The history of environmental ideas in Australia – scientific, Indigenous and non-Indigenous – reveals an entwining of purpose over the past six decades, as the extent of the local and global ecological crises have become more apparent. Some Aboriginal elders indeed comment that 'the future is behind us'.[6]

I want to start with an event from recent history to highlight the range of issues that Australian conservation biologists are grappling with simultaneously, and then work back from there, to unpack the history of their dilemmas.

A parable of our times

On 22 July 2006, David Lindenmayer, a prominent conservation biologist, was awarded the inaugural Daimler-Chrysler Australian Environmental Research Award for leadership in environmental research at a black-tie dinner in Melbourne. As he walked up to the podium to accept his award and the cheque, he made a quick decision. Lindenmayer declared he was 'honoured to be recognised as a leader in outstanding environmental research', but he immediately

---

[4] Jackson, Storrs and Morrison (2005), 105–9.
[5] Sörlin and Warde (2007), esp. 118.
[6] Deborah Rose, *pers. com.*, 24 March 2006.

gave away the money. He publicly presented it to the government to maintain a traineeship program for Darren Brown, a senior Wandandian man from the Wreck Bay Aboriginal Community, who was working with Lindenmayer in a fire ecology research project.

'We're not doing enough . . . to use the incredible environmental knowledge and skills of Indigenous people in this country', Lindenmayer declared.[7] Darren Brown's work had involved radio-tracking bandicoots and diamond pythons as part of Lindenmayer's team at Booderee National Park (Jervis Bay) on the coast south of Sydney. The project was to close prematurely because the government had cut its funding.[8] Lindenmayer had been urging the government officers to reconsider this decision, and felt that by presenting the $30,000 (€18,000 or $US24,000, approximately) associated with such a prominent and prestigious award, he could show the value of Aboriginal traineeships to conservation biology.

Lindenmayer's action catapulted him and Darren Brown into the media. A partnership between conservation biology and Indigenous knowledge was news. 'People like Darren are the future', Lindenmayer said. 'The environment is their passion and if they get the right opportunities, conservation in this country will really take off in some exciting new directions.'[9] Environmental supporters, including many key government decision makers, sitting at a $100/head (€60/$US 80) dinner needed reminding that his Aboriginal training program offered employment and scientific skills to local people, as well as Indigenous knowledge for practical science. Darren Brown was, Lindenmayer observed, younger than he was. At just 34 years old, he was an elder in an Aboriginal Community where people, particularly men, die young. Life expectancies for Indigenous people continue to be very low in Australia – more than 20 years lower than for non-Indigenous people.

Working with Aboriginal people 'cuts two ways', Lindenmayer commented. Understanding the Australian environment is a complex and unfolding task: science and Aboriginal knowledge can learn much from each other. Darren Brown agreed. He said that conservation biology could enrich his cultural perspectives by 'bringing the science to broaden out the picture'.[10] Aboriginal health issues are a major problem in Australia's otherwise first-world society. 'Caring for country' and having a passion for the bush are good for both health and environment, but while this connection is acknowledged by Indigenous

---

[7] Quoted in Beeby (2006).
[8] The responsible group was the Commonwealth Department of Transport and Regional services because Jervis Bay (where Booderee National Park is located) is a Federal Capital Territory, but discontinuous with the land that surrounds the nation's capital, Canberra.
[9] Beeby (2006).
[10] ibid.

people, non-Aboriginal environmental managers have yet to make the link. Conservation work and Aboriginal health are in different ministerial portfolios. People themselves maintain that understanding the connection between health and 'country' is essential for Indigenous futures. Conservation partnerships between science and Aboriginal knowledge are part of a set of revolutions that have the potential to transform conservation biology and sustainability science into something that really address Australia's exceptional biophysical conditions.

One of the other revolutions captured in this moment is the recognition that biodiversity is important *beyond* national parks. It is important for culture, Aboriginal and settler, alike. The idea that biodiversity itself depends on country outside and beyond reserves is an international one. *Benefits Beyond Boundaries* was, for example, the theme of the Vth IUCN World Parks Conference held in Durban, South Africa in September 2003. The idea of biodiversity conservation *alongside* people (in economically productive landscapes) was crucial to the politics of post-Apartheid South Africa, where 'protected areas' had been perceived as benefiting rich visiting game shooters at the expense of disadvantaged local people.[11] But the idea of considering the biodiversity values of land outside parks is politically important in different ways in Australia, where there has been traditionally little research undertaken on the 'nature' that has survived under farming and forestry regimes. David Lindenmayer has been a pioneering exponent of the study of landscape change in the context of habitat fragmentation.[12]

Lindenmayer has increasingly embraced people (Indigenous and non-Indigenous alike) in the practice of biodiversity conservation. Thinking beyond and outside park boundaries has enabled him and others to recognise that a national park like Booderee is also a 'production landscape' for Indigenous peoples. Reserves are cultural and not just biophysical places. Blurring the natural/cultural divide in science is enabling for collaborative work.

Lindenmayer himself was surprised about the other trend revealed by his generosity. The government flatly refused to reconsider the program. His gift bought time, but not much. In the brave new world of 'small government' Australia, there was no longer room for bureaucratic concessions to good ideas. The Australian tradition of public sponsorship of science had evaporated as the twenty-first century began, sacrificed in the ideology of small government. What used to be regarded as science for the 'public good' was no longer to be funded as such. Simultaneously, federal and state governments (of both major political persuasions) have divested themselves of the tasks of managing

---

[11] Hall-Martin and Carruthers (2003); Child (2004); Carruthers (1995); Carruthers (2006).
[12] Lindenmayer and Fischer (2006).

reserves for conservation. Biodiversity is no longer regarded as a resource worthy of public funding. Lindenmayer felt that putting his own money on the table would coax or at least embarrass the government department to match the funds to maintain the program he had developed to complement his Australian Research Council funded scientific research. But despite enthusiastic media coverage, the government department did not budge.

Partnerships for science and sustainability

The global idea of sustainability has distinctive local manifestations in Australia, the most obvious of which is 'ecologically sustainable development' (ESD). ESD was a uniquely Australian response to the declaration of the United Nations Conference on Environment and Development at Rio de Janeiro in 1992, that 'environmental protection shall constitute an integral part of the development process and cannot be considered in isolation from it'. By the end of a decade ruled by a *National Strategy for Ecologically Sustainable Development*, ESD was enshrined in 123 pieces of state and federal legislation.[13] Australia's decision to call its path to sustainable development 'ecological' was at odds with all the other 178 signatories to the Rio Convention. 'Ecology' had a political ring in Australia, different from other places. ESD endorsed the notion that science, particularly ecology, should be paradigmatically central to development and environmental management.

Conservation biology, an applied science 'concerned with the management of biological diversity (or biodiversity)' is also prominent in Australia, a legacy of the close relations between science and government over most of the twentieth century.[14] This has come at a cost. The strongest incentives for researchers have always been to undertake science of 'policy relevance'. Historically, the lack of support for 'pure science' dated right back to the colonial period, long before Federation, when a pragmatic attitude also prevailed: 'Zoology, Mineralogy and Astronomy and Botany, and other sciences are all very good things', commented the *Sydney Monitor* of 1833, 'but we have no intention of an infantile people being taxed to promote them'.[15]

The science of conservation biology emerged internationally contemporaneously with the forces for sustainability, defined most prominently in the Brundtland Report in 1987, which explicitly linked science with practical outcomes for development.[16] Biodiversity has had a high political profile in Australia since the late 1960s, but this has accelerated rapidly since the 1990s

---

[13] Stein (2000).
[14] Lindenmayer and Burgman (2005), 5.
[15] Moyal (1986), 88.
[16] World Commission on Environment and Development (WCED) (1987).

as Australia's exceptional extinction rates have become better documented and managers of conservation reserves have sought guidance for stemming the tide.

An exceptional place

David Lindenmayer and Mark Burgman's textbook, *Practical Conservation Biology* (2005) opened with the idea of Australia's physical exceptionalism:

> Australia is the oldest, flattest and most isolated continent in the world. It supports some of the oldest living life forms on the planet...the largest area of coral reef, the longest fringing reef, the largest areas of seagrass meadows and the most species-rich marine fish, mangrove and algal assemblages in the world...Less than 0.2% of [its] land surface [is] subject to snowfall...Australia also has the smallest area of wetland of any continent...Climatic conditions over much of Australia are highly variable over time – more variable than anywhere else...Australia's soils, and its freshwater and marine systems, are nutrient-depleted. All of these factors directly influence the ecology and dynamics of the species that inhabit the continent and the marine systems surrounding it.[17]

Flora and fauna (terrestrial and marine) are 'markedly different' from elsewhere on the planet. In the jargon, 'most taxonomic groups are species-rich and highly endemic', which means that within Australia animals and plants generally have many different varieties of each type (rich in species), but often no closely related forms outside Australia (endemic). This makes Australia 'megadiverse', one of only 17 countries that together harbour some 70% of the earth's species. Among the 17 megadiverse countries, Australia is exceptional too, because there is only one other, the United States, which has a developed, industrialised economy.

It is not just biodiversity and climate that are exceptional in this country. Even the way water flows defies logic. Dictionary definitions of 'rivers' do not account for Australian reality, as lexical cartographer Jay Arthur has argued. The Lachlan River, for example, is not 'a considerable natural stream of water flowing in a definite course or channel', but rather a 'series of meandering channels and marshes' that loses itself 'in a maze of swamps';[18] the Lachlan's behaviour confused explorers seeking inland seas. Unpredictability also limits the ways farmers can work in these wetlands. Irregular and intermittent watercourses are vulnerable, and need understandings beyond hydrology and engineering. Aboriginal people on the river Darling near Bourke in New South Wales told

---

[17] Lindenmayer and Burgman (2005), 1.
[18] Arthur (2003), 8–19.

historian Heather Goodall that too much management can make a river run 'backwards'. The story goes that when the pumps were turned on in the Darling for the irrigation of cotton, the flow reversed. For Aboriginal people in the Bourke district, a river is all about 'flow' and the failure to flow is a sign of a problem for the river and for the country.[19]

Such knowledge seems to fit uneasily with western understandings of rivers, but hydrologists also struggle to express the lack of reliability in 'annual flow' in Australian rivers as well. The United Nations Food and Agriculture Organisation 'Aquastat' statistics reveal orders of magnitude differences in levels of variability in the Murray–Darling river system when compared with other river systems. They are simply not on the same scale. For the Rhine in Switzerland or the Yangtze in China, the ratio between maximum and minimum annual flow is approximately 2, for the Amazon in Brazil, somewhat less at 1.3. The Murray is 15.5 (comparable with the Orange River in South Africa at 16.9). But the exceptionally variable Darling River is right off the graph – the figure is 4705.2![20] Aboriginal knowledge, practical farmers' traditions, and hydrological measures of water flows each grapple to express such variability of river flow.

Ephemeral wetlands – the 'occasional water places' of the desert country – are precious because they are rare (both in time and space). How they behave locally can matter for breeding populations of birds and animals from all over the continent.[21] How Australian scientists approach questions of wetlands and wildlife management has depended upon three factors: the exceptional biophysical conditions, the international historical odyssey of the science of ecology towards conservation biology and sustainability science and the particular demands to be policy relevant in Australia.

Historical strategies for protecting species

Land and natural resources are constitutionally the responsibility of state governments in Australia, and conservation activism has also tended to operate at this level. Scientists and activists began with a concern to reserve local 'land for nature' in national parks, reserves and other exclusionary zones, not a national vision. The first coordinated national view of efforts in nature conservation was that of the Australian Academy of Science, founded in 1954.[22] Its first publication, edited by John Turner in 1957, considered the environmental condition of the Australian Alpine regions straddling the Australian Capital

---

[19] Goodall (2002). This story is a parable, rather than a first-person witness account, but it had wide currency in the community.
[20] http://www.fao.org/AG/AGL/aglw/aquastat/background/index.stm
[21] Robin (1998).
[22] Robin (1994), 1–24.

Territory, New South Wales and Victoria.[23] The foundation of the Ecological Society of Australia (ESA) in 1961 added a further 'national science' perspective. In 1965, when the Australian Conservation Foundation was formally established, scientists from both the Academy and the ESA were prominent in defining its aims. Australia's national perspective emerged quietly in large part, through prominent senior scientists with international profiles and requests for information from international organisations like UNESCO.[24] Conservation in Australia became national because of international imperatives, but still largely operates through the filter of state instrumentalities.

The Academy's National Parks Committee worked closely with the new ESA, which sought the opportunity to 'make useful comments on the proposals before your committee'.[25] Max Day, the Chair of the Academy's committee, responded with the central 'complex ecological question' facing the Academy of Science:

> What is the minimum size necessary to ensure that a reserve is a self-perpetuating unit in various types of environment in Australia?...We are constantly embarrassed by lack of knowledge of the nomadic habits of macropods, for example, the size of an area necessary to recover from the effects of fire and so on.[26]

This question has dogged conservation biologists ever since, although few now would believe that a large park alone could save a species. In 1961, it was well beyond the scope of a brand new small professional organisation working in a country where so little was documented about the varieties of fauna, let alone the habits of particular species and responses to common major events such as fire and flood.

Gap analysis: a scientific basis for surveys

The Academy's work began with its scientific strengths. Australia's plant ecology was better documented, and vegetation did not move as rapidly and unpredictably as fauna. In an ecological era when vegetation was seen to stabilise in a 'climax' state, vegetation-mapping provided a firm basis for the 'gap analysis' approach to biological resources in reserves. The scientific community was reluctant to use its limited resources on simple descriptive surveys

---

[23] Turner (1957); Robin (1998).

[24] Robin (1994).

[25] John Calaby, Secretary of the ESA, to Max Day, Chair of the Academy's National Parks Committee 16 May 1961 [Australian Academy of Science (AAS) file 1004].

[26] Day to Calaby, 3 August 1961, AAS file 1004.

although it was clear that there was a severe shortage of baseline information. Gap analysis redefined a descriptive survey as a part of an overarching system amenable to scientific insight. The nature reserves of the country were taken together as a single 'system' preserving as many of the nation's ecosystems as possible (This notion continues to the present through what is now called a National Reserves System). Each local survey contributed to a larger 'national' project to determine what plant ecosystems were present in the region and absent from existing reserves. A set of surveys together identified 'gaps' in the reserves system, indicating the places where the missing ecosystems that could fill the gaps could be reserved. The gap analysis approach was useful because it was exhaustive and scientifically justifiable. It also provided a means to prioritise reserve selection. The need to advise governments on policy was never far away.

Gap analysis was one of the first attempts to unite the cause of the resource manager with that of the amateur field naturalist, who in this era was the typical national parks activist. The scientific approach cooled emotional pleas for saving bushland, and provided a way to make pragmatic choices. Biological resource potential was evaluated by experts in the same way that the mineral resources had been identified since before the nineteenth-century gold rushes. Nature reserves were no longer just a matter of personal taste or aesthetics, nor would they be limited to the 'leftover lands', deemed worthless after every possible economic use had been tried first.[27] This approach valued biological resources in economic terms, a precursor of what is now recognised as 'ecosystem services'.[28] Gap analysis also marked the beginnings of the 'science of crisis', conservation biology, although it would be some years before it was known by that name.

The fact that potential reserves were identified on the basis of biological criteria more or less independent of existing land tenure patterns made this system seem rather different from lobbying to save the 'last of' and the 'remnants'. The presence or absence of an ecological community depended on how that area had been used in the past, but not on the present title-holder. Filling in the gaps made better sense in small, well-documented states like Victoria than in the poorly documented central deserts or northern tropics. Identifying places spatially, while essential for a 'reserve system', did not allow for dynamics and natural change. The reserves themselves were frozen in time, legally 'permanent' – in line with the Victorian National Parks Association's first motto in 1952, 'for all people for all time'. Yet what happened to ecosystems within their boundaries could change according to what happened outside.

---

[27] The idea of national parks as favouring 'worthless lands' was developed for the United States case by Alfred Runte (1977).
[28] Daily (1997).

The science behind the 'gap analysis' approach was based firmly on the idea of climax ecosystems and steady-state climate conditions. It begged Day's original concerns about the effects of fire or the need for a range for nomadic animals beyond the immediate reserve. In 1981, the historian Jim Davidson, looking back on this period, mocked the short-sightedness of the technique adopted by reserves managers in most states in his wry comment that the National Park Service of Victoria displayed 'an almost philatelic concern... to complete its set of parks drawn from the 62 major habitat types to be found in the State'.[29] Stamp collecting was hardly great science, nor could it deal with the dynamics of change. But gap analysis shifted biological diversity from an abstraction of science to an imperative of management. The next step was to alert the politicians to the 'crisis'.

Conservation biology: international science of crisis in the Australian context

In 1980, one of the earliest textbooks to carry the title *Conservation Biology* hit the streets. Its subtitle was *An Evolutionary-Ecological Perspective*, and it was edited by Michael Soulé and B.A. Wilcox. It followed the first international conference on conservation biology, held at the University of San Diego in 1978. Six years later Soulé (without his co-editor) produced another *Conservation Biology* with a completely different subtitle: *The Science of Scarcity and Diversity*. The 1986 book was 25% longer, with 25 new chapters, and about 90% of its authors were new.[30] Between 1980 and 1986 the 'process' of conservation biology (built on evolutionary-ecological principles) had become a 'science of scarcity and diversity', with the principle of diversity integral to its status as a science. If 'diversity and rarity are synonyms for "everything" in ecology', as Soulé observed, the practical science of conservation biology needed to start with these fundamental issues.[31] The popular shorthand for biological diversity officially became 'BioDiversity' (biodiversity) at a National Academy of Sciences/ Smithsonian Institution forum in Washington, D.C. in September 1986, according to Edward Wilson.[32] The International Society for Conservation Biology also formed at this time (8 May 1985), with its journal *Conservation Biology* appearing from May 1987.[33]

---

[29] Davidson (1981), 3–7.
[30] Soulé (1986), ix.
[31] Soulé (1986), 117.
[32] Wilson (1997), 1.
[33] Society for Conservation Biology website: http://www.conbio.com/AboutUs/History; Fazey, Fischer and Lindenmayer (2005).

The theoretical origins for conservation biology and the idea of biodiversity were based on autecology (studies of the relation of a single species to all aspects of its environment), but since the late 1980s there has been a growing counter-argument that *populations* rather than species should be the 'operational unit of conservation evaluation'.[34] Graeme Caughley and Ann Gunn in the mid-1990s were still advocating the species as the unit of conservation because species are more measurable than populations. Caughley, sensitive to the work-ings of conservation bureaucracies, recommended 'species' because they were amenable to the sorts of quantitative statistics useful to managers. But the fact that the species is also the unit of natural selection made his choice attractive to evolutionary ecologists beyond the bureaucracy as well.

Soulé identified a number of problems with a 'science of crisis'. A 'mission-oriented crisis discipline' is constantly required to cross a frontier between the 'real world' of management situations, which are, in a sense, the 'ulti-mate test' of the theories of conservation, and the 'private world' (the world of pure science). Conservation biology demanded a constant dialogue of ideas, guidelines and empirical results travelling from science into management, and issues, problems, criticism, constraints and changed conditions feeding back into the science.[35] This was familiar territory for Australian scientists, whose funding since the 1920s had depended on such 'policy relevance'. And legisla-tors were willing to assume a high level of general scientific literacy. By 1999, Australia's environmental legislation defined biodiversity for management pur-poses as 'the variability among living organisms from all sources . . . and includes diversity within species and between species and diversity of ecosystems'.[36] Legislators were perhaps more concerned to be inclusive than being scientifi-cally rigorous. The pragmatic craft of 'ought-ecology' differed significantly from its scientifically exact progenitor (autecology). Conservation management is not strictly about biological communities.

Where to begin?

Measuring whole populations and diversity within and between species is often beyond the first concern in a biodiversity crisis. Steve Morton, a senior Australian arid-zone ecologist, commented that in much of his ecological work he felt like an ambulance officer arriving at the scene of a horrific accident. Triage was the only option remaining.[37] 'Threats' are often related to human behaviour (for example, land clearing), but they tend to be listed in a report about the species threatened, rather than as comments on the society. The

---

[34] Mackey et al. (1998), 14.
[35] Soulé (1986), 2–3.
[36] Australia *(91/1999)* S528.
[37] Morton (2000).

expertise for an analysis of society is seldom included in the Triage Team. A state of panic can inspire action, but it also tends to entrench certain ways of thinking, and to limit input from experts beyond scientific disciplines.

The first conference on Conservation Biology in Australia and Oceania was held at the University of Queensland in 1991. It professed not just an interest in data, but also in 'the uses to which the data are put'. The editors of the conference proceedings, Craig Moritz and Jiro Kikkawa, articulated the responsibilities of scientists thus: 'biologists are the representatives of the natural world... We can see very clearly what is happening, what will be the irreversible consequences for biology and humanity, and how the solutions must be constructed.'[38] But rather than describing biological communities, most of the chapters in the book in fact reviewed policy-making and environmental management. Authors recognised that practical outcomes had an important part in justifying the continuation of scientific programs, and generally shared Soulé's view that crisis management demanded action ahead of complete knowledge. 'The sheer number of undescribed taxa and the threats they face call for new ideas and approaches to taxonomic description and genetic studies,' botanist Steve Hopper, one of the presenters, commented.[39]

Australian conservation biology has quickly become a major player in sustainability science, testing – and often disproving – international theories in distinctive Australian conditions. David Lindenmayer and Mark Burgman had an experience that paralleled Michael Soulé's in the 1980s. They prepared a textbook entitled *Conservation Biology for the Australian Environment* in 1998 that, just seven years later, was completely rewritten as *Practical Conservation Biology*.[40] It went from 380 pages to over 600. Four major new chapters were added on Harvesting natural populations (including native and plantation forestry, kangaroos and fisheries), Vegetation loss and degradation (clearing, mining and urbanisation), Landscapes and habitat fragmentation and Fire and biodiversity. The treatment of traditional Aboriginal land use, although brief in both versions, was revised significantly in the later book to include more on ongoing Aboriginal management, and the views of Indigenous managers themselves, especially Marcia Langton, a prominent Indigenous spokesperson on fire and land rights.[41] In the first textbook, most of Aboriginal management was 'Pleistocene history'. Co-management partnerships between traditional owners and conservation biologists, and ongoing traditional caring for 'country', have been increasingly important in recent years, and the later version emphasised this.

---

[38] Moritz and Kikkawa (1994), v.
[39] Hopper (1994), 269.
[40] Burgman and Lindenmayer (1998); Lindenmayer and Burgman (2005).
[41] Langton (1998); Langton et al. (2004).

The original book dealt with philosophies of conservation and scientific methods, in terms of managing and monitoring species, populations and genetics, and in selecting and designing reserves. The great expansions in the new book concern conservation *outside* reserves, and the value to biodiversity conservation of production landscapes (places where people live, and economy and ecology actively compete).

Michael Soulé's legacy is acknowledged warmly by the Australian authors, but local case studies have tested and shown some of the limitations of the international theory of 'island biogeography' in reserve design, promoted by Jared Diamond and others.[42] Diamond posited in 1990 that reserves in developed landscapes function rather like oceanic islands surrounded by the sea. The shape of the reserves makes a difference to their potential success in 'saving' species. In general, bigger, less-fragmented reserve shapes allow fewer threats to the species in the reserve. Such a theory went some way to answering the question posed by Max Day of the Australian Academy of Science in the 1960s, but increasingly conservation biologists argue that management cannot simply end with setting up a reserve. As theories of ecology have become more accommodating of uncertainty, the ordered and regulated steady-state ecological systems that were assumed to apply to reserves have become more problematic.

Daniel Botkin and others have argued that 'wherever we seek to find constancy ... we discover change ... We see a landscape that is always in flux, changing over many scales of time and space.'[43] Finnish ecologist Yrjö Haila observed that comparing landscape fragments with oceanic islands, gave rise to two problematic assumptions: the first was that habitats surrounding fragments were like a 'hostile sea' that did not support most of the organisms; the second was that natural pre-fragmentation conditions were uniform (or had no history).[44]

David Lindenmayer and his students Joern Fischer and Adrian Manning have found empirical support in the Australian landscape for Haila's critique. Their work has shown that even small reserves may be useful to certain species and that the landscape that surrounds 'habitat patches' is often very important to species' survival. Habitats are no longer defined by what the human eye appreciates – nor can they be assumed to be similar for all species.

A 'habitat contour' model can capture ecological complexity, including the perspectives of different species. Fischer and his colleagues have defined contour maps of suitability for different species (defined by how the species uses the space), that may overlay the same area differently for different species. This idea moves beyond mosaics and patches to multiple definitions of the

[42] Soulé (1986); see also Diamond (1975); and Shafer (1990).
[43] Botkin (1990); Bocking (2005), 69.
[44] Haila (2002), 321–34.

same area of land, recognising that areas of high suitability for one species may not suit others. One of the important implications of the habitat contour models for management is to acknowledge that the sorts of habitats that suit less charismatic animals are often overlooked in 'patch' reserves. For example, endangered lizards depend heavily on native grasslands, but patch reserves are seldom designed to preserve grasslands.[45]

Far from being a 'hostile sea', the landscape that surrounds reserves represents a positive 'heterogeneous matrix' constructed by its varied history. This proves to be important to biodiversity conservation. The animal may look very differently at the matrix from the conservation manager, and may value things that humans do not consider or cannot even observe. The new conservation biology harks back to the early twentieth-century ideas writings of Estonian biologist-philosopher Jacob van Uexküll. Uexküll's subjective biology was built around the notion of organisms (from ticks to human beings) living within an *Umwelt* or sphere of meaning that they create for themselves. The *Umwelt* is much more than habitat, a space or a plot on a map. It allows the animal a subjective 'sense of place'.[46] The idea of managing for the place as it is valued by the endangered lizard or the rare butterfly adds another layer of complexity (possibly one that is impossible to accommodate fully) to the task of conservation management. But it suggests a precautionary principle that even unlikely areas have the potential to be valuable to some species, and that confident predictions of 'ideal habitat' based on a few colourful or prominent species may not provide best for a full range of species. Biodiversity management has to allow for the places beyond the 'patches', at least some of the time. *Practical Conservation Biology* is one of a number of newer books that counters the earlier emphasis in Australian and international research on conservation in 'nature reserves' and draws attention to the important role the 'matrix' and other management actions can play in conserving biodiversity.

Despite this new work on places outside reserves, Australia is still best known as a leader in 'reserve selection algorithms' (systems for prioritising choice amongst possible reserves) and in systematic conservation planning. Conservation of biodiversity still begins in reserves, but they no longer simply sample the biodiversity, but must also 'separate the biodiversity from processes that threaten its persistence'. This latest model for reserves includes the objectives of the original gap analysis for 'representativeness', and adding 'persistence' (maintaining natural processes and viable populations, and excluding threats). Biodiversity now needs more than 'space'. Nature is in constant flux and needs

---

[45] Fischer, Lindenmayer and Fazey (2004).
[46] Manning, Lindenmayer and Nix (2004). Jacob van Uexküll's work is assessed in an anthropological context in Roepstorff, Bubandt and Kull (2003).

optimal conditions to allow for evolving processes. There is no singular climax state, and therefore no singular management formula or reserves algorithm.[47]

Conservation Biology in the twenty-first century de-emphasises reserves, and has turned with a renewed concern to the ecological integrity of agricultural landscapes. Australian ecologists have been prominent in promoting what is now an international trend. The 'wilderness reserve' is no longer a sole goal of conservation, but rather the more complex idea of preserving present and future biodiversity, and maximising nature's opportunities for evolution.[48] Scientific principles for selecting reserves separate conservation planning from the aesthetic preferences for remote, rugged and beautiful scenery promoted by wilderness activists and green political groups, but they have been further nuanced by other developments in the science of conservation biology – and its political context. Scientists working closely with reserves managers are increasingly becoming conscious of the need to understand reserves and other places for biodiversity in cultural as well as biophysical contexts.

Since 2000, an increasing number of new conservation reserves in Australia have been identified and managed by Aboriginal people through a program of Indigenous Protected Areas (IPAs). Unlike the programs that involve Aboriginal people in co-management of National Parks and World Heritage Areas such as Uluru and Kakadu, IPAs provide support for *private* conservation initiatives by Aboriginal communities. The Program provides incentives to Indigenous landowners to establish and manage IPAs on their lands as part of the National Reserve System, and thereby to facilitate the promotion and integration of 'Indigenous ecological and cultural knowledge into contemporary land management practices', as well as significantly increasing the extent of lands held in the National Reserve System.[49]

Conservation biology and geographical inequity

Despite the fact that a very significant portion of the world's biodiversity is located in developing countries, conservation biologists tend not to study these places. The United States and Australia are two of the 17 most 'megadiverse' nations, but all the other 15 nations in this group have developing economies. In a recent study, Ioan Fazey, Joern Fischer and David Lindenmayer analysed the articles published in the world's three leading conservation biology journals in 2001: *Biological Conservation* (established 1968), *Conservation Biology* (established 1987) and *Biodiversity and Conservation* (established 1992). Less than a

---

[47] Margules and Pressey (2000).
[48] Holling, Gunderson and Ludwig (2002).
[49] 'What are the goals of the Indigenous Protected Areas Program?' http://www.environment.gov.au/indigenous/publications/ipa-newsletter-1.html#first

third (28%) of the studies concerned lower income countries, and only 15% had a primary author located in such a country.[50]

Conservation biology research typically focused on researchers' home countries (Europe, North America and Australasia). Within these well-resourced places, research is strongly biased towards areas designated as 'reserves' or national parks, according to a recent survey by Fazey et al.[51] The survey notes that conservation biologists assume (not always explicitly) that diversity will be richer where it is 'undisturbed' (that is, in reserves) that makes researchers favour these areas. However, Fazey et al. observe that the typical threats posed by human–environment interactions are in fact at their minimum in a biodiversity reserve. This effect is magnified in places like United States and Australia, where reserves themselves have been historically designated because of their 'wilderness' values. If conservation biology is a 'science of crisis', then more attention should be paid to how diversity works in production landscapes, where people's interactions with the environment are high. It is in these landscapes where the principles of sustainable development, including that elusive 'triple bottom line', will be most severely tested.

Australia is very unusual in that it has a national capacity to invest in science and a megadiverse continent at home to study. It has a legacy of 'wilderness' politics – where people and nature are sharply divided, and national parks have been established as 'people-free' places. Lands transferred to Aboriginal ownership since the 1980s typically have a history of being 'unimproved' Crown land. These often include places designated 'national parks'. In the last decade there have been some major critiques of wilderness by Indigenous leader Marcia Langton and many others, who argue that people are good for biodiversity.[52] 'Wilderness is a whitefella word', declared one bumper sticker.[53] Aboriginal co-management and representation on boards of management from the 1980s have raised the level of political debate about Aboriginal involvement in national parks, and the program for IPAs has taken these ideas beyond parks in the twenty-first century.[54] These management decisions have in turn shaped

---

[50] Fazey, Fischer and Lindenmayer (2005a). It is not just conservation biology that is neglectful of the non-western world. A two-year study by Raja and Singer (2004), showed that less than 15% of the papers published in 2002–2003 in leading medical journals (*New England Journal of Medicine, Journal of the American Medical Association, Lancet* and *British Medical Journal*) concerned the diseases and illnesses of developing countries, despite the much greater rate of suffering in these places. See: http://bmj.bmjjournals.com/cgi/content/full/329/7480/1429?etoc.

[51] Fazey, Fischer and Lindenmayer (2005).

[52] Langton (1995); Langton (1998).

[53] Head (1997), 19; Head (2000).

[54] Lawrence (2000), 181–92; Fleming (2000); Rose (2004b); Rose, James and Watson (2003).

ecological science. The Ecological Society of Australia's journal *Ecological Management and Restoration* (established April 2000), which has a motto of 'Linking science and practice', features at least one article on Indigenous management for conservation objectives every issue.

An increasing number of Indigenous land managers, working inside and outside parks, are finding that programs designed to improve cultural appreciation are advantageous to biodiversity, and vice versa. The land manager of Ikuntji-Haasts Bluff, Scott McConnell, worked with his community to establish a conservation area that also 'keeps culture strong' in the remote Cleland Hills in the western Northern Territory.[55] There are important synergies in the Indigenous ideas of 'country' as a 'place that gives and receives life', as anthropologist Deborah Rose puts it, and the conservation biologists' 'matrix that matters'.[56]

Yrjö Haila's critique of island biogeography models for reserves noted that they depended on an unfounded assumption that human-influenced environments were essentially different from 'natural' environments. David Lindenmayer calls this a 'Land Apartheid'. The Stanford ecologist Gretchen C. Daily argued recently in *Nature* that 'humanity has always been, and always will be a part of nature'. Daily described 'countryside', a concept of unbuilt landscape where 'ecosystem qualities are strongly influenced by humanity'.[57] This revived and gave an international scientific profile to a word popular in the post-war years of English and Welsh countryside planning. A nature–culture divide could never be maintained in a country where 350,000 people lived in national parks.[58] But in the twenty-first century, the nature–culture divide was being challenged in *Nature*, suggesting that it was no longer helping conservation biologists or managers make sense of environments, even in places like the United States, where wilderness had a positive political profile.[59]

Sustainability and history

The imperatives of sustainability have influenced biology and other disciplines such as economics (in particular, ecological economics), but they have yet to engage seriously with some branches of knowledge. Take history, for example.

---

[55] The parallels between nurturing culture and biodiversity were noted by Ikuntji-Haasts Bluff land manager, Scott McConnell, who worked with the community to establish a conservation area in the remote Cleland Hills in the western Northern Territory (pers. com. May 2004). See also Martin, Robin and Smith (2005).

[56] Rose (1996); Rose (2004a).

[57] Daily (2001), 245.

[58] Countryside Commission (1989), 3. The term is evident in the (English and Welsh) *Access to the Countryside Act 1949* and the later *Countryside Act 1968*, but has been used less commonly in the United States.

[59] Haila (2002).

Despite the fact that sustainability is a concern in management over time – 'for the long haul' – and that the timescales offered by history might inform the state of an ecological community, there has been only limited historical interest in such questions.

What are the obstacles to engaging with history? The first is the legacy of 'wilderness' thinking, where a place is valued for its lack of human interference, its pristine 'natural' state. Looking for wilderness and 'pristine country' has screened out the history in the landscape. This is a form of *terra nullius* that Aboriginal activists have been fighting against for decades. But settler history also has the potential to be overlooked in landscape analysis, particularly if the focus is on 'climax state' nature in reserves (which typically are based on land originally zoned as undeveloped crown land). Biologists no longer see 'climax states' as the ideal concept, and this has directed their work to consider biodiversity beyond reserves.

Conservation biology is reloading the 'matrix' – the areas outside reserves – with significance for biodiversity. Ecologists Ian Lunt and Peter Spooner have alerted their discipline to the value of historical studies, because they want to view ecosystems as '*historically* and spatially influenced non-equilibrium systems that are complex and open to human inputs'.[60] They argue that history is essential to understanding how the landscape is organised, why the remnant woodlands are where they are, and not elsewhere. Other biologists are concerned about the dynamics of change, understanding the way landscape alteration and disturbance has unfolded. As Sue McIntyre and Richard Hobbs noted, landscapes classified as 'intact, variegated, fragmented or relictual' are in different points on a *time*-continuum that shows a decrease in available habitat and an increase in disturbance.[61] The more 'edges' in the fragments, the greater the vulnerability of the species contained within them. 'Edge effects', as these are called in the technical literature, magnify with increasing fragmentation. These scientists are concerned that mapping in geographical space omits the element of time, but realise that history defines 'edges' as surely as a Global Positioning System. Few historians, in Australia or elsewhere, have written histories in a form useful for the sort of landscape reconstruction that ecologists seek to undertake, so ecologists are often left to write this sort of history themselves, as Lunt and Spooner did.

Donald Worster observed that one of the critical differences between science and history is the assumption in science that the present subsumes and includes all pasts. The date of the paper is 'an index to truth' for a scientist – 'the more recent the date, the more truthful the paper'.[62] Young scientists are trained to

---

[60] Lunt and Spooner (2005), 1860, my emphasis.
[61] McIntyre and Hobbs (1999); Also Fischer et al. (2004).
[62] Worster (1996), 11.

feel that there is no point to read anything except the latest paper, as that is where knowledge is *now*, and, by implication, is the likeliest basis for the future discoveries. Reading is urgent as science marches forward briskly. A historian of science, by contrast, needs to spend time reading all the other papers in order to unpack the intellectual journey, to look at the evolution of the ideas and the sequence of practices, and consider their politics and social context. The historian must 'waste time' considering the possibility of intellectual journeys not taken. *Environmental* history adds the dimension of watchfulness about place: how local ecologies and the land itself can shape the journey and its politics, and this is a major element in the story of conservation biology and sustainability science in Australia.

There is another role for history. The ideas of conservation biology have a history of constructing different understandings of biodiversity, and of the sustainability science that strives to preserve it. The Australian story is distinctive from elsewhere because the exceptional flat land with its ancient life forms also has such a long human history. International conservation biology begins with a global environmental history that started in the late Pleistocene. The first paper in the first issue of *Biodiversity and Conservation* in 1992 was entitled 'The Sinking Ark'. Jeffrey A. McNeely offered a 'global view' from his Swiss office of the International Union for Conservation of Nature (IUCN). His first two sentences show immediately how Australia's history is not the same as the global story:

> The period following the withdrawal of the last great glaciers of the late Pleistocene saw the emergence of *Homo sapiens* as a dominant force in ecosystems throughout the world. A relatively benign climate returned to Europe after tens of thousands of years of ice; people learned how to domesticate plants and animals and to exert increasing control over the landscape.[63]

Such 'initial conditions' have resonance in North America and Europe, both of which were highly glaciated in the Pleistocene. But the interplay between biodiversity and human settlement was very different in Australia where there was no glaciation – and of course, none of the scouring effects of glaciation that released nutrients into the soils of those northern climes. What George Seddon calls our 'uneventful Pleistocene' is one of the key factors in understanding our nutrient-poor soils.[64] Assumptions about 'more and less benign' times sit uneasily in Australia and are still controversial. Sixty thousand years of human land management and fire practices, followed (abruptly) by 200 years

---

[63] McNeely (1992), 2.
[64] Seddon (1996); Seddon (1997); Seddon (2005).

of European agriculture and pastoralism leave different signatures on the continental land mass and surrounding seas. Such legacies make development on the European model problematic, and inevitably shape the emerging ideas about sustainability itself in this country.

Epilogue

In an international forum like this book, it is possible to explore the conversations between the local and the global that have played out historically very differently in different places. It is useful to get beyond the 'Leagues Tables' of biophysical comparisons that define Australia simply as say 'megadiverse', or 'flat'. While the statistics of exceptionalism – the flattest, driest, highest levels of extinction and so forth – are important to the local politics of motivating governments to take short-term action, so is an understanding of the complex human history of each place. And social contexts are also dynamic, and may become part of new solutions in times of crisis.

I want to return now to the Jervis Bay story, where David Lindenmayer's generosity was shunned by government. Late in 2006, in the very nick of time, a very new player in the Australian conservation scene, the environmental philanthropist, stepped into the breach. The Thomas Foundation, in partnership with Bush Heritage Australia, established the Rick Farley Memorial Scholarship, honouring a farmers' advocate who worked actively for Aboriginal rights. On 14 December 2006, it was announced that the inaugural winner of this Scholarship was Darren Brown, the Aboriginal elder at the centre of Lindenmayer's research training scheme. The government failed to find 'matching funds', but the Thomas Foundation provided them with grace and positive publicity.[65] The Thomas Foundation is a private philanthropic group that has, since 1998, supported a range of conservation and art initiatives involving Aboriginal people, and is one of several new patrons of conservation and science for environmental management.[66]

---

[65] http://www1.bushheritage.org/default.aspx?MenuID=625;http://www.nature.org/where wework/asiapacific/australia/files/david_thomas_bio.pdf

[66] Although the Australian Bush Heritage Fund has been in operation since 1990, when prominent Green politician Bob Brown created it using an environmental prize to buy two small reserves in Tasmania, it has expanded rapidly. It moved to Melbourne (on the mainland) in 2004, closer to large philanthropic groups and other conservation groups including the US group, The Nature Conservancy's Australian office, and Trust for Nature. In 2000, it boasted 12 properties, in 2005, 23 – and they are larger. New partnerships with other groups, such as the 2007 Gondwana Link project in southern West Australia (http://www.gondwanalink.org/) mean that it now has a very significant set of holdings or 'land for conservation'. Another organisation, the Australian Wildlife Conservancy, has shown similar strong growth trends since its listing as a public charitable organisation in 2001. (see: http://www.australianwildlife.org/aboutAWC.asp)

This is Australia's newest and quietest revolution. In a country where private funding for science, education and conservation has been traditionally weak, there are signs that this sector, particularly the philanthropic (as opposed to the industry) sector, is growing fast and becoming significantly more influential.[67] The science of conservation biology finds itself an important player in a rather new 'biodiversity' market place.

Environmental history is important as it can add another layer of understanding to such a new market place and more broadly, to the ways that cultural shifts can affect environmental options. Environmental action that extends beyond national government programs is essential to engaging with current and emerging issues such as climate change, which have both local and global dimensions. New players, as the Jervis Bay story shows, can lead governments. What remains at stake are the fundamental questions of what constitutes the common social good and the health of the planet. And the future might well be behind us.

References

Arthur, J.M. (2003), *The Default Country: A Lexical Cartography of Twentieth-century Australia*, Sydney, UNSW Press.

Australia, *Environment Protection and Biodiversity Conservation Act (91/1999)* S528.

Beeby, R. (2006), 'Giving Back to the Land', *Canberra Times* 30 September. See http://www.ausbushfoods.com/articles/giving_back_to_land.htm

Bocking, S. (2005), *Nature's Experts: Science, Politics and the Environment*, New Brunswick, Rutgers University Press.

Botkin, D. (1990), *Discordant Harmonies: A New Ecology for the Twenty-first Century*, New York, Oxford University Press.

Burgman, M. and Lindenmayer, D. (1998), *Conservation Biology for the Australian Environment*, Chipping Norton, Sydney, Surrey Beatty and Sons.

Carruthers, J. (1995), *The Kruger National Park: A Social and Political History*, Pietermaritzburg, University of Natal Press.

Carruthers, J. (2006), 'From "Land" to "Place": Environmental Activism in the Magaliesberg, South Africa, and Coopers Creek, Australia', In Mauch, C., Stoltzfus, N. and Weiner, D.R. (eds), *Shades of Green: Environmental Activism around the Globe*, Lanham MD, Rowman and Littlefield, 69–99.

Child, B. (ed.) (2004), *Parks in Transition: Biodiversity, Rural Development and the Bottom Line*, IUCN South Africa, Pretoria.

---

[67] Scientific research in Australia has been very much 'government science', funded by state and federal agencies, CSIRO and through universities until the most recent period. The private sector (both philanthropy and industrial research and development) have been largely absent until the last decade, where the ideological commitment of politicians of all persuasions to 'small government' has forced an increasing reliance on other sources of funding. The historic case for ecological science is discussed in Robin (2005); and also in Robin (2007), 191, 201–04.

Countryside Commission (1989), *National Parks Review: A Discussion Document*, Countryside Commission.

Daily, G. (ed.) (1997), *Nature's Services: Societal Dependence on Natural Ecosystems*, Washington D.C., Island Press.

Daily, G.C. (2001), 'Ecological Forecasts', *Nature*, 411, 17 May, 245.

Davidson, J. (1981), 'Victoria', In Australian Heritage Commission, *The Heritage of Australia*, 3–7.

Diamond, J. (1975), 'The Island Dilemma', *Biological Conservation*, Vol. 7, 129–45.

Fazey, I., Fischer, J. and Lindenmayer, D.B. (2005), 'What do Conservation Biologists publish?', *Biological Conservation*, 124, 63–73.

Fazey, I., Fischer, J. and Lindenmayer, D.B. (2005a), 'Who Does all the Research in Conservation Biology?', *Biodiversity and Conservation*, 14, 917–34.

Fischer, J., Lindenmayer, D.B. and Fazey, I. (2004), 'Appreciating Ecological Complexity', *Conservation Biology*, 18(5), 1245–53.

Fleming, T. (2000), 'New Directions for Parks and Wildlife in South-eastern NSW', Environment Institute of Australia Seminar, Canberra, 17 May.

Goodall, H. (2002), 'The River Runs Backwards', In Bonyhady, T. and Griffiths, T. (eds), *Words for Country: Landscape and Language in Australia*, Sydney, UNSW Press, 30–51.

Haila, Y. (2002), 'A Conceptual Genealogy of Fragmentation Research', *Ecological Applications*, 12(2), 321–34.

Hall-Martin, A. and Carruthers, J. (2003), *South African National Parks: A Celebration*, Johannesburg: Horst Klemm.

Head, L. (1997), 'Second Nature. Imaging Australia as Aboriginal Landscape', In Unlocking Museums, Proceedings, 4th National Conference of Museums Australia Inc., Darwin, 6–12 September 1997, 19.

Head, L. (2000), *Second Nature: The History and Implications of Australia as Aboriginal Landscape*, Syracuse, NY: Syracuse University Press.

Holling, C.S. (Buzz), Gunderson, L.H. and Ludwig, D. (2002), 'In Quest of a Theory of Adaptive Change', In Gunderson L.H. and Holling, C.S. (eds), *Panarchy: Understanding Transformations in Human and Natural Systems*, Washington, Island Press, 10–20.

Hopper, S. (1994), 'Plant Taxonomy and Genetic Resources', In Moritz, C. and Kikkawa, J., *Conservation Biology in Australia*, Chipping Norton, Surrey Beaty.

Jackson, S., Storrs, M. and Morrison, J. (2005), 'Recognition of Aboriginal Rights, Interests and Values in River Research and Management', *Ecological Management and Restoration*, 6(2), 105–9.

Kenchington, R.A. (1994), 'Conservation and Coastal Zone Management', In Moritz, C., and Kikkawa, J. (eds), *Conservation Biology in Australia*, Chipping Norton, Surrey Beaty.

Langton, M. (1995), 'The European Construction of Wilderness', *Wilderness News* (TWS), summer 95/96.

Langton, M. (1998), *Burning Questions*, Centre for Indigenous Natural and Cultural Resource Management, Northern Territory University.

Langton, M., Tehan, M., Palmer, L. and Shain, K. (eds) (2004), *Honour among Nations: Treaties and Agreements with Indigenous People*, Carlton, Melbourne University Press.

Lawrence, D. (2000), *Kakadu: The Making of a National Park*, Carlton South, Miegunyah.

Lindenmayer, D. and Burgman, M. (2005), *Practical Conservation Biology*, Melbourne, CSIRO Publishing.

Lindenmayer, D.B. and Fischer, J. (2006), *Landscape Change and Habitat Fragmentation*, Washington D.C., Island Press.

Lunt, I.D. and Spooner, P.G. (2005), 'Using Historical Ecology to Understand Patterns of Biodiversity in Fragmented Agricultural Landscapes', *Journal of Biogeography*, 32, 1859–1873.

Mackey, B.G., Lesslie, R.G., Lindenmayer, D.B., Nix, H.A. and Incoll, R.D. (1998), *The Role of Wilderness in Nature Conservation*, Report to the Australian and World Heritage Group Environment Australia.

Manning, A., Lindenmayer, D.B. and Nix, H.A. (2004), 'Continua and *Umwelt*', *Oikos*, 104(3), 621–8.

Margules, C.R. and Pressey, R.L. (2000), 'Systematic Conservation Planning', *Nature*, 405, 243–53.

Martin, M., Robin, L. and Smith, M. (eds) (2005), *Strata: Deserts Past, Present and Future. An Environmental Art Project about a Significant Cultural Place*, Mandurama: Mandy Martin.

McIntyre, S. and Hobbs, R. (1999), 'A Framework for Conceptualizing Human Effects on Landscapes', *Conservation Biology*, 13(6), 1282–92.

McNeely, J.A. (1992), 'The Sinking Ark: Pollution and the Worldwide Loss of Biodiversity', *Biodiversity and Conservation*, vol. 1, 2.

Moritz, C. and Kikkawa, J. (1994), 'Introduction', In Moritz, C. and Kikkawa, J. (eds.), *Conservation Biology in Australia*, Chipping Norton, Surrey Beaty.

Morton, S.R. (2000), 'Ecology' (Workshop presentation). *Methods in Environmental History*, Canberra, Australian National University, 5 April.

Moyal, A. (1986), '*A Bright and Savage Land*': *Scientists in Colonial Australia*, Sydney, Collins.

Raja, A.J. and Singer, P.A. (2004), 'Transatlantic Divide in Publication of Content Relevant to Developing Countries', *British Medical Journal*, Vol. 329, 18 December, 1429–30.

Robin, L. (1994), 'Nature Conservation as a National Concern: The Role of the Australian Academy of Science', *Historical Records of Australian Science*, 10(1), 1–24.

Robin, L. (1997), 'Ecology: A Science of Empire?', In Griffiths, T. and Robin, L. (eds), *Ecology and Empire: Environmental History of Settler Societies*, Edinburgh, Keele University Press, 63–75.

Robin, L. (1998), 'Radical Ecology and Conservation Science: An Australian Perspective' *Environment and History*, 4(2), 191–208.

Robin, L. (2005), 'Migrants and Nomads', In Sherratt, T., Griffiths, T. and Robin, L. (eds), *A Change in the Weather: Climate and Culture in Australia*, Canberra, National Museum of Australia Press, 42–53.

Robin, L. (2007), *How a Continent Created a Nation*, Sydney, UNSW Press.

Roepstorff, A., Bubandt, N. and Kull, K. (eds) (2003), *Imagining Nature: Practices of Cosmology and Identity*, Aarhus, Aarhus University Press.

Rose, D.B. (1996), *Nourishing Terrains: Australian Aboriginal Views of Landscape and Wilderness*, Canberra, Australian Heritage Commission.

Rose, D., James, D. and Watson, C. (2003), *Indigenous Kinship with the Natural World*, Sydney, National Parks and Wildlife Service of NSW.

Rose, D.B. (2004a), *Sharing Kinship: How Reconciliation Is Transforming the NSW National Parks and Wildlife Service*, Sydney, National Parks and Wildlife Service of NSW.

Rose, D.B. (2004b), *Reports from a Wild Country: Ethics for Decolonisation*, Sydney, University of New South Wales Press.

Runte, A. (1977), 'The National Park Idea', *Journal of Forest History*, 21(2), 65–75.

Seddon, G. (1996), 'Thinking Like a Geologist: The Culture of Geology', Mawson Lecture 1996, *Australian Journal of Earth Sciences*, 43, 487–95.

Seddon, G. (1997), *Landprints: Reflections on Place and Landscape*, Melbourne, Cambridge University Press.

Seddon, G. (2005), *The Old Country: Australian Landscapes, Plants and People*, Melbourne, Cambridge University Press.

Shafer, C.L. (1990), *Nature Reserves: Island Theory and Conservation Practice*. Smithsonian Institution Press, Washington.

Sörlin, S. and Warde, P. (2007), 'The Problem of the Problem of Environmental History: A Re-reading of the Field', *Environmental History*, 12(1), 107–30.

Soulé, M.E. (1985), 'What is Conservation Biology?', *Bioscience*, 35(11), December, 727.

Soulé, M. (1986), *Conservation Biology: The Science of Scarcity and Diversity*, Sunderland Massachusetts, Sinauer Associates.

Stein, P.L. (2000), 'Are Decision-makers too Cautious with the Precautionary Principle?', *Environmental and Planning Law Journal*, 17(1), 3–23.

Turner, J.S. (ed.) (1957), *A Report on the Condition of the High Mountain Catchments of New South Wales and Victoria*, Canberra, AAS Report No. 1, May.

Wilson, E.O. (1997), 'Introduction', In Reaka-Kudla, M.L., Wilson, D.E. and Wilson, E.O. (eds), *Biodiversity II: Understanding and Protecting Our Biological Resources*, Washington D.C., Joseph Henry Press, 1.

World Commission on Environment and Development (WCED) [Chair: Gro Harlem Brundtland]. (1987), *Our Common Future*, Oxford University Press.

Worster, D. (1996), 'The Two Cultures Revisited: Environmental History and the Environmental Sciences', *Environment and History*, 2(1), 3–14.

Part III

Making Space: Environments and Their Contexts

9

54, 40 or Fight: Writing Within and Across Borders in North American Environmental History

Matthew Evenden and Graeme Wynn

The tongue-in-cheek title of this chapter invokes a slogan popular among American residents of the Oregon Country and associated with Democratic candidate James K. Polk in the US Presidential campaign of 1844. The numbers refer to the northern latitudinal limit of territory then held jointly by the United States and Canada (an area beyond the boundary along the 49th parallel agreed to in 1846), that Polk claimed for the United States and promised to fight for if necessary. Polk's demands emerged out of a complex domestic political scene, and were twinned with a plan to annex Texas, all part of a grand policy of manifest destiny.[1] The slogan has been largely forgotten in Britain and the United States, but it remains in play in Canada, invoked occasionally by nationalists fearful of American influence, and adopted by a successful rock band, that has saved the numbers (54, 40) but lost the fight.

This slogan reminds us how borders are claimed and contested, how regions like the Oregon country came to be re-territorialized as national and imperial space, and how the places we write about as environmental historians must be thought about partly in terms of their political geography and history. It also points to the problems of writing within and across borders in North American environmental history. Political boundaries are malleable and ephemeral by comparison with the age-old, seemingly immutable lineaments of continents and oceans. Critics of efforts to forge a Canadian nation across the northern reaches of North America, who styled themselves 'Continentalists,' spoke often, in the late nineteenth century, of the absurdity of constructing a country athwart the north–south 'grain' of the continent, and some would find vindication of their views in the establishment, a hundred years later, of a free-trade area encompassing the enormous triangle of territory between Chiapas,

---

[1] On Polk in the context of Manifest Destiny, see: Stephanson (1995), 35.

Nunavut and the north slope of Alaska. Should *environmental* historians shape their studies according to geopolitical templates, and confine and constrain their inquiries within the narrow, perhaps evanescent and often 'artificial,' envelopes of national boundaries? Several American environmental historians, and others who call themselves bioregionalists, suggest not. From this perspective, 'natural' regions – drainage basins, ecological zones, vegetation complexes – are more useful and appropriate spheres of inquiry than political units. Lakes and rivers (think of the Great Lakes and the Rio Grande) are ecological units. Upstream actions have downstream effects. Traditionally and historically water bodies have unified peoples and interests more than they have divided them. Why should environmental (and social) histories be put asunder by political fiat?

These questions go to the heart of the enterprise of environmental history, within North America and beyond. The ways in which they are answered will shape the kinds of work that environmental historians do and affect the influence of that work upon both scholarly practice and popular conceptions of the past. Ultimately, they also force practitioners to ponder the purpose of their endeavors, to ask 'Why and for whom do we write?' We come to these matters as two Canadian scholars. Our work focuses on Canadian regions, but we find a good deal of inspiration and interest in comparative and transnational research problems and we frequently situate our work in the international literature. Our concern here is to engage these questions from a Canadian perspective. This forces us to confront the dominance of American environmental historiography in Canada and elsewhere, and to contemplate the existence of, prospects for and possible value of a distinctive Canadian environmental history. To put this another way, we wonder whether there is, was, or should be anything beyond simple location (within the territorial bounds of the present Canadian nation state) that distinguishes environmental historical writing on northern North America and whether perhaps the 'continentalists' if not the bioregionalists have effectively won the day. Our anxieties on this score are perhaps given added point by the rather limited visibility of Canadian environmental history in the international realm – a fact which allowed the distinguished world historian John McNeill to write (we believe mistakenly) in 2003 that 'Canadianists have almost entirely ignored the genre' of environmental history.[2] In response, we reflect on the institutional and intellectual growth of the field in Canada over the last half century and provide a rough, introductory sketch of its current emphases and intellectual content before grappling with the larger challenges of writing environmental history within and across borders.

---

[2] McNeill (2003), 18. McNeill is not alone in this view, see also Coates (2004), 422.

Roots and branches

Use of the phrase 'environmental history' to characterize a particular field of inquiry is relatively new. In North America it began, by common consensus, with the rise of the environmental movement in the 1960s, and the name was soon applied to an increasingly diverse body of work concerned with the relations between people and nature through time. As a relatively new 'coinage,' environmental history promised historians new perspectives on the past. Its explicit interest in the environment as an influence upon (some said actor in) history offered scholars new questions to answer and posed new challenges to their conceptions of the past. It provided, reflected one recent American commentator, 'a locus for exploration and intellectual adventure,' and offered humans a spirited, reflexive understanding of themselves and their world.[3] Yet it is as well to recognize that disciples of the new are often fervent believers in the distinctiveness of their cause, and to remember how central 'the environment' was for an earlier generation of Canadian historians. For the Ontario-born, Harvard-educated historian and public intellectual Arthur R.M. Lower (1889–1988), after all, the history of Canada, as of the new world more generally, was inescapably environmental.[4] In this view, the North American past was 'largely the story of man's struggle with nature.' North to south and east to west, pioneer peoples, anxious to better their circumstances, sacrificed the environment to material ends. They had little sense of limits, either of resources or of possibilities. Wasteful plunder marked their advance across the continent. Half a century after Lower wrote, Ramsay Cook, one of Canada's leading historians, offered a refined echo of his point in a lecture pointing the way toward an ecological interpretation of early Canadian history. 'In a broad sense,' he observed in 1990, 'the writing of Canadian history has always been concerned with the environment. The story of opening up and developing a new country had, in significant measure, to be an account of the reshaping of the environment by fur traders, mining companies, the assault on the forest, as well as the furrowing and fencing of the land.'[5]

In the broad sense, this argument is unassailable. Much work, especially in English–Canada, has turned on the challenge of settling and surviving in a difficult environment. By the 1920s there was plenty of evidence of a strong geographical or environmental dimension in the writing of Canadian history. Many of Canada's first generation of professional historians recognized the importance of 'geographical' considerations (such as distance, space, and climatic marginality) to their inquiries. Early in the 1920s, the pages of the

---

[3] Weiner (2005).
[4] Lower (1938), 1.
[5] Cook (1990–91).

newly established *Canadian Historical Review* announced that geography held an important place in shaping the national consciousness of Canadians, and that history, the study of the completed, unalterable past, should be linked with geography, the study of 'the forces of nature which will mould the yet unformed future.'[6] A few years earlier, Emile Miller, a Professor in the Ecole des Arts et Manufactures in Montreal had embraced the teachings of France's Jean Brunhes, and endorsed the sentiments of Victor Cousin when he proclaimed 'Donnez-moi la géographie d'un pays et je vous trouverai son histoire.'[7] A few years later, Arthur Lower's accounts of the assault on the forest, Harold Innis's detailed studies of the fur trade and the cod fisheries, Innis and Lower together on the settlement of the forest and mining frontiers, as well as their contemporary Donald Creighton on the St. Lawrence, paid considerable attention to the environments in which the activities that were at the forefront of their analyses unfolded.[8]

Nature held an important place in histories that adopted the staples approach to Canadian history. Innis (1894–1952, the economic historian who is most widely associated with the idea that the development of new world societies turned around the exploitation of a staple product or 'natural resource,' and who argued forcefully against a continentalist view of Canadian history), began his books on the fishery and the fur trade with extended discussions of the cod and the beaver. He recognized that fish and fur-bearing animals had their 'geographies' (they were found in certain places), and that these were influenced by ecological considerations (water temperatures, habitat, growing seasons). Lower did much the same for pine trees. The story of Prairie wheat is in large part a story of triumph over environmental obstacles – swamps in Manitoba, drought in Palliser's triangle, short growing seasons almost everywhere. By late twentieth century intellectual historian Carl Berger's reading of Innis's work (and particularly of *The Cod Fisheries*), 'the net impression…was…of human helplessness in the face of brutal limitations imposed by nature and the intervention of blind, uncontrollable forces.'[9] By this account, at least, Innis was attempting 'to define the limits which material [environmental] conditions imposed upon the range of human activity' – and less-intentionally perhaps carving a furrow that would often be revisited by students of the Canadian past.

These are difficult waters. Their exploration can slide quickly into the dark swamp of debate about 'environmental determinism' that preoccupied

---

[6] Wallace (1920) and Wrong (1924), both cited in Shore (1995), 419; Shore (2002).
[7] Miller (1915).
[8] Lower (1938); Innis (1930) and (1954); Lower (1936); Creighton (1937).
[9] Berger (1976), 85–111, quote 102. Innis was born in Otterville, Ontario in 1894, and died in 1952. References in this paragraph are to Innis (1923); (1956 [1930]); and (1954 [1940]). See also Dunbar (1985); Parker (1988); Barnes and Hayter (1990).

a generation of geographers in the first quarter of the twentieth century and held the attention of some into the 1950s. Few Canadians embraced full-fledged environmental determinism during these years. Their interests were firmly focused – as the works of Innis and his contemporaries indicate – on human uses of the environment and upon the limitations that environments imposed on human will. Most historians emphasized the achievements of those who subdued the wilderness. In their accounts, nature was a barrier to progress, advance, improvement. It played scant part, other than as obstacle, in stories of the settlement and development of the nation. Even those who ordered their work around staple products and staple trades assumed that these products, like the land itself, were commodities for human exploitation. Environmental conditions affected the distribution of staple products and their locations shaped patterns of development, economies, and even societies, but there was little concern in the work of the staples school for the reciprocity and mutuality of relations between people and nature that proponents of environmental history now regard as central to their enterprise.

However, a somewhat different formulation appeared in the work of the Manitoba-born and Oxford-educated W.L. Morton (1908–1980), who became one of Canada's most eminent and respected historians. In the 1940s, when Morton first began to reflect in print upon the relationship between nature and culture in the western interior, he sensed that Prairie history should have a strong ecological dimension but found it difficult to escape the 'possibilist' mould that shaped the thinking of Creighton, Innis and dozens of other contemporaries.[10] 'Here,' he wrote of his native province, 'the environment is ruthless and not to be denied. Adaptation is the price of survival.' The story of the western interior was the story of 'the long clash of hereditary culture and a dominating environment,' and in the aftermath of the Depression and the dust-bowl it was by no means clear 'whether or not the traditional European culture can survive by adaptation to the great plains.' Might the settlement of the West, Morton wondered, be the 'high-water mark of the expansion of Europe, "the fading trace of a culturally important people...forced to forget the traditions and customs of a richer past" and the affiliations that went with them,' by a hostile environment.[11]

Through a quarter century of thinking about the West and Canada, however, Morton developed a less somber view of the prairie past. Writing of his

---

[10] W.L. Morton, b. Gladstone, Manitoba 1908, d. 1980. Berger (1976), 241–243, 245–250. Morton reflected on his childhood in Morton (1970a), 1–10, reprinted in McKillop (1980), 15–25.

[11] Morton (1946), 26 and 31, reprinted in McKillop (1980), 41–47. Archeologists no longer accept the interpretation of indigenous culture-history that Morton used as a springboard for his argument, viz. Vickers (1945).

'unliterary' native landscape, he recognized that three or four generations had brought a new environment into being. The 'great lone land' described by William Francis Butler in 1870 – a 'prairie-ocean' without a past, through which 'men had come and gone, leaving behind them no track, no vestige of their presence'...where 'nature alone tilled the earth and the unaided sun brought forth the flowers' – had been 'remade by the farm culture of North America.'[12] This was an important realization. The transformation of the prairies, wrote Morton, 'was at once a material reshaping of the land, and also a firm and confident expression of a way of life.' People had possessed this landscape, 'not only with their hands but with their whole being. They had made it, not only into farms, but into townships, school districts, electoral districts, town lots, church yards, cemeteries. It...was a cultural as well as a material landscape,' shaped reflexively over time 'to men's purposes by men's hands... [and] illuminated [albeit slowly and faintly] by men's minds.'[13]

There is much in these words that seems familiar. Consider that John McNeill defines environmental history, in a world survey of the field, as history that attends to the 'mutual relations between humankind and the rest of nature,' and that one of the three main varieties of the subject he identifies – 'material environmental history' – 'emphasizes the economic and technological in human affairs and focuses upon changes in biophysical environments.'[14] In his quest to illuminate the remaking of the prairie landscape in 'the terms of those who knew and lived it,' Morton provided an early model of environmental historical scholarship in Canada, and one not radically different, in impulse, from later work taken to lie near the center of the new field of environmental history. But Morton stood very much alone. In 1970, he lamented that the impulses that drove his scholarship in this direction were no longer much heeded by Canadian historians.[15] Some of them, he observed regretfully, condemned 'his respect for environment in history.' They had what he called a 'pavement mentality' – city-bred, they knew 'neither the revolution of the seasons nor the relevance of time and place, but live[d] contained and self-impelled lives.' This judgment was harsh. Histories of an increasingly urban people might say more of strikes and immigrant neighborhoods than of 'islanded farmsteads' and 'drifting buffalo herds,' but they were not necessarily less sensitive to time and place. Still, the rueful point was clear. The development of urban, social, labor, ethnic, and women's history as distinct and vigorous subfields of inquiry among Canadian historians in the late 1960s and early 1970s turned the weight of scholarship away from specific engagement

---

[12] Butler (1873).
[13] Morton (1970a).
[14] McNeill (2003), 6.
[15] Morton (1970a).

with ecological and environmental questions. The environment did not disappear from Canadian history, but it was, increasingly, portrayed and analyzed as a backdrop to human affairs, a platform upon which the concerns that captivated the current generation of scholars was played out.

Time has shown much of this work to be of considerable interest to the generation of environmental historians that has risen in its wake. Thus Viv Nelles's study of Ontario, accurately titled *The Politics of Development*, led a resurgence of interest in the venerable Canadian political economy tradition. Its focus was firmly upon the role of the state in natural resource development and regulation before World War II, but in tracing this story as it pertained to the 'new staples' of pulpwood, mining and hydro electric power, it also attended to some of the environmental consequences of their exploitation. In a very different vein, Paul Voisey's history of Vulcan, Alberta paid close attention to agricultural practices and the techniques farmers employed in adapting to the particular environmental challenges in this corner of the Prairie west. In Quebec, historians also produced some remarkable work with a strong environmental cast during these years. For example, René Hardy and Normand Séguin's *Forêt et société en Mauricie: La formation de la région de Trois-Rivières, 1830–1930* is a model of focused scholarship sensitive to the close connections between 'conditions de vie' and environmental circumstances in the nineteenth century.[16]

Nor can the developing strength of historical geography in Canada be ignored in considering the place of the environment in Canadian historical scholarship during these years. When Cole Harris (recently appointed to the Department of Geography in the University of Toronto) produced the first survey of geographical writing on the Canadian past in the mid-1960s, he noted that there were 'few trained historical geographers' in the country, and that they had done little research.[17] Thanks in substantial part to the appointment to Canadian universities of several students trained by the Canadian–American historical geographer Andrew Clark at the University of Wisconsin, and the development of vigorous graduate programs in a number of institutions, the field flourished in the 1970s. Efforts to trace its pedigree almost invariably noticed the strong commitment of Clark's mentor, the Berkeley geographer Carl Ortwin Sauer, to work on the human–environment interface, a commitment

---

[16] Nelles (1974); Voisey (1988); Hardy et Séguin (1984).

[17] Harris (1967). The main themes of this diffuse 'under-researched and poorly-written' literature were, in Harris's view, migration, settlement patterns, cultural transfer, survey systems, landscape description, distributions, and locales. J.D. Wood had voiced a similar lament a few years earlier when he reflected that the rich bounty of HBC records relating to Alberta had yielded an historical geography 'measured in fragments'. See Wood (1964–65).

that had shaped Clark's first book and produced a good deal of influential work with a strong environmental element before the World War II. But this invigorated Canadian historical geography had not escaped the influences running through the wider discipline, which focused on patterns of regional differentiation and, increasingly, spatial analysis.[18] Surprisingly little of the work done by historical geographers after 1960 was directly concerned with environmental change. Reflecting its strong intellectual pedigree in the United States, most Canadian historical geography of the 1970s and 1980s was more economic than environmental in focus. It emphasized spatial patterns, social collectives (ethnic groups), and landscape forms more than it explored processes, interactions, and ecologies. There were, of course, exceptions. Conrad Heidenreich wrote brilliantly on the ecology of seventeenth-century Huron settlement. Kenneth Kelly, interested in early Ontario agriculture, had a clear eye for the environmental circumstances with which settlers grappled. And Arthur J. Ray on his way to becoming a leading figure in the study of native–newcomer interactions in the western interior, wrote of disease epidemics that decimated native populations and early conservation schemes of the fur-trading Hudson's Bay Company.[19] Overall, however, relatively little scholarship explored the ways in which people made sense of their settings and conducted their lives within them.[20] Issues of ecological and environmental change received short shrift, especially when one considers their magnitude, Geography's long-standing claims of interest in human modifications of the natural environment, and the rise in public concern over environmental issues since the 1960s. The dedication in the first volume of the *Historical Atlas of Canada* traced the main lines of the field's intellectual pedigree at that juncture back to Harold Innis and Andrew Clark, for whom the environment was essentially an inert and rather featureless stage upon which patterns of human activity were traced and imprinted through time.[21]

Against this backdrop, the contributions of J. Gordon Nelson are particularly noteworthy.[22] Appointed to the Department of Geography in the University of Calgary at the beginning of the 1960s, Nelson brought the skills and insights developed in the course of a diverse education in physical and human geography to bear on understanding changing human impacts on the landscapes

---

[18] Wynn (1993).

[19] Examples of work by all three are included in Wynn (1990), viz Heidenreich (1990); Ray (1990); Kelly (1990).

[20] The characteristic emphases are well reflected in Harris (1987) which represents the culmination of scholarship by historical geographers during this period.

[21] Harris (1987).

[22] The substance of what follows is largely derived from Nelson (1976), which reprints the pivotal essays discussed below and includes an introduction by Nelson, commenting on the development of his work.

of western Canada. Late in the decade he wrote a series of landmark papers. The first was a broad survey of 'man and landscape in the Western Plains'; the second offered 'some comments on the causes and effects of fire' between 1750 and 1900 in the northern grasslands straddling the international boundary; and the third, which paid homage to James Malin's 'neglected' *The Grass-lands of North America*, focused on the changing fauna of the northern plains, combining the theoretical and empirical findings of scientists with historical records to understand – in a way that would have been entirely congenial to W.L. Morton – how immigrant peoples 'perceived, managed and changed the northern plains landscape.'[23] Taken together, these papers (and related works that included a book-length 1973 study of the exploration and land-use history of the southern Alberta–northern Montana border country centered on the Cypress Hills) reflect an interest with the intertwined history of physical and cultural processes.[24]

Even this proved insufficient, however. Early in the 1970s, Nelson began a series of ecological studies in Canada's National Parks, and his expertise soon generated requests for advice on the management of these areas. This meant 'exposure to the theory, methodology and findings of biologists, psychologists, lawyers, political scientists, economists,' and others. Looking back from the throes of engagement with these concerns, he concluded that his earlier work had erred in treating humans as distinct from 'animals, trees and the world around' them and separating their effects from those of other (wrongly considered 'independent') forces or processes. Yet the plethora of perspectives with which he was now confronted posed 'serious organizational and analytical problems.' Looking forward, it seemed that 'some type of integrative organizational framework or model' was necessary to make sense of the complexities he faced. As this model was elaborated, it entailed, at the most abstract level, a move toward the explicit use of systems analysis. Expressed less formally, it implied an emphasis on ecosystems instead of, or alongside, the earlier concern for landscape (although the terms 'landscape,' 'environment' and 'ecosystem' were, and are, often used in overlapping, non-exclusive ways). In practice, the model came down to four basic parts, each of which was considered sequentially. They were 'ecology; strategies and institutional arrangements; perceptions, attitudes and values; and technology.'[25] Through the early 1970s, Nelson wrestled with the challenges involved in merging 'historic studies of [hu]man's effects on landscape with the current high level of concern for better management of [the] environment.'

---

[23] These papers are Nelson (1967); Nelson and England (1971); Nelson and Byrne (1966).

[24] Nelson (1973).

[25] As exemplified in Roe and Nelson (1972) and Nelson (1976), 103–139, and 140–164.

In our judgment, questions of environmental impact assessment and management and policy issues gradually drew Nelson's focus away from the environmental historical concerns with which his research had begun. By his own retrospective account, however, there was less rupture than progression in this shift. He was wrestling with the issue of what should be encompassed in environmental history. Just as past, present and future are a continuum, he concluded, research was arrayed along a spectrum, and individual studies differed in the degree to which they emphasized 'analysis,' 'assessment,' or the development of normative principles. According to this conceptualization, shared with his students in Geography and Planning at the University of Waterloo, Ontario, Nelson regarded analytical work as 'pure research'; mid-spectrum studies focused on (environmental) assessment drew implications – or raised questions and suggested directions – for those engaged in policy and planning work; and normative reports made specific recommendations to managers, politicians, and others. On this reckoning, studies of the second and especially the third type were more applied than those in the first category but all should rightly attend to the past in some degree. Increasingly engaged with questions of contemporary policy and practice, Nelson concluded that the academic/policy and practice interface was the most opportune place to work, but never entirely forsook his interest in environmental history. In sum, Nelson's contributions to understanding the environmental historical geography of the Canadian west, to demonstrating the importance of the past to present and future concerns, and to the practice of a broad form of environmental history were highly original. Yet they spawned little analogous work at the 'pure' rather than the policy end of the research spectrum.[26]

In aggregate then it may be said that although some important Canadian scholarship acknowledged the environmental challenges and opportunities that faced early Canadians, before 1980 the country's historians and geographers wrote relatively little about the ways in which humans were 'affected by their natural environment through time' or to the manner in which they 'affected that environment and with what results.' As products of their times, these scholars reflected and responded to intellectual contexts quite different from those that prevailed in the last quarter of the twentieth century, and (with the possible exceptions of Morton and Nelson) they pursued agendas that were

---

[26] The argument here is underpinned by the work of Nelson's many graduate students at the University of Waterloo. See for example: Neugebauer (1975); Battin (1975); Mann (1978); Heffernan (1978); Bastedo-Hans (1983); Neufeld (1984); Honderich (1991); Cruik-shank (1991); Skibicki (1992); Beazley (1993); Sportza (1997); Lawrence (1996). Thanks to Gordon Nelson (personal communication) for his response to an earlier version of this paper.

rather different from those which animate current environmental-historical scholarship.

Florescence

Since 1980, and particularly since the beginning of the millennium, Canadian scholars have shown increasing interest in human relations with the rest of nature. Adapting a widespread current usage, one might say that Canadian historical scholarship has embarked on an 'environmental turn' – albeit a somewhat gentle one – in recent years. New work affirming the importance of human interactions with the environment in the Canadian experience has ranged over thought and policy, over cultural attitudes to the environment, and over the ways in which they and the laws and regulations that have flowed from them have influenced environmental change. It has striven to recognize the political, social, intellectual, and economic contexts in which environmental change occurs. And it has sought to understand nature's ecosystems as well as the systems by which people seek to comprehend nature. These developments have been driven by contemporary concerns about resource depletion, limits to growth, and environmental despoliation, as well as by scholarly arguments over the social construction of knowledge and its implications. At the same time 'the environment' has become an important focus of political concern.[27] This has led some to see the past as a valuable source of perspective on urgent issues. In this view, environmental history places present concerns in context and allows careful, considered judgment of completing claims. Others regard the past in more instrumental terms as a repository of lessons that we ignore at our collective peril as long-entrenched patterns of profligacy and resource consumption threaten the future of life on earth. Just how quickly these changes seemed to occur was captured with characteristic wit by the Federal Cabinet Minister and prominent Newfoundlander, John Crosby, in the 1980s, when he said, 'Ten years ago we didn't know about the environment – but now it is all around us.'[28]

In the past five or ten years, a growing number of Canadian scholars appear to have recognized environmental history's 'potential for changing the way we conceive of the past.'[29] A wide range of Canadian studies have appeared in international journals such as *Environmental History, Environment*

---

[27] Bassand (1995). The volume in which this appeared emerged from the efforts of the Social Sciences and Humanities Research Council of Canada to play a role in the formulation of Canada's Green Plan, and the development of the Eco-Research initiative is itself a useful marker of the broad concerns identified in this paragraph.

[28] Quoted in Rowe (1990), p. 29.

[29] Worster (1988), viii.

and History, and the *Journal of Historical Geography* as well as the international book series in environmental history. Several Canadian journals have published work in Canadian environmental history, including the *Canadian Historical Review*, the *Journal of the Canadian Historical Association*, and the *Journal of Canadian Studies*. In the past few years five journals have published special issues in environmental history (*BC Studies, Urban History Review, Revue d'Historie de l'Amérique Française, Globe. Revue internationale d'études québécoises*, and *Environmental History*). All of the major Canadian university presses have published work in environmental history and UBC press has recently launched a new series Nature/ History/ Society under the general editorship of Graeme Wynn.

Although no environmental history journal, formal association, or regular meeting of environmental historians exists in Canada, a wide range of what might be called occasional activity does. This activity suggests another measure of the field's consolidation. In British Columbia, an interdisciplinary and multi-institutional group has met for several years during the academic term at Green College, UBC under the banner Nature, History and Society to hear speakers and for occasional field trips and workshops. In Central Canada, the Quelques Arpents de Neige group operates a 'migrating workshop' in Quebec and Ontario that draws scholars from both provinces and beyond to debate set readings and to foster research-in-progress (http://www.arpents.ca/). In September 2005, this group co-sponsored with the Quebec Studies Centre at McGill a conference aimed at situating Quebec environmental history in its global contexts. A national network in Canadian environmental history or NiCHE also emerged in 2004–2005 funded by an SSHRC clusters grant and lead by Alan MacEachern and Bill Turkel at the University of Western Ontario. This network aims to draw together a widely dispersed constituency of scholars across the country engaged in environmental history to provide a platform for discussion, debate, and research. This network grew in part from a successful breakfast meeting of Canadian environmental historians organized by Alan MacEachern with support from SFU history and UBC geography at the American Society of Environmental History Meetings in Victoria (2004). NiCHE developed a community workspace using network software, launched several publication projects and organized field trips and summer schools. In 2007, its efforts were recognized by the award of new funding under the Social Science and Humanities Research Council's Strategic Knowledge Clusters Program, that has created large opportunities for the promotion and dissemination of research in environmental history, including the establishment of new regional 'workshops' in the Atlantic and Prairie regions. (See http://www.ssc.uwo.ca/history/niche/.)

All of this recent activity has built on earlier foundations. The Nature, History, and Society group emerged from a graduate reading group in the UBC geography department. The Quelques Arpents group includes members such as Colin Duncan who organized an environmental history reading group at Queen's University in the early 1990s and echoes a Toronto environmental history group spearheaded by Elinor Melville that operated for a time in the mid-1990s. The highly successful ASEH meeting held in Victoria and organized by Lorne Hammond was the second environmental history conference in that city within a decade, following the Environmental Cultures conference organized by Hammond and Richard Rajala in 1995. A year later, Richard Hoffman and Elinor Melville hosted a conference on 'Humans and Ecosystems Before Global Development' at York University, Ontario. It included research presentations by medieval and early modern scholars and reflections from modern scholars, including several Canadianists. Two years later, Laurel MacDowell organized a North American environmental history conference at the University of Toronto. The increasing number of publications and organizations in the field needs to be viewed as the outcome of a longer process of institutional and intellectual development.

In their recent spate of activity, Canadian environmental historians have addressed an array of important problems that defy simple classification and spill beyond our capacity to inventory in full. Thus we offer a crude categorization of recent work that reflects our shared sense of the main emphases evident in a disparate literature. We attempt no more than a tentative imposition of order (aware that even this is likely to be contentious) and suggest but a few exemplars of the kinds of work that continue, month by month, to flesh out our categorization. We are aware of many more works than those mentioned, as well as of differences between works within categories. We are also conscious that the table does a rather poor job of suggesting the analytical and conceptual aspects of the work that it catalogues. Subjects fit in tables; arguments and subtleties do not. In any event, our reading of the recent literature leads us to suggest that Canadian environmental historians focus primarily on ten problems. In no particular order, they are:

Topical and analytical emphases of recent writing in Canadian environmental history

Aboriginal life and colonialism: treating the environmental contexts of aboriginal experience before and after contact. Ted Binnema's *Common and Contested Ground* offers the most sweeping analysis of pre-contact processes, while Doug Harris's *Fish, Law and Colonialism* analyzes salmon fisheries as a case study of resource dispossession and the law. Michael Thoms' recently completed PhD thesis takes up related problems in Ontario.[30]

---

[30] Binnema (2001); Harris (2001); Thoms (2004) and (2002).

(Continued)

Disease diffusion: analyzing the origin and course of introduced disease in various sections of northern North America with particular attention to epidemiological aspects and social consequences. The work of Arthur Ray, Cole Harris, Bob Galois, Jody Decker, Paul Hackett, Mary-Ellen Kelm, and Robert Boyd, none of whom identify as environmental historians, has done much to bring one key aspect in the study of ecological imperialism, a central subject in the international field, to bear on Canadian space.[31]

Resettlement and environmental change: analyzing the environmental aspects of the settlement process in particular regional settings. Neil Forkey's work on the Trent River region, Matthew Hatvany's environmental historical geography of salt marshes, David Wood's historical geography of settlement in Ontario, Clint Evans's *The War on Weeds*, Barry Potyondi's *In Palliser's Triangle*, and James Murton's *Creating a Modern Countryside* all offer regional explorations of these problems.[32]

Science, technology, and environment: considering the place of nature in the history of science and technology and the role of science and technology in constructing ideas of nature. Suzanne Zeller's *Inventing Canada* (and various of her essays), Stephen Bocking's work on the history of ecology, John Varty's analysis of the refashioning of wheat, and Stéphane Castonguay's history of entomology all contribute to the history of Canadian science, technology, and environment.[33]

Places and Place-making: treating the significance of places to peoples and the imaginative fashioning of particular areas. Claire Campbell's *Shaped by the West Wind*, Laura Cameron's *Openings*, and William Turkel's *The Archive of Place* approach this problem most directly, but so too do other works such as Colin Coates' *Metamorphoses of Landscape and Community in Early Quebec*, and Alan MacEachern's *Natural Selections* on the creation of national parks in Atlantic Canada.[34]

Wilderness and Wildlife politics: concerning the making of law, expertise, and authority over wilderness and wildlife. Kurk Dorsey's work on

---

[31] Ray (1974); Harris (1994); Galois (1996); Decker (1996); Hackett (2002); Kelm (1999); Boyd (1983) and (1999).

[32] Forkey (2003); Hatvany (2004); Wood (2000); Evans (2002); Potyondi (1995); Murton (2007).

[33] Zeller (1987); Zeller (1996); Bocking (1997); Varty (2004); Castonguay (2004).

[34] Campbell (2005); Cameron (1997); Turkel (2007); Coates (2000); Murton (2007); MacEachern (2001).

wildlife diplomacy between Canada and the United States, Bill Parenteau's work on east coast sport fishing, Tina Loo's essay on wildlife policy and her book on wildlife conservation between the wars (*States of Nature*), Janet Foster's re-released *Working for Wildlife*, George Colpitt's *Game in the Garden*, John Sandlos' work on parks and game in the north (*Hunters at the Margin*), Jean Manore and Dale Miner's edited collection on *The Culture of Hunting in Canada*, and George Warecki and Gerald Killan's studies of wilderness preservation in Ontario all deal with the problems of wilderness and wildlife policy formation and their social and environmental contexts.[35]

Gender and environment: treating the social construction of nature and gender as intertwined problems. Tina Loo's discussions of the gendered aspects of game-hunting and Cate Sandiland's work on nature and gender experiences in national parks develop a subject that deserves wider attention.[36]

Resources, conflict, and environmental change: dealing primarily with the environmental aspects of primary resource development in peripheral regions and social conflicts associated with their development. Building on an earlier Canadian political economy tradition as well as Canadian historical geography, these works include Richard Rajala's *Clearcutting the Pacific Rainforest*, Jean Manore's *Cross Currents*, Matthew Evenden's *Fish versus Power*, and Jennifer Read's work on the International Joint Commission and pollution in the Great Lakes.[37]

Environmental perception: dealing primarily with the cultural history of nature. We think in this category of Joy Parr's essays exploring a more sensuous history, Patricia Jasen's cultural history of wilderness travel and Colin Coates' writings on landscape in early Quebec.[38]

Urban, class, and environmental justice: dealing with the development of cities and the emergence of environmental issues therein, ranging from the development of parks, through the challenges of water supply and sewage disposal to discussions of the uneven distribution (across society and space) of the impacts of urban and industrial development. Here we

---

[35] Dorsey (1998); Parenteau (1998); Loo (2001a) and (2006); Foster (1998); Colpitts (2002); Sandlos (2001) and (2007); Warecki (2000); Killan (1993).
[36] Loo (2001b); Sandilands (2004).
[37] Rajala (1998); Manore (1999); Evenden (2004); Read (1999).
[38] Parr (2001); Coates (2000); Jasen (1995).

> **(Continued)**
>
> think of the work among others of Arn Keeling, Ken Cruikshank, and
> Nancy Bouchier, Joanna Dean and others, examples of which can be found
> in a special issue of the *Urban History Review*, 34, 1 (Fall 2005).[39]

Every reader and every author will quibble with a table like this. It contents
and choices reflect our reading and pre-occupations. Other table makers would
make different tables. They might have a different number of boxes and more
or less extensive registers of work within them. Almost all would, like this one,
reveal an abundance and diversity of output. But what more can be made of
these broad brush strokes? Might the table help us to identify lacunae in topical
emphases and the analytical and scholarly traditions that have shaped recent
writing in Canadian environmental history?

Certainly our classification points up some significant lacunae: it suggests
that very little has been written on the methodological/ theoretical issues confronting environmental historians in Canada; that Canada's place in the global
history of hydrocarbon development and use remains to be examined; and
that besides the work on disease diffusion few studies explore the problems of
ecological imperialism across northern North America.[40] In addition, the view
enshrined in this table indicates that there have been few efforts to situate
canonical events and problems in Canadian history within an environmental context. What do environmental historians have to say about the building
of the railroad, the growth of the welfare state or Quebec nationalism? Further, the works that populate our taxonomy suggest that writing in this vein
has been at local, regional, and trans-national scales. The unit of analysis has
tended to be a river basin, park, or region. Revealingly, and characteristically,
the 2007 special issue of *Environmental History* takes Canada as its focus but
operates primarily at local, regional, and trans-national scales. Few studies have
been conducted entirely at a single scale – multi-scalar analyses have been the
norm – but work at the national scale has been rare indeed. In September 2007,
there is only one 'Canadian environmental history', Graeme Wynn's *Canada
and Arctic North America* (which emphasizes the ways in which human actions
have shaped and reshaped northern North American environments)[41] for Canadians to put alongside John Opie's *Nature's Nation* or Ted Steinberg's *Down to
Earth* on the United States, T.C. Smout's environmental history of Scotland,

---

[39] Dean (2005); Nelles (2005); Keeling (2005); Hermansen and Wynn (2005); and Baldwin
and Duke (2005); Cruikshank and Bouchier (2004) and Keeling (2004).
[40] But see MacEachern and Turkel (2009) and Piper and Sandlos (2007).
[41] Wynn (1998) and (2007).

or Eric Pawson and Tom Brooking's *Environmental Histories of New Zealand.*[42] The first collection of readings in the Canadian field, now over a decade old, contains primarily local case studies.[43] So too does a recent collection focusing on methodological problems.[44] A few books deal with 'Canadian' subjects but they focus on particular themes or problems that fall primarily within federal jurisdiction, such as national parks policy or federal agricultural science.

Students of American (i.e. United States) environmental history would likely find little surprising about the ten pigeon-holes that frame our summary of the Canadian literature. Indeed they might readily fill our boxes with fine pieces of work pertaining to the United States and add a few more categories – such as Ecological Imperialism or Urban Metabolism – of their own. If so, this is surely to suggest that although there has been a good deal of work in and of Canada in the last quarter century, there may be nothing particularly distinctive about it. Reading beyond the titles of books and articles on, or deserving inclusion in, tabulations such as these soon reveals a good deal of trans-border influence and even a degree of territorial imperialism, although the explicit traffic in ideas and inspiration appears to flow more insistently in a northern direction, and the most obvious trans-border incursions are into Canadian space. This is to say that much Canadian writing is heavily influenced by American models, and that Canadian scholars typically acknowledge, draw upon, or connect their work to that done by their American cousins. By contrast, Canadian scholarship is essentially invisible – infrequently cited, rarely engaged, and largely ignored – in the American literature.

Consider the following somewhat attenuated intellectual thread that demonstrates many of these patterns succinctly. In 1968, as work in the fledgling field of environmental history began to proliferate, the intellectual historian Roderick Nash (whose *Wilderness and the American Mind* published in 1967 helped to frame an agenda for the first generation of environmental historians in the United States) arrived in Alberta to attend a conference in Calgary, and to pronounce (with certain conviction but little local knowledge) that Canadians lagged two generations behind Americans in wilderness appreciation. A few years later, graduate student Janet Foster wrote a doctoral dissertation at York University in Toronto (published in 1978 as *Working for Wildlife: The Beginning of Preservation in Canada*) that sought, in part, to correct this claim. Canada, she argued, may have lacked charismatic visionaries such as John Muir and Theodore Roosevelt (who figured large in the pages of Nash's *Wilderness*) but the work of a group of far-sighted civil servants (who were 'fifty years ["two

---

[42] Opie (1998); Smout (2000); Brooking and Pawson (2002).
[43] Gaffield and Gaffield (1995).
[44] Turkel and MacEachern (2009).

generations"?] ahead' of their time) nonetheless moved the Canadian conservationist agenda forward. In this, these Canadian bureaucrats were influenced only to some limited extent by American thinkers, such as Muir and Pinchot, whose developing views were convergent with conclusions that the Ottawa men had reached through 'their own experiences in nature' and in confronting, as administrators of parks and forests, the threats that modern civilization posed to wildlife populations.[45]

Here, reflected environmental historian Alan MacEachern some years later (and perhaps somewhat tongue-in-cheek), was the basis of 'a distinctly Canadian tale' with bureaucrats in starring roles.[46] Perhaps unsurprisingly, Foster's interpretation failed to sustain a great deal of scholarly excitement, but her interpretation stood, neither gathering nor generating much momentum, for a quarter century. Then, in a valuable recent book, Tina Loo offered a corrective to Foster's story. *States of Nature* readily concedes that Foster was 'right to highlight state involvement' but points out that much wildlife work was done at the provincial rather than the federal level, and argues forcefully that private individuals played significant roles in wildlife conservation in Canada.[47] Among them, insists Loo, 'Father Goose' Jack Miner 'was every bit as charismatic and, at the time, as well known as Muir or Roosevelt.' Yet in focusing attention on 'ordinary' Canadians who made important, and in some cases widely celebrated, contributions to the conservation of Canadian wildlife in the twentieth century, Loo also acknowledges (as many writing on other topics have done before) that 'her thinking about the subject' was strongly influenced by an American literature.

In *States of Nature* the debt is owed, particularly, to an armful of books dealing with wildlife and nature preservation, and highlighting the normative characteristics of environmentalism.[48] But Canadian and American stories are even more deeply intertwined than this. The first chapter of Loo's book begins with a quotation from one of Foster's prescient bureaucrats, to the effect that Canadians had opportunity to learn from the Americans' wholesale destruction of their wildlife, and continues, in Loo's words: 'As with much in Canadian history, the story of saving Canada's wildlife can be framed...with reference to the United States.' A short summary of Canadian wildlife conservation efforts before 1945 is then sketched out before Loo concludes that 'in a process broadly similar to that described by Louis S. Warren for the United States, the

---

[45] Nash (1967); Foster (1978).

[46] MacEachern (2002), 217–218.

[47] Loo (2006).

[48] The key critical works cited by Loo are Bullard (1994); Jacoby (2001); Sellers (1977); Spence (1999); Warren (1997); she also lists Dunlap (1988); Isenberg (2000); Reiger (1975); and Tober (1981) on wildlife conservation.

involvement of both levels of government in conservation work altered the nature of wildlife as common property.'[49] This is not the place to engage the details or the niceties of these arguments. They are encapsulated here simply to demonstrate the varied and pervasive, even seemingly inescapable, weight of American scholars, ideas and influences on the writing of Canadian environmental history. Even as they seek to tell Canadian tales, students of Canada appear hard pressed to escape the potent sway of American scholarship – or the perceived need to connect their accounts with, and render them relevant to, the larger corpus of American literature.

These tendencies are manifest in other ways. Viewed from some southern vantage points, borders and details almost vanish. Developing a series of books intended to treat the environmental histories of most of the major regions of the world, those commissioning volumes for ABC Clio Press divided North America among several volumes and authors. The southern states, the northeast and Midwest, the plains and intermontane west, and the Pacific coast were identified in approximately those terms. All of the area north of the conterminous states (including Alaska) was assigned a separate volume, to be titled *Arctic America*, although most Canadian settlement is well south of the Arctic Circle (66° 66' N), some Canadians live (within Canada) at latitude 42° N, and the climate of the southwestern-most part of the country is, technically at least, almost Mediterranean.[50]

All of this begs the questions with which we began even as it suggests ways in which answers to some of them may be shaded. Is there anything (other than location) that is particular to Canadian environmental history, or is it topically, methodologically, intellectually simply an imitation, of its more fully developed American counterpart? Is there room, in the shared confines of the North American continent for a distinctive Canadian perspective in environmental history, or is the field north of the border – like the Canadian feature-film industry unable to escape the long shadow of Hollywood – destined to win notice (if not box-office success) only when it produces work in the American vein? And finally, does any of this matter? Given the north–south grain of the continent, and the arguments of bioregionalists and others, why need environmental historians attend (for other than purely pragmatic reasons such as the availability, comparability or difficulty of obtaining data) to national, and more broadly political, boundaries? Brought to confront these mysteries once more, we turn in conclusion to reflect upon them, briefly and prospectively, more intent upon sparking a conversation about the future of our field than in asserting directions and defining bounds for future work. Our concerns, in a nutshell, are whether

---

[49] Loo (2006), 11–12.
[50] Wynn (2007), 257–265.

there can and should be a *Canadian* environmental history, whether this need be at a national scale, and whether such an endeavor, if valuable and possible, might possess any distinguishing characteristics.

Good fences and good neighbors, or multiple destinies manifest

Especially since 1994, when the American historian Dan Flores published his manifesto for bioregional history, scholars interested in human–environment interactions have had to wrestle with one or another version of the claim that 'the politically-derived boundaries of county, state and national borders are mostly useless in understanding nature.'[51] By Flores' reckoning, environmental historians should focus their inquiries on natural regions, areas defined by reference to ecological and topographical rather than political boundaries. Taking nature seriously entails recognition of its variable character; students should pay closer attention to its qualities and boundaries than to such 'artificial' social/cultural constructions as census districts and administrative jurisdictions. On this view, a *Canadian* environmental history makes little sense. The western section of the Canadian–American boundary line, often (though less and less accurately these days) described as the longest undefended border in the world, pays scant attention to nature as it tracks the forty-ninth parallel. Following a latitudinal calibration invented by humans to impose order upon global space and fixed upon through political machinations (in the far west at least it might equally logically have been located at 45° to place most of the Columbia River basin north of the line or at 54°, 40' to satisfy James Polk), it marches unerringly from Lake of the Woods to the Strait of Georgia dividing drainage basins and ecological regions between the two nations it helps to define. East of Manitoba, the border follows the Rainy River, and divides its basin as decisively as it does the waters of four of the Great Lakes, before tracing the course of the St Lawrence to the 45th parallel. It then tracks east across lakes and rivers to find and follow the height of land, more-or-less, until it joins and then leaves the St John River to reach the Atlantic via the St. Croix River. Contested historically along significant parts of its length, this convoluted eastern boundary is no more coherent, ecologically, than its straight-line counterpart west of Rainy River.

Well might bioregionalists conclude, with the vigorous, Continentalist critic of late nineteenth-century Canadian nation building Goldwin Smith, that the country is an 'artificial' creation. 'Whoever wishes to know what Canada is,' Smith proclaimed in 1891, 'should begin by turning from the political to the natural map.' For, by contrast with the impression of a vast and uniform territory conveyed by the former, the latter revealed 'four separate projections of the

---

[51] Flores (1994–95); and (2001).

cultivable and habitable part of the continent into arctic waste' set apart from each other by 'great barriers of nature, wide and irreclaimable wildernesses or manifold chains of mountains.' Smith's views were not shared by all. Where he saw the attempt to extend Canada east to west across the continent as an effort 'to wage a desperate war against nature,' the prominent nationalist Reverend George Grant regarded it as a magnificent commitment by people ready to bear the costs of realizing their vision. Canadians, said Grant, had agreed to 'rise up and build' their country, and this was not to be dismissed, in Smith's perfunctory way, as mere child's play among grown babies.[52]

Grant's indignation warrants the attention of environmental historians. Bioregional approaches draw attention to the north–south ecological grain of the continent, but they underplay the perseverance reflected in and the significance of human inscriptions on this terrain. Like Smith, bioregionalists are perhaps too fervent in defense of their particular perspective. Purportedly interested in human–environment interactions, they insist that boundaries be drawn by reference to environmental not human concerns. But 'just as environmental boundaries have had consequences for humanity, so human boundaries have had consequences for the environment.' Shannon Stunden Bower, whose sentence this is, has demonstrated this clearly and cogently in her recent study of wetlands, flooding and drainage in southern Manitoba. Ecological concerns are of paramount interest in this study of the Red River basin between the American border and Lake Winnipeg.[53] Stunden Bower traces human adaptations to and transformations of the marshes and wet prairie (periodically inundated grassland) that form so large a part of this section of southern Manitoba through the years between 1810 and 1980. She is well aware of the north–south grain of the continent and of the fact that 42 rivers cross (often more than once) the international boundary between Canada and the United States. She knows that Dan Flores sees watersheds as history's 'natural nations.' And she demonstrates that debates, developments, and drains in the lowlands cannot be understood without attention to the physical geography and hydrology of the larger watershed in which they are situated. Yet the international boundary is an important factor in her analysis. It is not impermeable. A tributary of the Red River, the Roseau, which rises in Minnesota and flows for some distance through Manitoba was the subject of hearings before the International Joint Commission established under the Boundary Waters treaty of 1909, intended to find means of cooperative trans-border river management, and Stunden Bower pays due attention to these developments. In the end, however, she declines to include the upper reaches of the Red River (which flows through

---

[52] Grant (1893), p. 26. Smith (1891).
[53] Stunden Bower (2006), 16.

Minnesota and North Dakota to the Canadian border) in her study area. Her work deliberately offers both more and less than a bioregional perspective on the environmental history of southern Manitoba. The challenges presented by this area, she insists, 'may be part of a larger bioregional problem,' but they are only properly comprehended by giving 'careful and extended attention' to 'the actions of Canadian governments (local, provincial and national).'

In sum, Stunden Bower would argue that however environmentally credible efforts to find unity in drainage basins and to see river catchments as natural nations might be, they are problematic in cultural (and historical) terms. The fact that rivers run through them does not mean that political, ethnic, social, or cultural boundaries can and should be ignored. From social and political vantage points, both watersheds and bioregions might well be entirely arbitrary and substantially irrelevant units of analysis. As Stunden Bower points out, taking the entire Red River valley as the focus of analysis would be 'inconsistent with how many Manitobans understood their lives and with how...political divisions became inscribed on the landscape.' Landscapes and environments are shaped by many forces. Some of them are 'natural.' Some are human. And the two are often tangled and intertwined in endlessly surprising ways to form 'hybrid landscapes,' products of both nature and society. In the Red River basin, 'the 49th parallel...had environmental consequences.'[54]

For all that environmental historians seek to insert nature into their accounts of the past, as historians they must surely remain mindful of Marc Bloch's assertion, that 'it is man that history seeks to grasp.' Passing by the gender insensitivity of words written many years ago, we read this claim as a concise and useful encapsulation of a fundamental point: that the writing of history is a human enterprise, undertaken to satisfy human curiosity and to provide humans with the bearings they need to navigate their ways through space and time – or land and life. As historians, environmental historians need to attend to the ways in which 'natural nations' have been shaped and sundered by human actions. They need to investigate and reflect upon the layered and competing human discourses that relate and give human significance to partic-ular places. But as students of the environment they also need to attend to the ways in which it has been affected by and responded to human actions.

All of this has implications for thinking about whether there can and should be a Canadian environmental history, about the scales at which it might be written, and the forms that it might take. On the one hand, although the grain of the continent may suggest that the international boundary (and thus 'Canada') is irrelevant from an environmental perspective, it is obvious that 'history' has sundered the natural nations of North America, politics has

---

[54] Stunden Bower (2006), 17 and 271–299.

divided them, opportunities (shaped by markets, capital, and location) have differentiated them and the experience of those who live north and south of the border has led people both to think about their places in different ways, and to attach diverse meanings to their environments. For all of these reasons, it seems to us that the argument for a Canadian environmental history is indeed a powerful one. For economic and environmental as well as political, social, and cultural reasons that reach back at least as far as the earliest European encounters with the northern part of the western hemisphere, human–environment interactions have been cast somewhat differently north and south of the 45th parallel or thereabouts (as Stephen Hornsby's recent *British Atlantic, American Frontier* demonstrates so effectively, despite its opening lament that 300 years of history before 1780 have been artificially contained within American and Canadian envelopes).[55] Since 1776, political and ideological differences (and all their entailments) have amplified divergent patterns of human–environment interaction in northern and central sections of the continent. To be sure, technology worked (increasingly powerfully) to obliterate some of these dissimilarities: similar techniques marked the fisheries of New England and Nova Scotia; lumbermen in Maine and New Brunswick followed almost identical practices; railroads devoured space and time on both sides of the 49th parallel, and patterns of suburban sprawl around burgeoning American and Canadian cities after 1950 were remarkably similar. Yet regulatory environments differed – in British North American colonies the Crown retained title to land and resources as it did not in the United States, for example – and this affected the ways in which resources were developed and how human–environment interactions proceeded. So too, different peoples held different values, embraced different ideas, envisioned different futures and made different choices. It is no accident that Vancouver's urban form (and all that that implies from an environmental perspective) differs so markedly from that of Seattle or Los Angeles. The environmental history of Canada cannot be 'read off' that of the United States. To know themselves, and to understand their particular place on earth, Canadians need to know the unique environmental history of their territory.

On the other hand, none of this is to argue the need for, or imply the possibility of, a single definitive environmental history of Canada *tout suite*. National and ecological scales are rarely (if ever) conformable, and cannot be mapped onto each other without remainder. Certainly this is the case with Canada, which encompasses tundra and prairie, Carolinian and Boreal forests, ancient bogs, newly vegetated alpine meadows, and so on. In this the bioregionalists have a point; political boundaries rarely coincide with ecological units. But

---

[55] Hornsby (2005).

how important is this to thinking about a Canadian environmental history. Generally watersheds can be delimited precisely. But the boundaries of eco-logical units are (frequently) indistinct. Most ecological communities are far from homogeneous, even in the mix of species that constitute them. Vege-tation complexes often blend, imperceptibly, into one another, in transition areas known as ecotones. National boundaries are also contingent and many have changed over time. What was 'Canada'? All of the land and water within the country's current territorial bounds (which were only extended to encom-pass Newfoundland in 1949)? The area included within the four colonies that became provinces of the new Confederation in 1867? Five provincial territories in 1870, six in 1871, seven in 1873 (notably, none of the three later additions shared a border with each other or with any of the original provinces)? Can we properly think, speak and write of 'Canada Before Confederation'? Where do indigenous people – the First Nations of Canada, present for millennia before the appellation 'Canada' (which probably meant village in one indige-nous language) was applied to a more expansive territory by Europeans – fit into all of this? From the perspective of environmental history they are cer-tainly important to the story of human–environment interactions across this northern realm, but much of their past is as 'pre-Canadians.'

These are real concerns, but they are hardly insuperable. With the emergence of a large battery of historical subfields in the last decades of the twentieth century, few 'national histories' now purport to offer THE story of a people or a country, as they once did. Then, the rise and fall of governments, politi-cians, and other great men (rarely women) provided the focus for a more or less orderly narrative parade of events and achievements loosely taken to encap-sulate the 'history' of a nation and/or its people (singular). Now, libraries of scholarship in social, race, women's, gender, regional, urban, ethnic and other branches of history beg attention and inclusion. But the coherent integration of fragments into a larger story, the synthesis of knowledge in each of these subfields, has proven far more challenging than writing political history as the national narrative. Research contributions in each of these new fields are typi-cally case or local studies: histories of widows, strikes, suffrage struggles, small groups of migrants, and so on abound. Each study is inflected by the particular time and place upon which it is focused. By the same token, the environmental historian's characteristic focus may be upon the proverbial 'acre in time,' and there are many thousand such, each constituting a very particular story, across a country as large as Canada.

Dealing simultaneously with change across space and through time is – as geographers recognized long ago – no easy task. Subsuming multiple case stud-ies into a larger account, or finding ways in which to integrate separate and distinct pieces into an incomplete puzzle without losing a sense of the whole is not easy, and the result may not be seamless, but there are ways in which these difficulties can be addressed, if not entirely banished. Whatever solutions are found, any large-scale synthesis must rest upon the revelations and insights of hundreds of more narrowly defined inquiries. Authors of sweeping surveys and insightful monographs must necessarily attend to questions of scale. By working at a range of levels, moving from the general to the particular, and seeking the broader import of specific instances, they can be encouraged by the prospect of writing more than 'parish pump' accounts while avoiding the pitfalls associated with finding the world in a grain of sand. Thus might environmental histories of Canada be envisaged and written alongside more territorially circumscribed and substantively focused studies.

In addressing both the monographic and the synthetic parts of this agenda, Canadian environmental historians might challenge themselves to consider whether there is anything unique about the n/Nature of Canada. Did the experience of settlers north and south of the international boundary differ because of the generally greater intractability of northern environments and, if so, might these differing experiences underpin consequential differences in attitudes to nature in the two countries? Many have suggested as much. Literary scholars, in particular, have made a great deal of the contrast between American abundance and Canadian dearth: newcomers to 'America' celebrated the fruitfulness of the new world, offered exuberant affirmations of the area's promise, thought of America as 'Nature's Nation'; immigrants to early Canada were 'swallowed by an alien continent,' the 'sinister and menacing' Nature of Canadian poetry evoked stark terror, northerners struggled for survival against an implacable antagonist, the Canadian environment. Historians and geographers have also trodden this path, finding the roots of Canadian distinctiveness in 'northern-ness' and in the contrast between the truncated possibilities for settlement and expansion and the broad and long-enduring potential of the American frontier. Such suggestive ideas have solidified into powerful interpretive metaphors that have been extended to account for differences in the relative importance of such things as individualism and co-operation in the two countries. But these bold extrapolations rest upon abstract and essentially deductive claims.[56]

Surely environmental historians might contribute to these and other similar discussions through their studies of human–environment interactions. Might prevailing ideas about nature and culture in the two countries reflect different political circumstances, and the ideological choices of societal leaders, rather than the everyday experiences of settlers in encounter with their local settings; the physical environments of Maine and New Brunswick, Michigan and

---

[56] Kline (1970); Atwood (1972); Frye (1965); Jones (1970); Morton (1970b); Harris and Warkentin (1974). Thanks to Jean Manore, personal communication for pointing our discussion in this direction.

Ontario, Montana and Saskatchewan are not, after all, sufficiently different to sustain a strictly materialist account of attitudinal differences north and south of the border. In similar vein, perhaps environmental historians should look to other northern countries – Scandinavia, Russia, even (embodying unique and potentially intriguing jurisdictional/locational considerations) Alaska – more than to the conterminous United States for comparative insights and the possible identification of Canadian distinctiveness in the development of northern environments.[57] So too might they ask whether environmental differences have contributed to the particular cast of Canadian regionalism and how the answer to this question might advance understanding of regionalism in other parts of the world.

None of these questions is intended to be agenda-setting. Rather they suggest, with much else covered in this broad, and thus necessarily incomplete, reflection on the writing of environmental history within and across borders in northern North America, the enormous scope for new, important, and invigorating work in the field. Only the evidential record and the ingenuity of researchers limit the scope of possibilities. With gathering momentum and a diverse array of innovative work in progress and in prospect, the way is open for studies on a range of topics at a variety of scales that encompass more or less of the vastness of Canada. Whether seemingly definitive or provisional and open-ended, ultimately their treatment of space, time, and interpretive themes will be uneven and their coverage incomplete. They will be subject to challenge and revision. But they will serve as route maps into and through little charted territory, and in doing so offer their readers a path to understanding, a springboard for further reconnaissance, and a new way of thinking about the Canadian past.[58]

References

Atwood, M. (1972), *Survival: A Thematic Guide to Canadian Literature*, Toronto: Anansi.

Baldwin, D.O. and Duke, D.F. (2005), ' "A Grey Wee Town": An Environmental History of Early Silver Mining at Cobalt, Ontario,' *Urban History Review* 34, 1, 71–87.

Barnes, T. and Hayter, R. (1990), 'Innis' Staple Theory, Exports, and Recession: British Columbia, 1981–86,' *Economic Geography* 66, 2, 156–73.

Bassand, M. (1995), 'L'environnnement programme,' In Quesnel (ed.) *Social Sciences and the Environment/ Les sciences sociale et l'environnemente*, Ottawa: University of Ottawa Press, 21–33.

---

[57] Morton (1970b).

[58] Though dealing with a far smaller and more obviously bounded territory, it was for essentially similar reasons that Brooking and Pawson (2002) called their carefully-shaped and well-integrated but hardly exhaustive or incontrovertible collection of essays *Environmental Histories of New Zealand*.

Bastedo-Hans, B. (1983), Cultural Aspects of Resource Surveys, A Human Ecological Approach: Aishihik, Yukon, The University of Waterloo (M.A).

Battin, J. (1975), Land Use History and Landscape Change, Point Pelee National Park, Ontario, The University of Western Ontario (M.A).

Beazley, K. (1993), Forested Areas of Long Point: Landscape History and Strategic Planning, Department of Geography, The University of Waterloo (M.A).

Berger, C.C. (1976), *The Writing of Canadian History: Aspects of English-Canadian Historical Writing since 1900*, Toronto: Oxford University Press.

Binnema, T. (2001), *Common and Contested Ground: A Human and Environmental History of the Northwest Plains*, University of Oklahoma Press.

Bocking, S. (1997), *Ecologists and Environmental Politics: A History of Contemporary Ecology*, New Haven: Yale University Press.

Boyd, R. (1983), 'Smallpox in the Pacific Northwest: The First Epidemics,' *BC Studies* 57, 68–85.

Boyd, R. (1999), *The Coming of the Spirit of Pestilence: Introduced Infectious Disease and Population Decline among the Northwest Coast Indians, 1774–1874*, Seattle: University of Washington Press.

Brooking, T. and Pawson, E. (2002), *Environmental Histories of New Zealand*, Melbourne: Oxford University Press.

Bullard, R.D. (1994), *Dumping in Dixie: Race, Class and Environmental Quality*, Boulder. CO: Westview Press.

Butler, W.F. (1873), *The Great Lone Land*, London: S. Low, Marston, Low & Searle.

Cameron, L. (1997), *Openings: A Meditation on History, Method and Sumas Lake*, Montreal and Kingston: McGill-Queen's University Press.

Campbell, C. (2005), *Shaped by the West Wind: Nature and History in Georgian Bay*, Vancouver: University of British Columbia Press.

Castonguay, S. (2004), *Protection des cultures, construction de la nature. L'entomologie economique au Canada*, Sillery: Septentrion.

Coates, C. (2000), *Metamorphoses of Landscape and Community in Early Quebec*, Montreal and Kingston: McGill-Queen's University Press.

Coates, P. (2004), 'Emerging from the Wilderness (or, from Redwoods to Bananas): Recent Environmental History in the United States and the Rest of the Americas,' *Environment and History* 10, 4, 407–38.

Colpitts, G. (2002), *Game in the Garden: A Human History of Wildlife in Western Canada to 1940*, Vancouver: University of British Columbia Press.

Cook, R. (1990–91), 'Cabbages not Kings: Towards an Ecological Interpretation of Early Canadian History,' *Journal of Canadian Studies*, 25, 4, 5–16.

Creighton, D. (1937), *The Commercial Empire of the St. Lawrence*, Toronto: Ryerson Press.

Cruikshank, R.W. (1991), Mountaineering in the St Elias Mountains, Kluane National Park Reserve: A Geographical Perspective, Department of Geography, The University of Waterloo (M.A.).

Cruikshank, K. and Bouchier, N.B. (2004), 'Blighted Areas and Obnoxious Industries: Constructing Environmental Inequality on an Industrial Waterfront, Hamilton, Ontario, 1890–1960,' *Environmental History*, 9, 3, 464–96.

Dean, J. (2005), 'Said Tree Is a Veritable Nuisance': Ottawa's Street Trees 1869–1939,' *Urban History Review*, 34, 1, 46–57.

Decker, J.F. (1996), 'Country Distempers: Deciphering Disease and Illness in Rupert's Land before 1870,' In Brown, J. and Vibert, E. (eds), *Reading Beyond Words: Contexts For Native History*, Broadview Press, 156–181.

Dorsey, K. (1998), *The Dawn of Conservation Diplomacy: US–Canadian Wildlife Protection Treaties in the Progressive Era*, Seattle: University of Washington Press.

Dunbar, G. (1985), 'Harold Innis and Canadian Geography,' *Canadian Geographer*, 29, 2, 159–64.

Dunlap, T. (1988), *Saving America's Wildlife*, Princeton: Princeton University Press.

Evans, C. (2002), *The War on Weeds in the Prairie West: An Environmental History*, Calgary: University of Calgary Press.

Evenden, M. (2004), *Fish versus Power: An Environmental History of the Fraser River*, Cambridge University Press.

Flores, D. (1994–95), 'Place: An Argument for Bioregional History,' *Environmental History Review* 18, 1–18.

Flores, D. (2001), *The Natural West: Environmental History in the Great Plains and Rocky Mountains*, Norman: University of Oklahoma Press.

Forkey, N. (2003), *Shaping the Upper Canadian Frontier: Environment, Society, and Culture in the Trent Valley*, Calgary: University of Calgary Press.

Foster, J. (1998), *Working for Wildlife*, Toronto: University of Toronto Press.

Frye, N. (1965), 'Conclusion,' In Klinck, C.F. (ed.), *Literary History of Canada: Canadian Literature in English*, Toronto: University of Toronto Press, 821–49.

Gaffield, C. and Gaffield, P. (1995), *Consuming Canada: Readings in Environmental History*, Toronto: Copp Clark Ltd.

Galois, R.T. (1996), 'Measles, 1847–1850: The First Modern Epidemic in British Columbia,' *BC Studies* 109, 31–46.

Grant, G.M. (1893), *Canada and the Canadian* Question, Toronto: C B Robinson.

Hackett, F.J.P. (2002), *'A Very Remarkable Sickness': Epidemics in the Petit Nord, 1670–1840*, Winnipeg: University of Manitoba Press.

Hardy, R. and Séguin, N. (1984), *Forêt et société en Mauricie: la formation de la région de Trois-Rivières, 1830–1930*, Montréal: Boréal Express.

Harris, D. (2001), *Fish, Law and Colonialism: The Legal Capture of Salmon in British Columbia*, Toronto: University of Toronto Press.

Harris, R.C. (1967), 'Historical Geography in Canada,' *The Canadian Geographer*, 11, 4, 235–50.

Harris, R.C. (ed.) (1987), *The Historical Atlas of Canada*, Volume 1, Toronto: University of Toronto Press.

Harris, R.C. (1994), 'Voices of Disaster: Smallpox Around the Strait of Georgia in 1782,' *Ethnohistory*, 42, 4, 591–626.

Harris, R.C. and Warkentin, J. (1974), *Canada Before Confederation: A Study in Historical Geography*, New York: Oxford University Press.

Hatvany, M. (2004), *Marshlands: Four Centuries of Environmental Changes on the Shores of the St. Lawrence*, Sainte-Foy: Les Presses de l'Université Laval.

Heidenreich, C.E. (1990), 'The Natural Environment of Huronia and Huron Seasonal Activities,' In Wynn, G. (ed.), *People Places Patterns Processes: Geographical Perspectives on the Canadian Past*, Toronto: Copp Clark Pitman, 42–55.

Heffernan, S. (1978), Long Point Ontario: Land Use, Landscape Change and Planning, The University of Waterloo (M.A.).

Hermansen, S. and Wynn, G. (2005), 'Reflections on the Nature of an Urban Bog,' *Urban History Review*, 34, 1, 9–27.

Honderich, J.E. (1991), Wildlife as a Hazardous Resource: An Analysis of the Historic Interaction of Humans and Polar Bears in the Canadian Arctic, 2000 B.C. to A.D. 1935, Department of Geography, The University of Waterloo (M.A.).

Hornsby, S.J. (2005), *British Atlantic, American Frontier: Spaces of Power in Early Modern British America*, Hanover, NH: University Press of New England.

Innis, H.A. (1923), *A History of the Canadian Pacific Railway*, Toronto: McClelland and Stewart, Ltd.

Innis, H.A. (1954[1940]), *The Cod Fisheries*, Toronto: University of Toronto Press.

Innis, H.A. (1956 [1930]), *The Fur Trade in Canada: An Introduction to Canadian Economic History.*

Isenberg, A. (2000), *The Destruction of the Bison: An Environmental History*, New York: Cambridge University Press.

Jacoby, K. (2001), *Crimes Against Nature: Squatters, Poachers, Thieves and the Hidden History of American Conservation*, Berkeley: University of California Press.

Jasen, P. (1995), *Wild Things: Nature, Culture and Tourism in Ontario, 1790–1914*, Toronto: University of Toronto Press, 1995.

Jones, D.G. (1970), *Butterfly on Rock: A Study of Themes and Images in Canadian Literature*, Toronto: University of Toronto Press.

Keeling, A.M. (2003[2004]), 'Sink or Swim: Water Pollution and Environmental Politics in Vancouver, 1889–1975,' *BC Studies*, 142/143, 69–101.

Keeling, A.M. (2005), 'Urban Waste Sinks as a Natural Resource: The Case of the Fraser River,' *Urban History Review*, 34, 1, 58–70.

Kelm, M.-E. (1999), 'British Columbia's First Nations and the Influenza Pandemic of 1918–1919,' *BC Studies* 122, 23–48.

Kelly, K. (1990), 'Damaged and Efficient Landscapes in Rural Southern Ontario, 1880–1900,' In Wynn, G. (ed.) *People Places Patterns Processes: Geographical Perspectives on the Canadian Past*, Toronto: Copp Clark Pitman, 213–27.

Killan, G. (1993), *Protected Places: A History of Ontario's Provincial Parks System*, Toronto: Dundurn Press.

Kline, M.B. (1970), *Beyond the Land Itself: Views of Nature in Canada and the United States*, Cambridge, MA: Harvard University Press.

Lawrence, P. (1996), Great Lakes Shoreline Flooding and Erosion Hazards: Towards A Strategy for Decision-Making in Ontario, Department of Geography, The University of Waterloo (Ph.D.).

Loo, T. (2001a), 'Making a Modern Wilderness: Wildlife Management in Canada, 1900–1950,' *Canadian Historical Review*, 82/1, 91–121.

Loo, T. (2001b), 'Of Moose and Men: Hunting for Masculinities in the Far West,' *Western Historical Quarterly*, 32, 296–319.

Loo, T. (2006), *States of Nature: Conserving Canada's Wildlife in the Twentieth Century*, Vancouver: University of British Columbia Press.

Lower, A.R.M. (1936), *Settlement and the Forest Frontier in Eastern Canada*, Toronto, Macmillan Company of Canada, Ltd.

Lower, A.R.M. (1938), *The North American Assault on the Canadian Forest: A History of the Lumber Trade between Canada and the United States*, Toronto: Ryerson Press, 1938.

MacEachern, A. (2001), *Natural Selections: National Parks in Atlantic Canada, 1935–1970*, Montreal and Kingston: McGill-Queen's University Press.

MacEachern, A. (2002), 'Voices Crying in the Wilderness: Recent Works in Canadian Environmental History,' *Acadiensis*, 31, 2, 217–18.

MacEachern, A. and W.J. Turkel (eds) (2009), *Method and Meaning in Canadian Environmental History*, Toronto: Nelson.

Mann, D. (1978), The Changing Rondeau Landscape, The University of Waterloo (M.A.).

Manore, J. (1999), *Cross-Currents: Hydroelectricity and the Engineering of Northern Ontario*, Waterloo: Wilfred Laurier Press.

McKillop, A.B. (ed.) (1980), *Contexts of Canada's Past: Selected Essays of W.L. Morton*, Toronto: The Macmillan Company of Canada.

McNeill, J.R. (2003), 'Observations on the Nature and Culture of Environmental History,' *History and Theory* 42/4, 5–43.

Miller, E. (1915), 'La géographie au service de l'histoire,' *Revue trimestrielle canadienne*, 1/1, 45–53.

Morton, W.L. (1946), 'Marginal,' *Manitoba Arts Review*, V (Spring), In McKillop, A.B. (ed.) *Contexts of Canada's Past: Selected Essays of W.L. Morton*, Toronto: The Macmillan Company of Canada.

Morton, W.L. (1970a), 'Seeing an Unliterary Landscape,' *Mosaic: A Journal for the Comparative Study of Literature and Ideas* (Manitoba Centennial Issue) III #3, 1–10.

Morton, W.L. (1970b), 'The "North" in Canadian Historiography,' *Transactions of the Royal Society of Canada*, NS, 4/8, 31–40.

Murton, J. (2007), *Creating a Modern Countryside: Liberalism and Land Resettlement in British Columbia*, Vancouver: University of British Columbia Press.

Nash, R. (1967), *Wilderness and the American Mind*, New Haven: Yale University Press.

Nelles, H.V. (1974), *The Politics of Development: Forests, Mines and Hydro-electric Power in Ontario, 1849–1941*, Toronto: Macmillan of Canada.

Nelles, H.V. (2005), 'How Did Calgary Get Its River Parks?,' *Urban History Review*, 34/1, 28–45.

Nelson, J.G. (1967), 'Man and Landscape in the Western Plains of Canada,' *The Canadian Geographer*, 11/4, 251–64.

Nelson, J.G. (1973), *The Last Refuge*, Montreal: Harvest House.

Nelson, J.G. (1976), *Man's Impact on the Western Canadian Landscape*, Toronto: McClelland and Stewart Ltd.

Nelson, J.G. and Byrne, R. (1966), 'Man as an Instrument of Landscape Change – Furs, Floods and National Parks in the Bow Valley, Alberta,' *Geographical Review*, 56/2, 226–38.

Nelson, J.G. and England, R.E. (1971), 'Some Comments on the Causes and Effects of Fire in the Northern Grasslands Area of Canada and the Nearby United States, Ca, 1750–1900,' *The Canadian Geographer* 15/4, 1971, 295–306.

Neufeld, K. (1984), Ranching and the Grasslands National Park: An Historic and Institutional Analysis, School of Urban and Regional Planning, The University of Waterloo (M.A.).

Neugebauer, P. (1975), Land Use History, Landscape Change and Resource Conflict in the Sandbanks Provincial Park Area, Prince Edward County, Ontario, The University of Western Ontario (M.A.).

Opie, J. (1998), *Nature's Nation: An Environmental History of the United States*, New York: Harcourt Brace.

Parenteau, W. (1998), '"Care, Control and Supervision": Native People in the Canadian Atlantic Salmon Fishery, 1867–1900,' *Canadian Historical Review*, 79/1, 1–36.

Parker, I. (1988), 'Harold Innis as a Canadian Geographer,' *Canadian Geographer*, 32/1, 63–69.

Parr, J. (2001), 'Notes for a More Sensuous History of Twentieth Century Canada: The timely, the tacit and the Material Body,' *Canadian Historical Review*, 82/4, 720–45.

Piper, L. and Sandlos, J. (2007), 'A Broken Frontier: Ecological Imperialism in the Canadian North,' *Environmental History*, 12, 759–95.

Potyondi, B. (1995), *In Palliser's Triangle: Living in the Grasslands, 1850–1930*, Calgary: Purich Publishers.

Rajala, R. (1998), *Clearcutting the Pacific Rain Forest*, Vancouver: University of British Columbia Press.

Ray, A.J. (1974) *Indians in the Fur Trade*, Toronto: University of Toronto Press.

Ray, A.J. (1990), 'Diffusion of Diseases in the Western Interior of Canada, 1830–1850,' In Wynn, G. (ed.), *People Places Patterns Processes: Geographical Perspectives on the Canadian Past*, Toronto: Copp Clark Pitman, 68–87.

Read, J. (1999), Addressing 'A Quiet Horror': the Evolution of Ontario Pollution Control Policy in the International Great Lakes, 1909–1972, University of Western Ontario, (Ph.D).

Reiger, J. (1975), *American Sportsmen and the Origins of Conservation*, New York: Winchester.

Roe, N.A. and Nelson, J.G. (1972), 'Man, Birds and Mammals of Pacific Rim National Park, B.C.: Past, Present and Future,' In Nelson J.G. and Cordes, L.D. (eds), Pacific Rim: An Ecological Approach to a New Canadian National Park, *Studies in Land Use History and Landscape Change*, No. 4, Calgary: University of Calgary.

Rowe, J.S. (1990), 'Wilderness as Home Place,' In *Home Place: Essays on Ecology*, Edmonton: NeWest Press, 29–34.

Sandilands, C. (2004), 'Where the Mountain Men Meet the Lesbian Rangers: Gender, Nation, and Nature in the Rocky Mountain National Parks,' In Hessing, M., Raglon, R. and Sandilands, C. (eds), *This Elusive Country: Women and the Canadian Environment*, Vancouver: University of British Columbia Press.

Sandlos, J. (2001), 'From the Outside Looking In: Aesthetics, Politics and Wildlife Conservation in the Canadian North,' *Environmental History*, 6/1, 6–31.

Sandlos, J. (2007), *Hunters at the Margin: Native People and Wildlife Conservation in the Northwest Territories*, Vancouver: University of British Columbia Press.

Sellers, R.W. (1977), *Preserving Nature in the National Parks: A History*, New Haven: Yale University Press.

Shore, M. (1995), '"Remember the Future": The *Canadian Historical Review* and the Discipline of History, 1920–95,' *Canadian Historical Review*, 76, 410–63.

Shore, M. (2002), *The Contested Past: Reading Canada's History*, Toronto: University of Toronto Press.

Skibicki, A. (1992), Land Use History and Landscape Change in the Grand River Forest, School of Urban and Regional Planning, The University of Waterloo (M.A.).

Smith, G. (1891), *Canada and the Canadian* Question, Toronto: Hunter Rose.

Smout, T.C. (2000), *Nature Contested: Environmental History in Scotland and Northern England since 1600*, Edinburgh: Edinburgh University Press.

Spence, M.D. (1999), *Dispossessing the Wilderness: Indian Removal and the Making of National Parks*, New York: Oxford University Press.

Sportza, L. (1997), Assessing the Evolution of Marsh Management in Protected Areas: With special reference to Point Pelee, Rondeau and Long Point, Lake Erie, Canada, School of Urban and Regional Planning, The University of Waterloo (M.A.).

Stephanson, A. (1995), *Manifest Destiny: American Expansionism and the Empire of Right*, New York: Hill and Wang.

Stunden Bower, S. (2006), Wet Prairie: An Environmental History of Wetlands, Flooding, and Drainage in Agricultural Manitoba, 1810–1980, PhD dissertation, University of British Columbia.

Thoms, J.M. (2002), 'A Place Called Pennask: Fly-fishing and Colonialism at a British Columbia Lake,' *BC Studies* 133, 69–98.

Thoms, J.M. (2004), 'Ojibwa Fishing Grounds: A History of Ontario Fisheries Law, Science, and the Sportsmen's Challenge to Aboriginal Fishing Rights, 1650–1900,' PhD thesis, University of British Columbia.

Tober, J.A. (1981), *Who Owns Wildlife? The Political Economy of Conservation in Nineteenth Century America*, Westport, CT: Greenwood Press.

Turkel, W.J. (2007), *The Archive of Place: Unearthing the Pasts of the Chilcotin Plateau*, Vancouver: University of British Columbia Press.

Varty, J. (2004), 'On Protein, Prairie Wheat, and Good Bread: Rationalizing Technologies and the Canadian State, 1912–1935,' *Canadian Historical Review*, 85/4, 721–53.

Vickers, C. (1945), 'Archeology in the Rock and Pelican Lake area of Southern Manitoba,' *Papers of the Historical and Scientific Society of Manitoba, 1944–1945*, Winnipeg, 1945, 14–24.

Voisey, P.L. (1988), *Vulcan: The Making of a Prairie Community*, Toronto: University of Toronto Press.

Wallace, W.S. (1920), 'The Growth of National Feeling,' *Canadian Historical Review*, 1, 138–40.

Warecki, G. (2000), *Protecting Ontario's Wilderness: A history of Changing Ideas and Preservation Politics, 1927–2000*, New York: Peter Lang.

Warren, L.S. (1997), *The Hunter's Game: Poachers and Conservationists in Twentieth Century America*, New Haven: Yale University Press.

Weiner, D.R. (2005), 'A Death-Defying Attempt to Articulate a Coherent Definition of Environmental History,' *Environmental History*, 10/3, 404–20.

Wood, J.D. (1964–65), 'Historical Geography in Alberta,' *Albertan Geographer*, 1, 17–19.

Wood, J.D. (2000), *Making Ontario*, Montreal and Kingston: McGill-Queen's University Press.

Worster, D. (1988), *The Ends of the Earth: Perspectives on Modern Environmental History*, Cambridge: Cambridge University Press.

Wrong, G.M. (1924), 'The Teaching of the History and Geography of the British Empire,' *Canadian Historical Review*, 4, 297–98.

Wynn, G. (ed.) (1990), *People Places Patterns Processes: Geographical Perspectives on the Canadian Past*, Toronto: Copp Clark Pitman.

Wynn, G. (1993), 'Geographical Writing on the Canadian Past,' In Conzen, M., Rumney, T. and Wynn, G. (eds.), *A Scholar's Guide to Geographical Writing on the American and Canadian Past*, Chicago: University of Chicago Press, 99–124.

Wynn, G. (1998), *Remaking the Land God Gave to Cain: A Brief Environmental History of Canada*, London: Canadian High Commission.

Wynn, G. (2007), *Canada and Arctic North America: An Environmental History*, Santa Barbara: ABC-CLIO.

Zeller, S. (1987), *Inventing Canada: Early Victorian Science and the Idea of a Transcontinental Nation*, Toronto: University of Toronto Press.

Zeller, S. (1996), *Land of Promise, Promised Land: The Culture of Victorian Science in Canada* (Ottawa: Canadian Historical Association, Booklet #56).

10
Modernity and the Politics of Waste in Britain*

Tim Cooper

Pollution rightly has a central place in environmental history. Urban environmental historians have focused a good deal of attention on industrial cities and the search for technological solutions to urban pollution problems.[1] Recently, scholars have also displayed an interest in the role played by pollution in early environmental organisation.[2] Ironically, one consequence of this focus on *pollution* has been that the significance of *waste* in the development of environmental concerns has been relatively underplayed. Waste and pollution were not simply two sides of the same, rather grubby, coin, although they were often closely associated. Waste can pollute, but the word itself carries a complex set of ethical meanings about the proper use of resources. As a category of thought 'waste' is a concept within which ideas about the right relationship between society and nature have been contested.

In contemporary discussion waste is commonly associated with environmental crisis, often expressed as a fear that affluent societies are running out of space

* In addition to the 'Uses of Environmental History' conference in Cambridge 2006, versions of this chapter have been presented to conferences at St Andrews University, and the ESEH conference in 2007 at the Vrije Universiteit, Amsterdam. I would like to thank the participants at those meetings, whose comments have helped to improve the chapter considerably. Special thanks are also due those who have read and commented on the chapter in detail, Dr John Clark, Tineke D'Haeseleer, and, of course, the editors of the present volume. Part of the research for this chapter was completed at the AHRC Centre for Environmental History at the University of St Andrews between 2004 and 2005. Information about the 'Waste' project can now be found online at http://www.st-andrews.ac.uk/envhist/ahrc.html
[1] Much of the environmental history literature is American in origin, reflecting its relative importance in the United States. For the most comprehensive treatments of pollution, waste and the technology of disposal, see Tarr (1997), Melosi (1981), (2000). Waste has also been widely treated in the context of public health and hygiene, for instance Hamlin (1985), (1998), Wohl (1983).
[2] See, for instance, Moore (2007), Mosley (1996), 41, (2001), Thorsheim (2006).

to dump their wastes.[3] The medical connotations associated with the use of 'crisis' in this context are not wholly coincidental. Waste is viewed as matter that is poisoning the environment and threatening society, as something in need of 'treatment' or 'cleansing'. Some sceptics have criticised the simplifying implications of these approaches to waste. Martin Melosi, for instance, has argued that the solid waste problem 'is too complex to regard... as a crisis', and that 'implied in this usage of "crisis" is the assumption that society has reached a point beyond which we can expect nothing less than a dangerous outcome'.[4] Similarly, Stephen Horton has claimed that focussing on the physical problems of waste disposal deflects attention from the political and ideological aspects of the problem.[5] These are important insights and it is undoubtedly true that the problem of waste *disposal* has been over-emphasised in environmental discourse at the expense of asking who actually benefits from waste creation in the first place. However, in pointing out these facts we must take care not to obscure the obvious historical significance of waste in the emergence of modern environmental politics. The fact that waste came to be seen as part of a systemic crisis and eventually took a central role in the development of environmentalism is itself in need of explanation.

In an important essay on the philosophy and sociology of garbage, John Scanlan has argued that waste was at the heart of modernity.[6] In post-Enlightenment Europe 'improvement' was conceived of as the detaching, or displacing, of the valueless from the valued; waste therefore became an inescapable by-product of progress. By seeking to unshackle the rational, scientific energies of industrial and agricultural progress, the enlightened propagandists of 'improvement' believed that they were opening up the prospect of the progressive and universal elimination of waste. That this utopian vision failed was largely due to the fact that the emergent capitalist order itself exhibited an unparalleled capacity for waste.[7] Waste consequently acquired an unexpected potential as a source of social and environmental critique.

---

[3] The press consistently employs this idea of a 'waste crisis'. See 'Dumped Mobiles Cause Waste Crisis', *The Observer*, Sunday March 26, 2006; 'The Alternative to the Waste Crisis', *Socialist Voice*, June 2003. There is a considerable market in popular books on waste and the waste crisis. See Royte (1999), Tammemagi (1999), Girling (2005).

[4] Melosi (2001), 90.

[5] Horton (1995).

[6] Scanlan (2005), 56–88. Scanlan demonstrates how the anti-waste rationale of modernity also affected the temporal and moral approach to poverty in the Victorian abjuration of idleness. Scanlan (2007).

[7] One useful means of comprehending how a social system based on market rationality and scientific progress could become so successful at generating waste is through Schumpeter's widely applied concept of 'creative-destruction'. See Schumpeter (1942). According to this perspective, economic growth is dependent upon innovation and obsolescence,

The wastes of capitalism were treacherous, exposing the entropic, degenerative reality underlying the capitalist conception of progress.

Scanlan demonstrates the necessity of understanding the history of waste within the story of the emergence of modernity. As well as suggesting utility of this perspective to environmental historians, this chapter attempts to open up lines for future research by sketching an outline of the evolution of waste as a critical element of environmental thought in modern Britain. The chapter is based on highly speculative reflections stemming from an ongoing project on the history of waste and recycling, and attempts to synthesise, as far as is possible, a broad and somewhat disparate secondary literature. It loosely addresses a number of phases in the making of the modern 'waste crisis' between roughly 1600 and the present. The first of these was rooted in the agricultural revolution of the seventeenth and eighteenth centuries and saw the meaning of 'waste' contested between advocates of agricultural improvement through enclosure and their opponents. A second phase opened in the early nineteenth century, when waste was established as an urgent challenge to the sustainability of urban-industrial society. The third phase saw apparent solutions to the problem of waste disposal found in what has been termed the 'refuse revolution', before a final, fourth phase began with the emergence of environmentalism and the re-articulation of waste as symptomatic of a systemic problem. This schema is a convenient means of addressing a complex subject, and should not necessarily be viewed as a 'hard-and-fast' narrative, but it reflects something of the trajectory of development.

It is hardly necessary to explain that one of the original meanings of the word 'waste' described the marginal lands in the possession of a mediaeval manor, and was sometimes also applied to uncultivated lands and to forests and wooded areas.[8] However, it is worth beginning with an outline of this conception of waste because it played an important, if latent, role as the starting point for later developments. The manorial waste was far from useless; it formed an important part of the fabric of the late-mediaeval village economy and usually composed scrub, woodland or roadside strips that were either not usually given over to cultivation or that had been deliberately taken out of use. There were various uses to which the waste could be put, including the grazing of cattle, the cutting of peat, wood or furze for heating and cooking, the

---

and economic progress on maintaining a constant rate of destruction of existing ideas and processes. Aghion and Howitt (1992), Horton (1997), Harvey (2003).

[8] Of course, there were other important meanings of waste as well, which I do not propose to analyse in detail. The *OED* gives the origin of the word in the Latin *vastum*, coming into modern English usage through the Old French *wast*. In both cases the meaning was that of a desolate, uncultivated wilderness. Another common usage was that relating to the devastation meted out in war.

collection of seasonal crops and the hunting of animals.[9] In suitable areas sand and gravel might also be dug, and reeds and grasses provided roofing materials. The ecology of the waste fulfilled a wide variety of important human subsistence needs, and although the Statute of Merton (1235) made it subject to possible enclosure, in many areas the waste remained a significant part of the subsistence ecology of agricultural communities until the early nineteenth century.

While formally the property of the Lord of the Manor, the waste was nonetheless recognised as subject to common rights of usage by tenants, rights enforced through the manorial courts.[10] Consequently, waste and common lands were tightly woven into the moral economy of rural areas, and were subject to a complex set of customary practices designed to police access to common rights of grazing and avoid over-exploitation.[11] On the other hand the subversive potential of waste-land had long been apparent in the use made of moor, forest and fen as refuges for outlaws, gypsies, beggars, vagabonds and other such 'masterless men'. The communal principles underpinning both these variants of the economy of waste were inimical to nascent capitalist agriculture. In order to establish the conditions necessary for a commercially organised and scientifically advanced agriculture it was necessary to radically transform human relationships with the land.[12] The commons and wastes were among the first victims of this transformation as landowners sought the consolidation and rationalisation of estates around commercial tenant-run farms. The question of how the relationship between society and the land should be reformed was the main issue at stake during the long agricultural revolution of the sixteenth to nineteenth centuries. During this period, and particularly in the seventeenth and eighteenth centuries, the meaning of 'waste' came to be contested between two increasingly incompatible ecological visions.

This contestation of waste centred on whether it should be developed on the basis of communal principles or private property. Significant popular ideological coherence was given to this contest by the Digger project for a commonwealth of waste-lands. The appropriation and exploitation of the wastes formed the basis of what James Holstun has called the Digger's 'green millennialism'.[13] '[L]et the common people have their commons and waste lands set free to them from all Norman enslaving lords of manors', argued Gerrard Winstanley in 1649, 'those we call poor should dig and freely plant the waste and common land for a livelihood (seeing there is land enough, and more

[9] Neeson (1993), Neilson (1942).
[10] Hammond and Hammond (1978),1–13.
[11] Thompson (1991).
[12] Merchant (1980), 42–68.
[13] Holstun (2000), 367–433.

by half than is made use of), and not suffered to perish for want'.[14] While the Diggers envisaged agricultural improvement and technological advance, 'to lay hold upon, and as we stand in need, to cut and fell and make the best advantage we can of the woods and trees that grow upon the commons', they demanded that this improvement should be controlled on a popular basis, not by the owners of private property or a scientific cadre.[15] In Digger thought the very existence of under-utilised land offered a space within which the construction of an alternative society was possible that might come to triumph over both a defeated absolutism and an emerging capitalism.

With the suppression of this vision of agricultural progress the advocates of agricultural improvement on the basis of private property and enclosure gained the advantage in defining the meaning of waste. The late seventeenth and eighteenth centuries saw the repudiation of both the existing moral economy of waste and an important accompanying shift in the meaning of the word itself. For 'improvers' the extension of the rights of private property and the elimination of commons and wastes aimed to establish a more productive, and more profitable, ecology suited to the mercantilist objectives of national wealth and power. This was, after all, the period in which John Locke famously founded the very right to property on the preparedness to eliminate underdeveloped, waste land. In justification of the encloser, Locke observed that 'he who appropriates land to himself by his labour, does not lessen but increase the common stock of mankind. For the provisions serving to support of human life produced by one acre of enclosed and cultivated land are . . . ten times more than those which are yielded by an acre of land, of an equal richness, lying waste in common'.[16] In identifying right with the entrepreneurial improver, whether he was a commercial farmer or a reigning monarch exploiting the colonies, Locke expressed the growing conviction of his time that land not already under rational cultivation was merely awaiting appropriation and reclamation.

By disposing of waste, and the accompanying vestiges of the system of subsistence production, the agricultural revolution provided the key ecological turning point in modern British history, creating what John Bellamy Foster has termed a 'metabolic rift' in the relationship between man and the natural world that reinforced the subordination and dependence of both labour and nature.[17] The elimination of the wastes, after all, not only disposed of unproductive land, but also of idle people. One seventeenth-century English clergyman revealed these uses of enclosure when he observed that it 'will give the poor an interest

---

[14] *Winstanley*, 'An Appeal to the House of Commons', in Hill (1983), 115, 165.

[15] *Winstanley*, 'A Declaration from the Poor Oppressed People of England', in Hill (1983), 103.

[16] Locke (1993), 279.

[17] Foster (1999), 366–405, (2000).

in toiling, whom terror never yet could enure to travail'.[18] Busy hands were also a weapon against political radicalism. In 1794, another writer encouraged enclosure because it would ensure 'that subordination of the lower ranks which in the present times is so much wanted'.[19] However, as suggested above, the subordination was not just social but ecological. Improvers propounded a utopian ecology that contrasted wild, unimproved nature with the results of rational agriculture, reclaiming nature from a post-inundation ruin.[20] Newly enclosed waste was subject to drainage, deforestation, soil 'improvement', the planting of new crops and the replacement of an existing, unproductive flora and fauna with something more profitable. According to the account of one farmer in 1813, the wastes of Oxfordshire were 'overrun with ant-hills and coarse herbage', and subject to sheep-rot or 'moor evil'; therefore, like any worthy improving agriculturist he 'extirpated the aboriginal plants, carried on some new earth, and varied fruitful crops have succeeded'.[21] By the end of the early modern period, that distinctively modern meaning of waste, implying the unproductive or useless was becoming established as a justification of the necessity of progress. Gradually, the word 'waste' became capable of suggesting a reflection upon the moral character and progress of society as a whole.[22] At the beginning of the nineteenth century the common lands in the county of Sussex were described by Arthur Young as 'mere wastes' and the failure to improve them an 'unaccountable negligence'.[23] This moralising tendency also contributed to furthering the subordination of the cottager and the agricultural labourer, who were represented as incapable of using the waste either productively or sustainably.[24]

The ideas underpinning the agricultural revolution resulted in the remaking of waste as a threat to productivity and progress. However, the word had not yet acquired its association with physical excrescence, the excess of production or consumption. Nor had the moralising usage of waste, increasingly common in agricultural discourse, yet moved into discussions of the urban environment. Indeed, until well into the nineteenth century many waste products were

---

[18] Hill (1969), 151.

[19] Quoted in Hammond and Hammond (1978), 9.

[20] Womack (1989), 61–86.

[21] Farmer Davis quoted in Young (1969), 230.

[22] The evolution of waste may have paralleled that of the closely related idea of poverty. As Gertrude Himmelfarb has shown, the seventeenth and eighteenth centuries saw the making of poverty as a social and moral problem as opposed to an immutable state of nature. See Himmelfarb (1984).

[23] Young (1970), 187–188.

[24] One improving farmer observed that on Ottmoor Common, 'The abuses here…are very great, there being no regular stint, but each neighbouring householder turns out upon the moor what number he pleases'. Farmer Davis quoted in Young (1969), 229–230.

highly valued, and recycling was the unnoticed reality of everyday life. Agricultural improvement itself retained a dependence upon large amounts of organic waste, dung or rags for example, as fertiliser, especially in the era before foreign guano imports became widely available.[25] In the more modest domestic economy of the cottager household much was reused simply as a consequence of endemic poverty, 'conservation and recycling' as Jane Humphries has noted, 'made the cottager economy pay'.[26] It has even been argued that a good deal of general domestic consumption in early modern England was the result of some kind of recycling practice.[27] There was certainly little that could not be profitably employed; anything of value when either out-grown or worn-out could be sold to the itinerant rag merchant, or repaired by the tinker.

Naturally none of this should be taken as indicating the absence of pollution before the industrial era. Filth, dung and refuse defiled the pre-industrial landscape, and historians have demonstrated both the omnipresence of dirt and detritus, as well as widespread popular concern over its effects on health and well-being.[28] Yet, while there were plenty of reasons to be dismayed with urban *pollution* before the industrial era, contemporary language did not express the same ethical concerns with *waste* that would eventually be encapsulated in references to urban waste during the industrial and post-industrial periods. There was talk of dirt, filth, rubbish, muck, corruption and nuisance, but not of wasted natural resource.[29] Often the source of a nuisance was itself an attempt to utilise waste products.[30] The refuse, dust and rubbish of the pre- and early industrial city were in fact vital pre-requisites to its existence in an era when construction methods relied heavily upon the availability of dust and cinders from house-fires for the making of bricks.[31] Despite the fact that the meaning of waste was changing, down to the nineteenth century the conception of waste as a useless remainder of processes of production or consumption remained in a very real sense unthinkable.

Martin Melosi has observed that it was 'the saturation of cities and suburbs with air, water, refuse, and noise pollution [that] finally produced an

---

[25] W. Blith, 'Lime and Other Manures', in Thirsk and Cooper (1972), 126–130.

[26] Humphries (1990), 25.

[27] There is a relatively good historiography of pre-industrial recycling practices, reflecting its economic importance: See Ginsburg (1980), Lambert (2004), Lemire (1988), Woodward (1985).

[28] Mark Jenner has demonstrated the importance of discourses on cleanliness within wider political culture. See Jenner (1995), (2000), (2005). The pioneering work by E.L. Sabine is also still interesting. See Sabine (1933), (1934). Nor were mediaeval sanitary conditions wholly unpleasant. See Thorndyke (1928).

[29] Cockayne (2007), 181–205.

[30] King (1992).

[31] Clarke (1992), 94–105.

environmental consciousness among the complacent citizenry' of industrial cities.[32] This is a widely help assumption, and yet, given the observations made above, it is not entirely correct. It is, perhaps, more accurate to say that the saturation of cities and suburbs with pollutants had a new significance in the context of the industrial city. Waste was increasingly seen as a symptom of an unprecedented kind of economic and social failure, the inability to recycle waste. It was the declining capacity to utilise waste products that introduced a new element into the discussion of urban pollution, and in turn invented the modern conception of waste. It would be mistaken to exaggerate the speed of the transition. Even while traditional recycling practices were being overwhelmed by the sheer productive capabilities of the new economic order, the old ways of dealing with refuse remained surprisingly durable. For a time it was felt sufficient to leave urban waste disposal to the vagaries of the market and individual initiative. In its early phases industrial capitalism appeared capable of incorporating systems of widespread reuse and recycling. The scavenger, dustman and night-soil collector, who had served British cities for centuries, remained part of the social and economic landscape of 'shock cities' like London and Manchester until well into the nineteenth century.[33] The dust-contractor even became a symbol of the possibilities of private entrepreneurship, built on a myth of the great wealth contained within the dust heaps of London.[34]

Nonetheless, the idea of waste as a systemic problem became increasingly apparent in the nineteenth-century Britain. As cities grew and industry expanded, urban ecologies became increasingly unbalanced and the recycling of waste increasingly difficult. While human excrement, the ashes of thousands of fireplaces, animal and industrial waste remained subject to reuse, the profitability of recycling activity was undermined by the inexorably mounting volume of refuse.[35] In fact, one of the most enduring problems of creating an effective system of recycling was established in this period: while the market system dealt relatively well with industrial waste products of high value, concentration and relative scarcity (metals are a good example), it completely failed to cope with the glut of low-value, highly dispersed urban domestic and street waste. Mayhew's account of the changing situation of dust contractors in the mid-century London illustrates exactly this difficulty:

---

[32] Melosi (2001), 39.

[33] The dustman held an important position in the representational culture of the nineteenth century. See Maidment (2002).

[34] An obvious case is Dicken's *Our Mutual Friend*. See Sucksmith (1973), Metz (1979).

[35] For a detailed investigation of these developments in Victorian London, see Turner (2006).

Of late years, however, the demand [for dust] has fallen off greatly, while the supply has been progressively increasing, owing to the extension of the metropolis, so that the Contractors have not only declined paying anything for the liberty to collect it, but now stipulate to receive a certain sum for the removal of it. It need hardly be stated that the parishes always employ the man who requires the least money for the performance of what has now become a matter of duty rather than an object of desire.[36]

Mayhew was observing a critical juncture in the environmental history of the city. Until the middle of the nineteenth century cities had to some extent consume their own wastes by employing dust or manure to house or feed their growing populations. A tipping-point was reached sometime in the mid-nineteenth century when too much domestic waste began to chase too small a demand and recycling became uneconomic in many fields. This happened well in advance of the municipalisation of domestic waste disposal and the consequences for the urban environment were profound.

As the problem of urban detritus became increasingly apparent it fed into a wider sense of systemic crisis, and the ethical meanings of waste, which had been so carefully constructed in the first place, were increasingly employed to challenge capitalist claims of progress. A number of historians have recently suggested that the nineteenth-century's utilitarian obsession with metropolitan cleansing emerged from a political effort to remake the labouring poor in the liberal bourgeois image of the good citizen, clean and decent. Public cleansing and waste disposal were thus techniques of ideological hegemony as well as means of improving objective environmental problems.[37] No doubt there is some truth in this interpretation; however, it ignores the extent to which waste, rather than mere dirt, presented a parallel ideological problem for the bourgeoisie. While the 'liberal bourgeois' could accuse the poor of being filthy, it was possible to mount a counter-accusation that filth was the inescapable consequence, and therefore also a condemnation, of the capitalist city. Critics of capitalism often relied on the language of waste in descriptions of the detrimental environmental effects of capitalist society. In *The Condition of the Working Class in England*, Engels, a man who always had an acute eye for polemical advantage and a command of potent imagery, widely employed waste in a visual and olfactory critique of the conditions of urban life. In London, he noted, 'the streets are generally unpaved, rough, dirty, filled with vegetable and animal refuse, without sewers or gutters, but supplied with foul, stagnant pools instead'; similarly, in Bradford's 'lanes, alleys, and courts lie filth and debris

---

[36] Mayhew (1968), 167.
[37] Otter (2004), (2002).

in heaps'.[38] Across urban Britain it was possible to construct a litany of waste, nuisance and potential disease: 'filth and disgusting grime'; 'stagnant urine and excrement'; 'the stench of animal putrefaction'; foul streams in which 'slime and refuse accumulate and rot in thick masses'; 'Everywhere heaps of debris, refuse and offal'.[39]

The emergence of waste as an environmental threat was consequently related to the existence of a critique of the conditions endured by the working class under capitalism. James Winter has demonstrated how coping with the pollution caused by waste became a central plank to Charles Cochrane's idiosyncratic brand of metropolitan radicalism in the 1840s and 1850s.[40] Waste also played a significant role in the thought of romantic socialists. William Morris brought environmental and social critique together through waste to argue for a radically different future. 'I feel sure', Morris argued in *How We Live and How We Might Live* (1885), 'that the time will come when people will find it difficult to believe that a rich community such as ours, having such command over external Nature, could have submitted to live such a mean, shabby, dirty life as we do'. For Morris, profit was the cause of pollution:

> It is profit which draws men into enormous unmanageable aggregations called towns, for instance; profit which crowds them up when they are there into quarters without gardens or open spaces; profit which won't take the most ordinary precautions against wrapping a whole district in a cloud of sulphurous smoke; which turns beautiful rivers into filthy sewers.[41]

The solution to the polluted city was its dissolution into 'a garden, where nothing is wasted and nothing is spoilt'.[42] The experience of un-reclaimed waste in modern industrial society thus contributed to inspiring a new 'green millennialism'.[43]

By the mid-nineteenth century the failure to eliminate waste presented a serious environmental challenge to the proponents of *laissez faire* capitalism: one that demanded a response. The strategies employed to defend capitalism from criticism of its wasteful and polluting tendencies were diverse. The simplest was to present waste as a sign of progress and prosperity. Adam Rome has demonstrated that in early industrial cities the sight of smoke pouring from stacks was often represented as a sign of both productivity and economic success.[44] For as

---

[38] Engels (1999), 39, 53.
[39] *Ibid*, 61–63.
[40] Winter (1993), 118–229.
[41] Morris, 'How we Live and how we might Live' in Briggs (1980), 175.
[42] Morris quoted in De Gues (1999), 114.
[43] Gould (1988), Morris (1982).
[44] Rome (1996), Gugliotta (2000).

long as smoke was not equated with risk to health, this was a viable strategy, although one that was increasingly challenged by middle-class urban reformers. However, in the case of other forms of waste it was less useful, not least because miasmatic epidemiology suggested that organic waste was a severe health hazard.[45] Even apparently inorganic matter could be a source of suspicion; in 1878, a report for the Local Government Board warned luridly that, 'dust and dirt swept from houses and thrown in with other refuse may contain specifically infective matters, such as the cutaneous scales from a scarlatinal patient'.[46]

Another alternative was suburbanisation. Professor Rodger has argued that urban wastes and pollution were a factor in the making of the bourgeois suburb, as the British middle class sought to escape from 'the contagion, both moral and physical, of the degenerate city'.[47] In time the suburban trend also spread down to the working class as a result of a public policy deliberately aimed at improving the health and environmental conditions of the labouring population. It may not have been possible to build livable cities, but it was certainly possible to shift the working class out of a pernicious urban environment. One builder of working-class housing in the suburbs congratulated himself that he had 'brought many persons with their wives and families from the fog and smoke of London into a cleaner atmosphere'.[48] This solution had the additional advantage of being profit led, evidence that capitalist solutions to social and environmental problems could work. Charles Booth was a particularly vocal advocate of the suburb as a private enterprise solution to the ills of the city.[49]

Arguably, however, the public health movement was the most significant response to the challenge of waste. Built on miasmatic epidemiological theory, the sanitary movement reflected the fear that exposure to putrefying waste was a threat to the health of all, as well as the political reliability of the working class. Christopher Hamlin has demonstrated that the publication of Edwin Chadwick's *Sanitary Condition of the Labouring Population of Great Britain* (1842) and the subsequent campaign for sanitary reform identified waste as one of the main causes of disease and distress among the poor, as well a cause of such political infections as Chartism.[50] Chadwick's advocacy of sewer technology, and its eventual widespread adoption across Britain, sought to establish an urban circulatory system that would both cleanse the city and be proof

---

[45] Hamlin (1985), 382–383.
[46] See the *Eighth Annual Report of the Local Government Board, 1878–79*, Parl. Papers, 1878–1879, XXIX, 306.
[47] Rodger (2000), 244.
[48] Clarke (1861), vii.
[49] Cooper (2005).
[50] Hamlin (1998).

against political discontent.[51] However, an unforeseen consequence of the sanitary movement was that it further marginalised recycling: if proximity to waste always caused disease and political upheaval, then removal would always be preferable to utilisation. Bourgeois comment was revolted by the proximity in which the working class lived to waste, as well as by the uses to which it was put. As R. and W. Chambers argued, 'the existence of great masses of offensive matter creates a set of beings who pursue disgusting trades, and are as mentally impure as the garbage they handle'.[52] The exaggerated fears associated with the urban pig represent the way in which the sanitary idea prioritised the fear of filth over the rational use of waste.

> The houses of the poor sometimes surround a common area, into which the doors and windows open at the back of the dwelling. Porkers, who feed pigs in the town, often contract with the inhabitants to pay one small sum for the rent of their area, which is immediately covered with pigstys, and converted into a dung-heap and receptacle of the putrescent garbage, upon which the animals are fed, as also of the refuse which is now heedlessly flung into it from all the surrounding dwellings.[53]

Yet, even as towns progressively forbade 'unhygienic' practices such as pig-keeping, and the Metropolitan Sewers Commission began cleansing human filth from London's overflowing cesspits, the ideological contradictions provided by increasing reliance on disposal over re-use were apparent. As recycling declined so concerns arose over the ecological basis of capitalist society and the basic structure on which it was dependent: the town–country divide. Agriculturalists like J.J. Mechi doubted the wisdom of discarding previously valued urban wastes into rivers, especially when Malthusian population theory already emphasised the issues of scarcity and limits to growth.[54] Von Liebig's demonstration during the 1840s of the role played by nitrogen in plant growth raised fears of a shortage of organic fertiliser, as well as contributing to the development of a 'chemico-theological' view of the natural world as one vast recycling mechanism that man upset only at his own risk.[55] The challenge of these ideas at a time when waste was apparently already threatening the social and political viability of the city was profound. As the Health of Towns Commission observed in 1847, the whole idea of sewage *disposal* was wasteful:

---

[51] Adoption of new sanitary technologies was complex and owed much to local conditions. See Goddard and Sheail (2004), 93–95.

[52] Chambers and Chambers (1850), 290.

[53] Shuttleworth (1862), 21–22.

[54] Worster (1977).

[55] Mårald (2002), Foster (1998). See also Goddard (1996), Sheail (1996).

It is a striking example, on the grandest scale, of inefficiency and extravagance. We shall search in vain for a similar instance of the wholesale waste of resources. Not content with polluting our rivers, and heaping up upon their banks masses of offensive matter most injurious to health, we are constantly pouring into them treasures which cannot be estimated at a less sum than several millions a year.[56]

The concerns with agricultural sustainability reached a peak in the 1860s with the completion of London's system of intercepting sewers. In 1864 the *Select Committee on Metropolitan and Town Sewers* stated that 'the amount of artificial manures is even at present insufficient, and the sources whence some of the most important are obtained will, in a few years be exhausted. Other means of fertilizing land must therefore be resorted to'.[57] The essential problem was whether the need to make the urban environment sustainable could be reconciled with sustaining rural agriculture.

This was the context in which the 'sewage farm' and sewage irrigation emerged as a means of reunifying the divided ecologies of town and country. The concept of the sewage farm offered the prospect of the perfect reclamation of waste and improved agricultural fertility. Urban sewage would be fed out to the city's fringe where it would be applied to land to grow crops to maintain the city's populace. The practice proved much more difficult to apply than the principle, but for enthusiasts that was often less important than the excitement of apparently discovering a means of sustaining vast new urban agglomerations into the indefinite future. In a rhetoric that was later to be recycled in the twentieth century, the very idea of waste was challenged. Waste was no longer to be thought of as waste, it was resource. As Frederik Krepp explained, echoing the insights of 'chemico-theology', those who sought to dispose of London's wastes:

should know better than pronounce any matter, organic or inorganic, decomposing and offensive to the senses or not, a nuisance to be got rid of…In the sublime arrangements of Nature, not a single law is made in vain – that not a single atom is superfluous, but has its properly assigned function and place, for some good and wise purpose; that in fact no matter can be a 'nuisance' unless the law of Nature is violated, or its intention frustrated.[58]

---

[56] Health of Towns Commission (1847), 1.
[57] See the *Report of the Select Committee on Metropolitan and Town Sewers*, Parl. Papers, 1864, XIV, v.
[58] Krepp (1867), 42–43.

No nation could long endure that ignored the agricultural utility of its wastes.[59] Krepp admired the ancient empires of China and Japan, whose commitment to the recycling of waste provided not only agricultural success, but also political and social stability.[60]

Although the mid-Victorian obsession with sewage recycling waned during the 1870s, its popularity damaged by the fact that most experiments with it had proved either unworkable or unprofitable, the Sewage Question was nonetheless a turning point in the Victorian perception of waste.[61] The arguments over sewage and soil fertility invented the idea of recycling. This was significant because it came after the emergence of the modern idea of waste. When recycling had been a daily reality it had gone largely unnoticed and un-theorised. When waste finally presented a real problem, recycling emerged as a theoretical solution to polluting wastes that also reconciled capitalist progress with limited natural resources. In later periods recycling would re-emerge whenever waste seemed to challenge capitalism's sustainability, and would be consistently, and inaccurately, represented as an innovative technological solution to waste.

The first 'reinvention' of recycling took place in the last decades of the nineteenth century in the wake of waning enthusiasm with sewage recycling. Interest in industrial recycling was influenced by a growing awareness of the relationship between work and waste, which highlighted the entropic nature of all industrial activity.[62] In 1879, the *Scotsman* ran an article on 'The Utilisation of Waste Materials' which enthused over the economic and environmental advantages: 'The industrial progress of recent years has in no direction been more marked than in the utilisation of waste materials. New industries have thus been created, and old ones rendered more profitable, while grave public nuisances have in many cases been removed or abated.'[63] A number of authors took up the banner of industrial recycling as a means of reconciling industrial production with the cycle of nature. Peter Lund Simmonds, the author of *Waste Products and Undeveloped Substances* (1862), provides a good example.[64] Simmonds's work was strongly influenced by the 'chemico-theological' conception of the lessons of the natural economy for business. 'When we perceive in nature how nothing is wasted, but that every substance is reconverted, and again made to do duty in a changed and beautiful form, we have at least an example to stimulate us in economically applying the waste materials we

---

[59] Ibid, 9.

[60] On recycling in Japan, see Hanley (1983), (1997).

[61] Goddard (2005), 145.

[62] Norton Wise (1990), 94–96.

[63] *The Scotsman*, 17 March 1879, 2.

[64] Despite a tendency to overstate its achievements, Desrochers (2007) provides a nonetheless useful study both of Victorian industrial recycling and of Lund Simmonds.

make, or that lie around us in abundance, ready to be utilized.'[65] The same arguments may be discerned in the panegyrics of specific recycling industries. In his *History of the Shoddy-trade* (1860), Samuel Jubb noted that shoddy manufacture – the shredding and reweaving of old cotton and woolen cloth – was 'a source of national wealth, by utilizing materials of value, which were previously thrown away'.[66] It had obvious environmental benefits too, as, 'not a single thing belonging to the rag and shoddy system is valueless, or useless; there are no accumulations of mountains of debris to take up room, or disfigure the landscape; all – good, bad and indifferent – pass on and are beneficially appropriated'.[67]

While these texts reveal the strong intellectual interest of Victorian industrial and technological writers in recycling, it is important not to extrapolate from these any significant achievements. The discovery of recycling was, after all, in part a function of the increasing difficulty of achieving the perfect circulation and re-absorption of waste in industrial Britain. For all their optimism, these texts reveal the contradictions inherent in the promise to eliminate waste. The subtitle of *Waste Products* was, after all, *Hints for Enterprise in Neglected Fields*, a wording suggestive of the underlying concern that opportunities for profit in industrial recycling were themselves being wasted. While ostensibly writing a paean to *laissez faire* efficiency, Simmonds was nonetheless compelled to acknowledge that it was actually the 'improvements in arts and sciences . . . [and] . . . the greater demands hence made upon manufacturers and the continual waste occurring' that paradoxically created the necessity to invent new products made from waste. Capitalism's 'creative-destruction' ensured that waste could never be wholly prevented.

In addition to the limited achievements of Victorian recycling projects, other contradictions were also becoming apparent. One of these was the strong tendency in late-Victorian municipal administration towards the disposal of waste as the prime aim of public cleansing. From the Public Health Act 1875, local government was engaged in subjecting urban domestic, trade and street waste to increasingly sophisticated legislative and technocratic control, a process to which B. Luckin has aptly given the name the 'refuse revolution'.[68] The commissars of this revolution were the public health professionals whose influence within both local and central government was rapidly growing. From the 1870s medical officers of health, cleansing superintendents and municipal engineers developed a consensus that the best solution for urban waste, refuse and sewage

---

[65] Simmonds (1862), 2.

[66] Jubb (1860), 4.

[67] Ibid.

[68] On the idea of a 'refuse revolution', see Luckin (2000). A similar set of changes were also occurring across the Atlantic at this time. See Melosi (1973), (1980).

was not re-use, but rather rapid removal and, preferably, incineration. The fact that incineration was often termed 'destruction' at this period, despite the well-understood significance of the laws of thermodynamics, reflected the psychological significance of this tendency the consequences of which can be clearly seen in the fate of the Victorian dust-yard. For much of the nineteenth century this rudimentary form of recycling had recovered value from urban waste through the labour of some of the neediest of the urban poor. From the 1880s, however, the dust-yard was increasingly attacked as unhygienic by urban reformers, especially Progressives: the dust-yards, and their infamous sorting women, were displaced by 'clean' and 'efficient' municipal incinerators.[69]

One of the perceived advantages of the incinerator was that it enabled household waste to be dealt with on a regular, predictable basis, ensuring rapid removal of refuse and interrupting the life-cycles of urban pests such as rats and flies. The privileging of public health in decisions on how to treat urban waste was one of the key reasons why disposal triumphed over reuse in municipal waste management. Even well-established recycling industries, such as the 'shoddy' trade, now came under suspicion of permitting their raw materials to 'undergo no adequate process of disinfection in their progress from the gutter to the drawing room'.[70] The 'refuse revolution' and the triumph of disposal represented the logical consequences of the failure to fulfill the promise to absorb waste and make it productive. The 'throwaway society' began not in the post-war era of affluence, but amid the inability of late-Victorian society to sustain the widespread and effective system of waste recycling it had inherited.[71] The 'refuse revolution' also represented a significant change in the nature of environmental consciousness. Historians who have emphasised the role of 'preservationist' ideas in Britain from the 1860s have, perhaps, taken too little notice of waste.[72] The triumph of disposal, and the marginalisation of recycling, accompanied by apparent solutions to the problem of agricultural fertility, quashed debate over the sustainability of capitalist development until the twentieth century. Many of the issues of sustainability that were more or

---

[69] Clark (2007). In mid-Victorian England the hygienic state of the dust-yards was disputed, but nonetheless found defenders among medical experts and social observers, Guy (1848).

[70] See, for example, the pseudonymous pamphlet by Manufacturer, *The Reign of Shoddy* (1911). A similar process of 'marginalisation' of the waste trades was also occurring in the United States at this time see Zimring (2004), (2005).

[71] This is a different chronology from that presented by Susan Strasser in her fascinating history of the emergence of a throw-away culture in the United States, Strasser (1999).

[72] For a summary of the literature on the importance of 'preservationism' in English culture, see Burchardt (2002). The two most significant contributions to the debate are Weiner (1981) and Mandler (1997).

less explicit in nineteenth-century debates on waste were negated by the focus on disposal, public health and urban or rural amenity.[73]

One example of the way in which early twentieth-century environmental debate took an increasingly narrow view of waste was the debate on litter. The first half of the twentieth century saw intense discussion of the impact of litter on the landscape. Litter was, of course, partly a result of emerging mass consumerism, and especially the growth in pre-packaged foods.[74] However, the debate on litter did not lead to reflection on the compatibility of nature preservation with mass affluence or a campaign against packaging. Those societies in which anti-litter activism was most prevalent – such as the Society for Prevention of Disfigurement in Town and Country (SCAPA) or the Commons, Open Spaces and Footpaths Preservation Society – focused on the effects on landscape and amenity.[75] Controlling public behaviour was their prime concern; a SCAPA society pamphlet, *Litter*, ambitiously observed that, 'we have to induce each of forty-seven million odd people to add it to his or her social code that to throw a sticky wrapper on to the pavement or laneside is one of the things that should not be done'.[76] Few things illustrate better how deeply the idea of disposal had penetrated public consciousness, especially among those concerned with 'environmental issues', than the passing of the Litter Act 1958. This aimed to cajole people into throwing their waste into litter bins order to preserve aesthetic beauty. The question of what happened to this waste subsequently was forgotten.

The only significant challenge to the hegemony of disposal between 1900 and 1960 was provided by war.[77] The interruption of supplies of raw materials from Europe and the Empire created a strong, if temporary, imperative for government to increase recycling and also reignited consciousness of the finite quality of natural resources. As the *Scotsman* observed in 1940: 'In war-time it is necessary to exercise the strictest economy in the uses of our resources, and the careful collection of waste materials is a humble, but useful, way of contributing to the national effort.'[78] British wartime governments put pressure on local authorities to recover scrap and waste paper for the war effort, and successful salvage campaigns were run that mobilised national, patriotic sentiment to

---

[73] This was illustrated by the emergence of concerns among countryside preservation groups that rural facilities for waste disposal were insufficiently developed. Aesthetic concerns, not questions about the sustainability of a social and economic system, were what mattered. See Ashford and Baker (1933).

[74] See, for instance, the comments by E.A. Martin in *Journal of the Commons, Open Spaces and Footpaths Preservation Society*, March (1929), 114.

[75] Matless (1998), 67–70.

[76] See the article 'Litter' in *Scapa Society Quarterly Papers*, New Series, II (1930).

[77] Cooper (2007).

[78] *The Scotsman*, 3 February 1940, 8.

encourage participation. Wartime recycling even encouraged a few to question the desirability of disposal in peacetime.[79] H.J. Spooner's monograph *Wealth from Waste* (1918) indicated the impact the First World War had upon attitudes to waste.

> Before the war everything was so different; in the full enjoyment of unparalleled prosperity, extravagance and waste were rampant in the land; there was little or no thought of economy in any form ... The war, with its colossal requirements in men, munitions, material, and money has changed much of that, and the era of retrenchment, frugality, economy and thrift has dawned, whilst the gospel of the prevention of waste is being preached at every turn.[80]

In reality the lessons of war were quickly forgotten in peacetime and recycling never displaced disposal in municipal cleansing arrangements. Post-war affluence ensured that waste, litter, disposal costs and the ever-present difficulty of finding new dumping sites, all remained pressing problems, but at no stage did this provide the basis for a challenge to disposal or affluence. J.C. Wylie accurately perceived the reality of what the 'refuse revolution' had achieved when he stated that: 'The fact is that urban wastes have indeed been swept out of sight for the time being. But this has not been done without leaving nuisances; nor will it ever be done properly until the principle of mere disposal has been made to yield to the infinitely higher idea of use.'[81]

An enduring revival of the 'higher idea of use' eventually occurred in the 1970s under the shadow of the oil crisis. The revival of Malthusianism reproduced doubts over the sustainability of the existing economic and social systems and reworked aspects of Victorian 'limits-to-growth' concerns.[82] What was novel about the 1970s was the filtration of these concerns into an emerging popular environmental politics. As Professor Cotgrove has observed, the environmental movement that emerged from the 1960s was more radical than is often supposed, based on the belief that fundamental social change was necessary in order to avoid ecological catastrophe.[83] The new environmentalism articulated an economics of waste inimical to many capitalist assumptions. Nicholas Georgescu-Roegen, founder of an ecological economics based on the law of entropy, recognised that 'creative-destruction' was essential to the capitalist economy: 'the economic process consists of a continuous transformation of low entropy into high entropy, that is, into irrevocable waste or, with a

---

[79] Dawes (1942), 387.
[80] Spooner (1918), 3.
[81] Wylie (1959), 79.
[82] Schoijet (1999), Linnér (2003).
[83] Cotgrove and Duff (1981), 93.

topical term, into pollution'.[84] The critique of capitalism's profligacy with natural resources represented industrial–consumer society as a dangerous foot on the entropic accelerator pedal leading to disaster.

For the emerging British environmental movement waste provided both a *raison d'etre* and the basis of a radical practice that might construct the foundations of new society. Something of the politics of waste was revealed by a campaign led by the Camden Friends of the Earth in 1974 to establish a local municipal recycling scheme. 'We live in an era of shortages', they argued, 'this may sound more than obvious to those who have queued many hours for a gallon of petrol, or who have donated money after the failure of a harvest in some remote part of the world, or who have seen their newspaper shrink to the size of a brief handout'.[85] The politics of waste moved away from the aesthetic concerns of the 'anti-litter' campaigning of the early twentieth century and towards a concern with sustainability and resources. As an organisation Friends of the Earth (FOE) argued against allowing the litter problem to deflect attention from the true waste-makers, the packaging industry. In its formative years FOE focussed strongly on the issue of packaging, suggesting the potency of waste in highlighting the tension between affluence and finite natural resources.[86] When it launched its British section in the early seventies, FOE chose a publicity stunt that involved piling used drinks bottles outside Schweppes UK headquarters.[87] For radical environmentalists the solution to the problem of waste was for society to reassess its values, 'We must learn to value not merely how much we can move from hand to hand until we drop it, but how much we can manage to hold on to for our mutual long term benefit'.[88] Recycling, would play a political role to bring about this reordering. Camden FOE believed that a recycling scheme would mobilise popular desire for change, as well as challenging both the sense of political impotence and social alienation.[89] Recycling in the hands of radical environmentalists was more than a mere technological fix to the problem of finite resources; it also provided the basis for a political reformation that emphasised democratic, community-based responses to environmental problems.

However, despite the hopes, recycling retained the capacity to neutralise criticism; the idea was rapidly co-opted into technocratic responses to environmental critiques that were elitist and bureaucratic, and which underpinned the pursuit of economic growth. From a technocratic perspective the problem

---

[84] Georgescu-Roegen (1971), 281.
[85] Friends of the Earth (1974), 2.
[86] Lamb (1996), 54–56.
[87] Bate (1976), iv.
[88] Friends of the Earth (1973), 4.
[89] Friends of the Earth (1974), 8.

narrowed to one of waste as a 'source of raw materials', and presented waste as an opportunity rather than an environmental problem, reiterating tropes first developed by the Victorian recycling enthusiasts that there was no such thing as waste only raw material. As Andrew Porteous argued, 'It is important to realise that when refuse is tipped or incinerated a source of valuable resources is lost to the community and the national economy. Refuse is no longer a material to be disposed of. It is a mine of valuable resources from which raw materials may be salvaged.'[90]

Government action reinforced these tendencies. During the early 1970s the Labour Party faced significant pressure from a small but vocal left-wing minority within its own ranks for action on environmental issues. This clashed with the revisionist commitment among the party's leadership to the pursuit of affluence and economic growth, and demanded a response capable of deflecting the argument that compromise with consumerism was neither socialist nor sustainable. Pressure also came from without. A polemical attack by *The Ecologist* on Antony Crosland, who had placed the need for economic growth at the heart of revisionism in *A Social Democratic Britain*, was prefaced by a cartoon illustrating Crosland amid piles of industrial and consumer waste.[91] These political pressures contributed to the publication, in 1974, of the recently elected Labour government's green paper, *War on Waste*, which apparently promised dramatic action in promoting recycling and waste reduction.

> We all instinctively feel that there is something wrong in a society which wastes and discards resources on the scale which we do today...This squandering of resources will become more and more serious for us as consumption rises and with increasing uncertainty about world raw materials supplies.[92]

But despite admitting the popular sense of concern, and promising more direct government intervention in the secondary materials market, *War on Waste* repudiated the environmentalists' calls for radical change. There would be no significant action to compel reduction of packaging or increased recycling of industrial waste. Solutions to the problems of finite resources would be based on research, and the advice of a panel of technical experts (The Waste Management Advisory Council). Popular participation would be restricted to charitable endeavour and 'education'. The *Times* observed that 'industry fares quite well in the Government's new policy on waste reclamation and will be gratified it does

---

[90] See the article by A. Porteous, 'New Uses for Old Rubbish', *The Illustrated London News*, July, 1974, 51.
[91] *The Ecologist*, 1, 9 (1971).
[92] Department of the Environment (1974).

not emerge, as some would have hoped, as the ogre responsible for Britain's heading towards the "Throwaway Society"'.[93]

After the fanfare accompanying the launch of *War on Waste* there was relatively little administrative or legislative action. The new Waste Management Advisory Council, charged with advising the government on improving recycling performance, was dominated by manufacturing interests and headed by an industrialist, Robert Berry, which ensured that the interests of business came before those of environmental protection. After two years of relative inaction, Berry expressed the reality with blunt honesty: 'It is fine for pressure groups to worry about the earth's resources, but we have to look at economics.'[94] The failure of *War on Waste* to deliver any meaningful change meant that, during the 1970s and 1980s, British governments continued to rely primarily on voluntary and local government recycling initiatives, such as Oxfam's 'wastesaver' scheme, and there were no significant changes in either the trajectory of waste generation or levels of recycling. By the early 1990s many environmentalists had concluded that the idea of recycling, rather than offering a means of creating a sustainable economy, actually reinforced the idea that it was morally unproblematic to continue with voracious consumption.[95]

The aim of this chapter has been to set the British waste crisis into historical context. Despite recent denials, there has been a waste crisis in modern British history, which has played a crucial role in the emergence of modern environmental concerns. However, there was much more to waste than just the physical problem of its disposal or its tendency to despoil idealised landscapes. Waste presented a crisis character long before the advent of the throwaway society. Indeed, throughout the last four centuries waste was a central and contested issue at the heart of the capitalist vision of the road to progress and improvement. The elaboration of a new meaning of waste which emphasised capitalism's ability to achieve vastly improved, even the most perfect possible, efficiency was accompanied by the need to demonstrate that this vision could be sustained into an indefinite future. However, the omnipresence of waste, as refuse, or a host of other disorders such as disease, decay, unemployment or dereliction, always remained an implication of failure. It brought forth critiques of capitalism based on both the environmental and economic experience of waste. This use of waste to express concern over the sustainability of urban, industrial capitalism in the nineteenth century shared similarities with its later exploitation by twentieth-century environmentalism, a connection made possible by the shared Malthusianism of these movements. This suggests that, contrary to the arguments of those who represent environmentalism

---

[93] *The Times*, 12 September 1974, 25.
[94] *The Observer*, 12 December 1976, 13.
[95] *The Ecologist*, 22, 6 (1992). See also Luke (1993).

as a post-modern phenomenon, twentieth-century environmentalism actually derived from classically modern concerns with the relationship between universal improvement and its contradictory tendency to environmental degradation.[96] Recycling also developed in the nineteenth century as a response to fears that urban-industrial progress was destroying the natural basis of capitalism's success; in much the same way as conservation emerged as a response to the unsustainable exploitation of colonial resources.[97] Indeed, if one is searching for a way in which imperial interests in conservation may have fed back into the development of European environmentalism then the challenges, ecological and ideological, presented to modern capitalist societies by their own wastes are a possible place to begin.

References

Aghion, P. and Howitt, P. (1992), 'A Model of Growth through Creative Destruction', *Econometrica*, 60, 323–351.

Ashford, E.B. and Baker, H. (1933), *Rural Refuse and its Disposal*, London.

Bate, R.R. (1976), *Many Happy Returns: Glass Containers and the Environment*, London.

Briggs, A. (1980), *William Morris: News from Nowhere and other Writings*, London.

Burchardt, J. (2002), *Paradise Lost: Rural Idyll and Social Change since 1800*, London.

Chambers, W. and Chambers, R. (1850), *Sanitary Economy: Its Principles and Practice, and Its Moral Influence on the Progress of Civilisation*, Edinburgh.

Clark, J.F.M. (2007), '"The Incineration of Refuse is Beautiful": Torquay and the Introduction of Municipal Refuse Destructors', *Urban History*, 34, 255–277.

Clarke, E. (1861), *History of Walthamstow*, London.

Clarke, L. (1992), *Building Capitalism: Historical Change and the Labour Process in the Production of the Built Environment*, London.

Cockayne, E. (2007), *Hubbub: Filth, Noise and Stench in England, 1600–1770*, London.

Cooper, T. (2005), 'The Politics of Radicalism in Suburban Walthamstow, 1870–1914', unpublished Ph.D. thesis.

Cooper, T. (2007), 'Challenging the Refuse Revolution: War, Waste and Recycling, 1900–1945', *Historical Research* (Online Early Articles) doi:10.1111/j.1468-2281. 2007.00420.x

Cosgrove, D. (1990), 'Environmental Thought and Action: Pre–Modern and Post-Modern', *Transactions of the Institute of British Geographers*, New Series, 15, 344–358.

Cotgrove, S. and Duff, A. (1981), 'Environmentalism, Values and Social Change', *British Journal of Sociology*, 32, 93–111.

Dawes, J.C. (1942), 'Making Use of Waste Products', *Royal Society of Arts Journal*, 90, 387–408.

De Gues, M. (1999), *Ecological Utopias: Envisioning the Sustainable Society*, Utrecht.

Department of the Environment. (1974), *War on Waste: A Policy for Reclamation*, London.

Desrochers, P. (2007), 'How Did the Invisible Hand Handle Industrial Waste?: By-product Development before the Modern Environmental Era', *Enterprise and Society*, 8, 348–374.

---

[96] On environmentalism's post-modern qualities, see Cosgrove (1990).

[97] Drayton (2000), R. Grove (1990), (1995).

Drayton, R. (2000), *Nature's Government: Science, Imperial Britain and the Improvement of the World*, London.

Engels, F. (1999), *The Condition of the Working Class in England*, Oxford.

Friends of the Earth. (1973), *Packaging in Britain: A Policy for Containment*, London.

Friends of the Earth. (1974), *Waste Not: A Report Prepared for Camden Council on the Feasibility of Recycling Waste Paper*, London.

Foster, J.B. (1998), 'Liebig, Marx and the Depletion of Soil Fertility: Relevance for Today's Agriculture', *Monthly Review*, 50, 1–16.

Foster, J.B. (1999), 'Marx's Theory of Metabolic Rift: Classical Foundations for Environmental Sociology', *American Journal of Sociology*, 105, 366–405.

Foster, J.B. (2000), *Marx's Ecology: Materialism and Nature*, New York.

Georgescu-Roegen, N. (1971), *The Entropy Law and The Economic Process*, Cambridge, Mass.

Ginsburg, M. (1980), 'Rags to Riches: The Second-hand Clothes Trade, 1700–1978', *Costume*, 14, 121–135.

Girling, R. (2005), *Rubbish: Dirt on our Hands*, London.

Goddard, N. (1996), ' "A Mine of Wealth"? The Victorians and the Agricultural Value of Sewage', *Journal of Historical Geography*, 22, 274–290.

Goddard, N. (2005), ' *Sanitate Crescamus*: Water Supply, Sewage Disposal and Environmental Values in a Victorian Suburb', In Dieter-Schott, H. Luckin, B. Massard-Guilbaud. G. (eds), *Resources of the City: Contributions to an Environmental History of Modern* Europe, Aldershot, 132–148.

Goddard N. and Sheail, J. (2004), 'Victorian Sanitary Reform: Where Were the Innovators?', In Bernhardt C. (ed.), *Environmental Problems in European Cities in the Nineteenth and Twentieth Century*, Munster.

Gould, P. (1988), *Early Green Politics: Back to Nature, Back to the Land, and Socialism in Britain, 1880–1900*, Brighton.

Grove, R. (1990), 'The Origins of Environmentalism', *Nature*, 345, 11–15.

Grove R. (1995), *Green Imperialism: Colonial Expansion, Tropical Island Edens and the Origins of Environmentalism, 1600–1860*, Cambridge.

Gugliotta, A. (2000), 'Class, Gender and Coal Smoke: Gender and Environmental Injustice in Pittsburgh, 1868–1914', *Environmental History*, 5, 165–193.

Guy, W.A. (1848), 'On the Health of Nightmen, Scavengers and Dustmen', *Journal of the Statistical Society of London*, 11, 72–81.

Hamlin, C. (1985), 'Providence and Putrefaction: Victorian Sanitarians and the Natural Theology of Health and Disease', *Victorian Studies*, 28, 382–383.

Hamlin, C. (1998), *Public Health and Social Justice in the Age of Chadwick*, Cambridge.

Hammond, J.L. and Hammond, B. (1978), *The Village Labourer*, London.

Hanley, S.B. (1983), 'A High Standard of Living in Nineteenth-Century Japan: Fact or Fantasy?', *Journal of Economic History*, 43, 183–192.

Hanley, S.B. (1997), *Everyday Things in Pre-modern Japan: The Hidden Legacy of Material Culture*, Berkeley. Cal.

Harvey, D. (1989), *The Condition of Postmodernity*, Oxford.

Harvey, D. (2003), *Paris: Capital of Modernity*, London.

Health of Towns Commission. (1847), *Analysis of Evidence laid before the Health of Towns Commission and the Select Committee of the House of Commons on Metropolitan Sewage Manure*, London.

Hill, C. (1969), *Reformation to Industrial Revolution*, London.

Hill, C. (1983), *Winstanley, 'The Law of Freedom' and other Writings*, Cambridge.

Himmelfarb, G. (1984), *The Idea of Poverty*, London.

Holstun, J. (2000), *Ehud's Dagger: Class Struggle in the English Revolution*, London.

Horton, S. (1995), 'Rethinking Recycling: The Politics of the Waste Crisis', *Capitalism, Nature, Socialism*, 6, 1–19.

Horton, S. (1997), 'Value, Waste and the Built Environment', *Capitalism, Nature, Socialism*, 8, 127–139.

Humphries, J. (1990), 'Enclosures, Common Rights, and Women: The Proletarianisation of Families in the Late Eighteenth and Early Nineteenth Centuries', *Journal of Economic History*, 50, 17–42.

Jenner, M.S.R. (1995), 'The Politics of London Air: John Evelyn's *Fumifugium* and the Restoration', *Historical Journal*, 38, 535–551.

Jenner, M.S.R. (2000), 'Civilization and Deodorization? Smell in Early Modern English Culture', In Burke, P., Harrison, B., Slack, P. *Civil Histories: Essays Presented to Sir Keith Thomas*, Oxford, 127–144.

Jenner, M.S.R. (2005), 'Death, Decomposition and Dechristianisation? Public Health and Church Burial in Eighteenth Century England', *English Historical Review*, 120, 615–632.

Jubb, S. (1860), *The History of the Shoddy Trade: Its Rise, Progress and Present Position*, London.

King, W. (1992), 'How High is Too High? Disposing of Dung in Seventeenth-Century Prescot', *Sixteenth Century Journal*, 23, 443–457.

Krepp, J.C. (1867), *The Sewage Question*, London.

Lamb, R. (1996), *Promising the Earth*, London.

Lambert, M. (2004), ' "Cast-off Wearing Apparell": The Consumption and Distribution of Second-hand Clothing in Northern England during the Long Eighteenth Century', *Textile History*, 35, 1–26.

Lemire, B. (1988), 'Consumerism in Preindustrial and Early Industrial England: The Trade in Second Hand Clothes', *Journal of British Studies*, 27, 1–24.

Linnér, B. (2003), *The Return of Malthus: Environmentalism and Post-war Population-Resource Crises*, Isle of Harris.

Locke, J. (1993), *Political Writings*, London.

Luckin, B. (2000), 'Pollution in the City', In Daunton M. (ed.), *The Cambridge Urban History of Britain*, Vol. 3, Cambridge, 220–221.

Luke, T.W. (1993), 'Green Consumerism: Ecology and the Ruse of Recycling', In Bennet, J. and Chaloupka, W. (eds), *In the Nature of Things: Language, Politics and the Environment*, Minneapolis, 154–172.

Maidment, B. (2002), 'One Hundred and One Things to do with a Fantail Hat: Dustmen, Dirt and Dandyism, 1820–1860', *Textile History*, 33, 79–87.

Mandler, P. (1997), 'Against "Englishness": English Culture and the Limits to Rural Nostalgia, 1850–1940', *Transactions of the Royal Historical Society*, 6th Series, 7, 155–175.

Mårald, E. (2002), 'Everything Circulates: Agricultural Chemistry and Recycling Theories in the Second Half of the Nineteenth Century', *Environment and History*, 8, 65–84.

Matless, D. (1998), *Landscape and Englishness*, London.

Mayhew, H. (1968), *London Labour and the London Poor*, Vol. II, London.

Melosi, M. (1973), ' "Out of Sight, Out of Mind": The Environment and Disposal of Municipal Refuse, 1860–1920', *Historian*, 35, 621–640.

Melosi, M. (1980), 'Refuse Pollution and Municipal Reform: The Waste Problem in America, 1880–1917', In Melosi, M. (ed.), *Pollution and Reform in American Cities, 1870–1930*, Austin. Tex, 105–134.

Melosi, M. (1981), *Garbage in the Cities: Refuse, Reform and the Environment, 1880–1980*, Austin. Tex.

Melosi, M. (2000), *The Sanitary City: Urban Infrastructure in America from Colonial Times to the Present*, Baltimore.

Melosi, M. (2001), *Effluent America: Industry, Energy and the Environment*, Pittsburgh.

Merchant, C. (1980), *Death of Nature: Women, Ecology and the Death of Nature*, London.

Metz, N.A. (1979), 'The Artistic Reclamation of Waste in *Our Mutual Friend*', In *Nineteenth Century Fiction*, 34, 59–72.

Moore, T. (2007), 'Democratizing the Air: The Salt Lake Women's Chamber of Commerce and Air Pollution, 1936–1945', *Environmental History*, 12, 80–106.

Morris, J. (1982), *Back to the Land: The Pastoral Impulse in England from 1880 to 1914*, London.

Mosley, S. (1996), 'The "Smoke Nuisance" and Environmental Reformers in Late Victorian Manchester', *Manchester Region History Review*, 10, 40–47.

Mosley, S. (2001), *The Chimney of the World: A History of Smoke Pollution in Victorian and Edwardian*, Manchester, Cambridge.

Neeson, J.M. (1993), *Commoners: Common Right, Enclosure and Social Change in England, 1700–1820*, Cambridge.

Neilson, H. (1942), 'Early English Woodland and Waste', *Journal of Economic History*, 2, 54–62.

Norton Wise, M. (1990), 'Work and Waste: Political Economy and Natural Philosophy in Nineteenth Century Britain (III)', *History of Science*, 28, 221–261.

Otter, C. (2002), 'Making Liberalism Durable: Vision and Civility in the Late-Victorian City', *Social History*, 27, 1–15.

Otter, C. (2004), 'Cleansing and Clarifying: Technology and Perception in Nineteenth Century London', *Journal of British Studies*, 43, 40–64.

Rodger, R. (2000), 'Slums and Suburbs: The Persistence of Residential Apartheid', In Waller, P. *The English Urban Landscape*, Oxford, 233–268.

Rome, A. (1996), 'Coming to Terms with Pollution: The Language of Environmental Reform, 1865–1915', *Environmental History*, 1, 6–28.

Royte, E. (1999), *Garbage Land: On the Secret Trail of Trash*, New York.

Sabine, E.L. (1933), 'Butchering in Mediaeval London', *Speculum*, 8, 335–353.

Sabine, E.L. (1934), 'Latrines and Cesspools of Mediaeval London', *Speculum*, 9, 303–321.

Scanlan, J. (2005), *On Garbage*, London.

Scanlan, J. (2007), 'In Deadly Time: the *Lasting On* of Waste in Mayhew's London', *Time and Society*, 14, 205–222.

Schoijet, M. (1999), 'Limits to Growth and the Rise of Catastrophism', *Environmental History*, 4, 515–530.

Schumpeter, J.A. (1942), *Capitalism, Socialism and Democracy*, New York,

Sheail, J. (1996), 'Town Wastes, Agricultural Sustainability and Victorian Sewage', *Urban History*, 23, 189–210.

Shuttleworth, J.K. (1862), *Four Periods of Public Education*, London.

Simmonds, P.L. (1862), *Waste Products and Undeveloped Substances: Or Hints for Enterprise in Neglected Fields*, London.

Spooner, H.J. (1918), *Wealth from Waste*, London.

Strasser, S. (1999), *Waste and Want: A Social History of Trash*, New York.

Sucksmith, P.H. (1973), 'The Dust Heaps in *Our Mutual Friend*', *Essays in Criticism*, 23, 206–212.

Tammemagi, H. (1999), *The Waste Crisis: Landfills, Incinerators and the Search for a Sustainable Future*, Oxford.

Tarr, J.A. (1997), *The Search for the Ultimate Sink: Urban Pollution in Historical Perspective*, Akron, Ohio.

Thirsk, J. and Cooper J.P. (1972), *Seventeenth Century Economic Documents*, Oxford.

Thompson, E.P. (1991), 'Custom, Law and Common Right', In Thompson E.P., *Customs in Common*, New York, 97–184.

Thorndyke, L. (1928), 'Sanitation, Baths and Street-cleaning in the Middle-ages and Renaissance', *Speculum*, 3, 192–203.

Thorsheim, P. (2006), *Inventing Pollution: Coal, Smoke and Culture in Britain since 1800*, Athens, Ohio.

Turner, A. (2006), 'Dust-O! Rubbish in Victorian London, 1860–1900', *London Journal*, 31, 157–178.

Weiner, M.J. (1981), *English Culture and the Decline of the Industrial Spirit, 1850–1980*, London.

Winter, J. (1993), *London's Teeming Streets, 1830–1914*, London.

Wohl, A. (1983), *Endangered Lives: Public Health in Victorian Britain*, Cambridge, Mass.

Womack, P. (1989), *Improvement and Romance: Constructing the Myth of the Highlands*, Basingstoke.

Woodward, D. (1985), ' "Swords into Ploughshares": Recycling in Pre-industrial England', *Economic History Review*, Second Series, 38, 175–191.

Worster, D. (1977), *Nature's Economy: The Roots of Ecology*, San Francisco.

Wylie, J.C. (1959), *The Wastes of Civilization*, London.

Young, A. (1969), *General View of the Agriculture of the Oxfordshire*, Newton Abbott.

Young, A. (1970), *General View of the Agriculture of the County of Sussex, 1813*, Newton Abbott.

Zimring, C.A. (2004), 'Dirty Work: How Hygiene and Xenophobia Marginalized the American Waste Trades, 1870–1930', *Environmental History*, 9, 80–101.

Zimring, C. (2005), *Cash for Your Trash: Scrap Recycling in America*, New Brunswick, New Jersey.

11

Why Intensify? The Outline of a Theory of the Institutional Causes Driving Long-Term Changes in Chinese Farming and the Consequent Modifications to the Environment

Mark Elvin

The examination of a hitherto little studied fiscal aspect of a problem in Chinese environmental history suggests that, in cases like this, environmental history does not have a clearly defined separate character of its own. It is inextricably interwoven with other disciplinary perspectives. In turn it makes its own, often crucial, contribution to the comprehension of the subject-matter studied in these other perspectives. To the extent that the example presented here is representative, one could say that environmental history is thus both everywhere and nowhere.

The intensification of farming

Chinese farming during the last thousand years of the empire, and especially the last few hundred before the arrival of chemical fertilizers, has been famous as an example of fields farmed with a high input of hours of labour per hectare per year, and the subtlety of attention to detail that is normally associated with gardens. This is why it has sometimes been loosely described by the German term *Gartenbau*, 'garden-style farming' or some such equivalent. Wilhelm Wagner, the pre-eminent Western authority on the subject, preferred however to reserve 'Gartenbau' for agriculture that was predominantly irrigated and made minimal use of non-human labour, and to call farming that was mainly non-irrigated and made relatively greater use of animal-power *Ackerbau*, or 'field-farming.'[1]

---

[1] Wagner (1926), 636–637. A viewpoint that differs from Wagner's in being more convinced of the traditional system's long-term sustainability is Franklin King (1949; rev. ed. 1972). Note that the earliest version of King's book came out in 1911 from the University of Wisconsin: Madison WI. A summary that inclines more to Wagner's side of this argument may be found in Elvin (1982). The basic reference for the long-term history is Motonosuke (1962).

274

Table 11.1 Simplified version of the calendar for farmwork in South China from Bao Shichen

Food plants omitted in the simplified version: scallions, tea, yams, red hibiscus, ginger, melons, gourds, eggplant, edible greens, lettuce, sweet potatoes, lentils, buckwheat, cabbages, turnips, carrots, jujubes, lotus leaves, "fruit." Fibre plants omitted: hemp, ramie. Others: mulberry trees, indigo, linseed, bamboos, pond fish, silkworms, etc. The sowing time of barley and harvest time of buckwheat are not given in this part of the source, and are supplied from other parts.

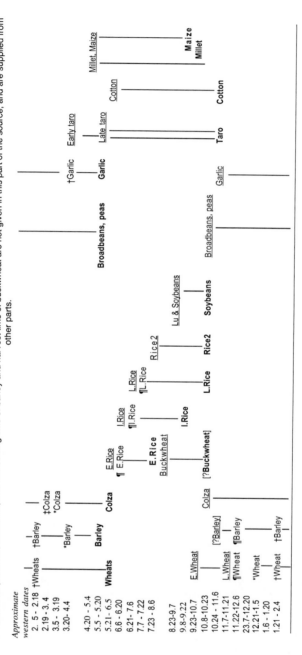

Approximate
western dates

2. 5 - 2.18 †Wheats †Barley
2.19 - 3. 4 ‡Colza
3.5 - 3.19 *Colza
3.20 - 4.4 *Barley

4.20 - 5.4 **Wheats** **Barley** **Colza** **E.Rice**
5.5 - 5.20 ¶ E.Rice
5.21- 6.5 I.Rice
6.6 - 6.20 ¶I.Rice
6.21- 7.6 **E.Rice** **I.Rice** **Broadbeans, peas** †Garlic Early taro Millet Maize
7.7 - 7.22 Buckwheat
7.23 - 8.6 L.Rice
               ¶L.Rice

8.23-9.7 E.Wheat Colza [?Buckwheat] **I.Rice** Rice2 Late taro Cotton
9.8-9.22 Lu & Soybeans
9.23-10.7
10.8-10.23 [?Barley] **L.Rice** **Rice2** **Garlic** **Taro** **Cotton** **Maize**
10.24 - 11.6 L.Wheat **Soybeans** **Millet**
11.7-11.21 ¶Wheat ¶Barley
11.22-12.6 *Wheat Broadbeans, peas Garlic
23.7-12.20
12.21-1.5 †Wheat †Barley
1.6 - 1.20
1.21 - 2.4

Key: † Hoe, * Manure, ‡ Pinch off flowers, ¶ Weed, ‡ Manure, ¶ Weed. <u>Plant in final place</u>, **Harvest** [*E.*' 'early', '*I*,''intermediate', 'L.' 'late']. *Note:* 'colza' = 'rape', 'Rice2' = 'interplanted rice', '[?]' = estimated. *Source:* Bao Shichen, *Qimin sishu* [Four techniques for ruling the common people] (1844. Reprinted, n.p. Zhuqingtang 1872), *j.* 25, 291-30.

In a very approximate sense the first term can be used to describe later imperial-age rice-farming in the Yangzi valley and further south, and the second, in implied contrast, the wheat, millet and barley farming of this time in the North-China plain. For many purposes it is a useful distinction, but for the purposes of this chapter it risks being misleading. The phenomenon of the historically ever more intensified use of labour per unit of cultivated area applied equally to *both* areas, and in terms of going *as close as possible* to what was technically and economically locally possible it is fair to say that it affected both regions roughly *equally*. No one, I think, would deny that there were environmental possibilities of several kinds in the South, such as crops that produced more per cropping acre than their northern counterparts, and more hours of insolation each year, and fewer days of frost, not to mention more abundant relatively reliable supplies of water from both rains and rivers, that enabled the South in the last millennium of the empire to exploit a wider range of possibilities of labour-intensification than the North (Table 11.1). But the North has also to be thought of as 'intensified'.

By premodern standards, the late-imperial mean yields per hectare *per year* for rice- and wheat-based farming, including multiple cropping and intercropping with secondary crops of various sorts, were indubitably high for the age *prior* to the use of chemical fertilizers when compared to most other countries that were broadly similar in terms of climate and soils. Important possible exceptions to this remark are Tokugawa Japan, and some parts of South India for rice, and eighteenth-century England for wheat. For the greater part of the thousand-year period there was no fallowing in China (a fact that boosted its overall average annual output per hectare relative to its European comparators). This meant an unremitting attention to the conservation of soil productivity. In addition to the use of rotations, a wide variety of fertilizers had to be constantly collected and matured, or purchased, and then applied (see Table 11.1 on southern farm-work schedules). These fertilizers included materials of vegetable origin, animal and human manures, and river-mud plus, where appropriate, lime. Put in different terms: the late-imperial farming system increased *labour* costs per unit output in return for increased output per unit *area*. In different terms again: it did in a *shorter time* by human agency the work (namely fertility restoration) that 'nature' could probably have done, if not in quite the same way, in a *longer time* (through fallowing).

Chinese farming was not always intensive. For example, around 763 CE a large new development in Jiaxing in the lower Yangzi region used the broad-casting of rice-seeds 'by means of the breezes,' rather than growing sprouts in seedbeds and then transplanting them in rectangular grids of a size to optimize the use of the land-area.[2] This was in one of the more advanced regions for

---

[2] Elvin (2004), 181.

rice-farming. Four centuries later, non-intensive rice-farming was retreating to the economic margins. In 1178 an official remarked contemptuously of Qinzhou in the far southwest:

> The farmers of Qinzhou are careless. When they work the soil with an ox they merely break up the clods; and when the time comes for sowing they go to the fields and broadcast the grain. They do not transplant seedlings. There is no more wasteful use of seed than this. After they have planted it, they neither hoe it nor irrigate it, but place their reliance on the forces of nature.[3]

But if land here was still plentiful, as seems likely, this style of farming may have been the best use of labour-power.

The simpler forms of intensive farming were however the foundation on which a significant proportion of China's population of perhaps 120 millions of people was already supported by around 1100 CE. A millennium and a half later, in its most developed form, circa 1850, it was the lifeline of a preponderant part of the late premodern population of perhaps 430 million people.

Outside the main domain of floodplain and valley-floor intensive farming there was also a periphery of semi-stable or shifting upland farming, often based on New World crops like maize and sweet potatoes, and using less-intensive methods. The difficulties of quantifying this peripheral but eventually significant sector, and incorporating it into meaningful aggregate historical statistics of arable land, farm labour, and per-hectare and per-worker output constitute something of an econometric headache.

Nonetheless, we may broadly say that the long-term relative success of intensive farming more than any other aspect of the economy made possible the massive weight of the demographic pressures bearing down in late premodern times and today on many aspects of China's environment. So what triggered, and then drove the millennial process of intensification in the first place?

I suggest in what follows that a combination of institutional factors, including first and foremost the structure of unit-area-based taxation on farmed land, and, in second place, social conventions shaping the levying of rents, and the status of those engaging in farm labour, with certain aspects of the technology of farming, plus, of course, as time went on, the growing pressure of population on farmable land, need to be looked at more deeply for their

---

[3] Elvin (1973), 114. Presumably there was suitable seasonal flooding here, as in much of southeast Asia.

possible interacting effects on the process of intensification. Pressures on my time during the last year means that it is inevitably still a sketch.

The core argument

The crux of the analysis presented here is that the key factor, in the sense that it could relatively easily have been different, was the practice of the Chinese government of basing taxes on farmers, including their labour-service obligations, on the area of the land that they owned. The ultimate origins of this go back to the sixth century BCE, but other bases were also used at various points, and it is simpler to look at the issues by taking the beginning of the Song dynasty (960 CE) as the starting-point of our discussion since it was close to the normal mode after that point. A complementary practice that had an analogous, though weaker, effect was that of often levying rents from tenants mainly on the basis of the main cereal crop, but not on supplementary crops.

When tax is levied primarily on an area, even if this is roughly graded by estimated quality, it usually pays to look first to increasing output per unit area, so far as this is possible, rather than farming a larger total area. There are also two 'ratchet' effects. When the normal yields have risen and the tax take adjusted upwards to match this, it is ruinous to let yields slide back again. When a larger population is being supported through increased unit-area productivity, if the productivity slides backwards, out-migration of a proportion of the population may be the only alternative to starvation.

A structural social shift that largely took place in the seventeenth century so far as the rice-farming region was concerned was the virtual ending of farming using unfree farm labour, the proportion of which varied markedly from one place to another, but was often quite significant. This is likely to have increased the proportion after this date of those engaged in farming who were motivated to drive themselves to the limit of the possible in improving yields on what were often very small units of operation. An interesting unresolved question is how far the pioneering of methods of intensification in earlier times was done by larger units of operation, which I have in the past referred to as 'manors,' using labour often under non-economic forms of constraint. That they seem to have done so is possibly simply the consequence of their being better covered by our sources, though non-trivial resources were probably needed in a majority of cases to open new land where new infrastructure in irrigation and drainage was required. In the eighteenth and nineteenth centuries, however, the smaller units of operation are usually thought to have had a higher productivity per hectare than the larger ones. A complementary explanation is that most large landowners had by this time long since moved their residences into the towns and cities, and were not as interested as before in farming technology and management.[4]

A formal and somewhat oversimple mathematical analysis of the economic logic of this core argument is presented at the end of this chapter just before the discussion. It seems likely that most readers will find it easier to understand at this point, once the fiscal and rent structures are more familiar. But the foregoing is, in simplified and non-formalized form, the heart of the theory that I am suggesting.

There is, however, an immediate objection that will occur to most of those already well acquainted with the economic history of late-imperial China, and which needs to be tackled at the outset. Was the burden of taxation and rents on farmers working relatively small areas of land heavy enough to have been a possible long-term influence of any important kind on their style of economic behaviour? In spite of some fine work on late-imperial Chinese tax history, such as that of Wang Yeh-chien and Ray Huang,[5] the real nature of the tax 'burden' has not been well understood by historians of China. This is in good measure because the essential source materials are lacking until the early twentieth century. In spite of my admiration for the pioneering work of Wang Yeh-chien, it is essential to point out that even his most sophisticated formula

$$\text{Real land tax burden} = \frac{\text{Land tax quota} \cdot \text{rate of collection (in money)}}{\text{Cultivated} \cdot \text{acreage} \cdot \text{land yield} \cdot \text{price}}$$

does not give adequate insight into all the factors to which one needs to be sensitive.[6] Essentially he leaves out costs, both actual and implied. Nor does he deal adequately with the horrifyingly difficult, but vital, issue of how far the official figures are usable at all for aggregate analysis.

Though it breaks the flow of the logic, it seems essential at this point to insert two excursuses, one on the nature of the tax burden and the second on the reliability of the late-imperial figures.

Excursus A: The nature of the tax burden on farmers

If we use the data from the early years of the twentieth century provided by Wilhelm Wagner,[7] it is possible to establish the theoretical foundations of a

---

[4] For a more detailed discussion and documentation of all these changes, see Elvin (1973), especially ch. 15, 'The disappearance of serfdom.' In comparison to that of the south, the agrarian history of north China under the Ming dynasty (1368–1644) is opaque, but it seems that unfree labour had diminished to a low level here very considerably earlier.
[5] For example, Wang (1973); Huang (1974).
[6] Wang (1973), 111.
[7] Wagner (1926), 636–657.

more realistic approach to the question of the possible impact of taxation on intensification. Wagner's material was obtained through his students and relates only to Shandong, hence to northern dry-farming. None the less the principles emerge clearly enough. We will consider as illustrations two of his 'representative' examples, one an owner-operated farm that was free of debt, and the other a holding worked by tenants.

The first was a property of 42 *mou*, which were each locally 0.09 hectares, hence with a total farmed area of 3.78 hectares. Its capital value at market prices was 864 *taels* [ounces of silver], made up of the following:

Farmland at 12 taels per mou	504
Buildings	250
Livestock [1 mule, 1 donkey, 1 ox]	70
Other inventory [probably tools etc.]	40
Total capital in taels	864

The clayey soil each year produced winter crops of winter wheat, winter barley, and winter peas, summer crops of millet, gaoliang, summer peas, sweet potatoes, and peanuts, and 'intercrops'[8] of beans (probably mainly soy), maize, sweet potatoes, and turnips, plus straw and leaves. The value at current market prices of this total output was 480.08 taels; and the farm also raised 6 pigs worth a total 42 taels (one of which was usually eaten by the family), poultry worth about 2 taels, and over 100 eggs. The household also produced straw braid worth about 5 taels.

This output had to support the owner, his wife, 3 sons and their wives, and four grandchildren. In this regard, Wagner counts females as 0.75 of a male, and children as 0.25 of an adult. There were thus the equivalent of 8 person-years to be looked after.

Except for a few days each year of hired help with the harvesting, no monetary wages were paid. The cost-equivalent in market prices of supporting the humans and animals from the unsold portion of the total produce was 327.98 taels. From this 70 taels as the annual cost of shoeing the three animals needs to be deducted. Thus supporting the adult labour force plus the 4 children required 258.0 taels. It cost approximately 8.06 taels a year to support one child, 24.19 taels to support an adult female, and 32.25 taels to support an adult male. The males had also annually to supply the state with a certain amount of unpaid labour services, but we have to leave this on one side in the estimation for lack of quantified information.

---

[8] *Zwischenfruchte.* See Table 1 on the farming calendar for an illustration of what this term presumably referred to.

The calculable residue was just over 201 taels.[9] When the land tax of 30 taels had been deducted this left about 171 taels. What of the remaining factor of production, the cost of the labour? Without the work done by the four adult men and women, or substitutes hired to replace them, the land would have produced nothing. The wages paid locally to a man, in addition to providing him with his food, were 20 taels a year. For women it was 8 taels. Adding in the additional harvest workers to the four men and women in the household, the total value of the year's labour was 117 taels, or about 2.8 taels per mou. This left a net profit of 54 taels. In other words, a return on the capital of 6.25%.

The local annual rate of interest was on the order of 10%. Thus the notional payment that would have been necessary to supply the non-land component of the farm capital, which amounted to 360 taels, was 36 taels. The total yield-based notional ground rent for the farmland (if it had had to be rented) can thus be estimated as (54–36). Dividing by 42, this gives 0.43 taels per mou per year. Capitalized over 10 years, this points to a yield-based land price of 4.3 taels per mou, rather than the market value of 12 taels.

Several conclusions emerge from this example, assuming that it is reasonably representative for owner-operated properties at this time and place. Farmland was 2.8 times the price that the value of its output warranted, given prevailing interest rates and agricultural commodity prices. Further intensification relative to area and time-density of cropping, where technically possible, would have made sense. The tax burden took away 36% of what would have been the net profit if there had been no tax. According to Wagner's calculations, the Chinese tax on farmland per hectare at the end of the Qing dynasty and start of the Republic was 15 times heavier than that in Prussia in the mid-nineteenth century calculated in the same way; and the better access that Prussian farmers had to woodland and pasture land, close to lacking in most of the core areas of China by this date, has not been factored in to this assessment.

In the second case, that of a tenant holding of 20 mou of 0.09 hectares per mou, that is 1.8 hectares, but with the buildings, livestock, and tools owned by the tenant, the net profit, calculated on the same lines as in the case just presented, was just negative, by the slenderest of margins. (This can be seen as equivalent to accepting imputed wages below the current local mean value.) The family, consisting of a father and his wife plus his son and the son's wife, with five children, must have relied to a critical extent on their small but presumably non-rented garden and perhaps some other unlisted employments, as there were no domestic handicrafts recorded. They had one ox and one donkey, and normally fattened five pigs a year,

---

[9] There are uncertainties, such as the income from eggs. Wagner gives a precise figure is 201.10 taels.

of whom they sold three. The normal tax levy on farmland in the area was reported to be twice the legal level, but their landlord was an influential merchant who, exceptionally, only paid at the legal rate of 5 taels a year for the property. Their total annual rent was 34.60 taels, paid in fixed amounts[10] of two different grains, plus 5.10 taels in copper cash. The landlord in this particular case gained each year 1.48 taels per mou after tax. The ordinary landlord who lacked such political pull would have made 1.23 which capitalizes the land to a value per mou of 12.2 taels, very close to that in the first example. The pressure on the tenants to intensify in any technically and economically feasible way was even greater than in the first case.

Excursus B: The formal level and the actual level of taxes

Wagner marshalls a sample of evidence from informed foreigners long-resident in China during the later nineteenth and early twentieth century to suggest that the amounts of land tax collected were often over twice the legal amount, and in extreme cases (in Sichuan) up to ten times more. He also provides evidence for a prima facie case that the rates levied on farmland of comparable quality and nearby location sometimes varied significantly.[11]

I long ago noted Chinese evidence from slightly earlier times suggesting that the formal burden might at times be misleadingly low compared to what was actually levied. A passage from the Qing-dynasty *Veritable Records* for 1806 notes that:

> When the tribute grain is collected, the local officials in the provinces collect more than the amount sanctioned by law. They make arrangements to have gentry of bad character act as their agents.... They bribe them in advance, granting them the right to contract for a certain portion of the tribute grain. The rustics and the poor have a redoubled burden because these persons can levy and excess amount from them just as they please.[12]

A substantial number of other items from the same source confirm that this kind of malpractice could and did happen. Thus an entry from 1735 reads as follows: 'In one place after another reports of tax grain resulting from the opening of new land have added nothing to the state's [actual] quota of taxes. Not only has it not proved possible to do the land-surveying [required], but the clauses

---

[10] Although it is hard to say for sure, some variety of crop-sharing, at least as regards the main crop, seems to have been somewhat more common in late-imperial China as a whole, but this question is hard to resolve.

[11] In one case of two small properties near Shangahi that he records the ratio of the difference between them as of the order of 2:3 per mou. Wagner (1926), 140.

[12] Elvin (1973), 267.

in the demand for the first reporting [of liability] have been the occasion for intimidatory and fraudulent assignations [to pay], the high officials [seeking] a reputation for zeal for the public good [by claiming to be raising more taxes] and the lesser officials having the additional motive of seeking for profits.'[13] A number of other passages from official sources quoted later show that taxes and labour-services (which were also based pro rata on the acreage owned) were seen as a serious burden in various places at various times because of official malpractice. The Emperor Kangxi in 1668 thought that official cheating of the taxpayers had been rife.[14]

Pressure from the land tax only affected small and perhaps some medium owners. Powerful local people had the political and social clout to defend themselves against extortion. This was why substantial numbers of free peasant landowners are reported as having 'commended' themselves as rent-paying tenants to such powerful local people.[15] Those commending their own land may have paid lower than usual rents (the sources do not tell us such details), but ordinary rents were substantially higher than formal taxes, and tenants also risked losing tenure of their land by eviction after the passage of time had removed the memory of the commendation. That there were good reasons for paying more than the formal tax in order to avoid the actual tax, even taking this risk of losing the use of the land as well, seems reasonably plausible. This sort of behaviour is hard to explain if the total tax burden was not felt to be onerous.

Talking in terms of tax as a percentage of some such total as national product or income is likely to be an unsatisfactory guide to real-life economic behaviour. What mattered in the context of the analysis given in the Excursus A above was the impact on the family budgets of the smallholders, who were the ones who, more than others, intensified their farming. A Jesuit report from the later eighteenth century noted that '[The lands] of peasant proprietors are of an astonishing fertility, in contrast with the large estates.'[16]

Further, we need to think in terms of vulnerability to variability in the weather and to other hazards, like locusts. The per-area land tax was relatively inflexible, though bureaucratic mechanisms did exist for reducing it somewhat in circumstances of generally harsh local weather. It is unclear how far they were effective.[17] Thus a prudent peasant would have wanted as big a buffer

---

[13] Nankai University Department of History (1959), *Qing shilu jingji ziliao jiyao* [Digest of materials on economics from the Qing-dynasty *Veritable Records*, hereafter *QSLJJ-ZLJY*], 75.

[14] *QSLJJZLJY*, 664.

[15] Example: *QSLJJZLJY*, 660.

[16] Elvin (1982), 14.

[17] There is a good discussion of them by Robert Marks for the far south of China in his *Tigers, Rice, Silk, and Silt* (1998).

as possible against uncertainty due to variability (which for rain in northern China was around 30% a year). What might be easily borne in an average year might be disastrous in a bad one.

The land-based levies included not only the formal tax but a variable number of surcharges like that for 'wastage,' miscellaneous fees, and, above all, obligatory conscripted labour. Some of the latter was outside the formal quota, though only imposed as needed.[18] Some of the surcharges were legal, others illegal. The add-on in Jiaxing prefecture before 1672 at one stroke put the land tax up by 80%. In Hunan province the unauthorized levies were reported to be several times the official tax.[19]

It is easy to multiply citations of the type just given. Bluntly, it is not possible to calculate useful statistics in this sort of situation, especially when one is concerned with the impact on peasant behaviour.[20]

If we turn now to qualitative materials, the Qing-dynasty poems on aspects of everyday life reveal a view on the part of a number of different poets that the evil done by tax-collectors was at times literally murderous. These poems were somewhat like newspaper headlines – depicting extreme and shocking cases – but they would not have been anthologized if they had not spoken of something that was quite possible to imagine occurring.

In "The Ballad of Date-Tree Lane" by Li Fuqing on the decision of parents to sell their son because of the pressure of taxes we read that

The time for rearing silkworms had not run its course in full
When the red warrants demanding tax were sent to the countryside.
In the morning, people were pressed to pay, no hen or pig overlooked.
They were pressured again, late in the day, and the weak and the elderly died.

A poverty-stricken married couple, then living in Date-Tree Lane,
Set off to the county capital market, to dispose of their only son.
To feel affection for one's child is only human nature,
But when taxes have to be paid to the State, what else is there to be done? . . . .

In due course, having used her last pennies to buy food for her son and husband, the mother hangs herself from a tree, but the father and son vow at once to follow her in death, and hang themselves on a branch beside her. So far

---

[18] See Elvin (1973), 265, for an example.

[19] *QSLJJZLJY*, 667.

[20] This remark is of course without prejudice to the massive efforts made by Angus Maddison in his *Chinese Economic Performance in the Long Run* (2007) to produce usable, albeit very rough-grained, estimates for China's overall economic performance in premodern times.

from indulging in Hardyesque morbidity, the poet then honours their mutual devotion:

Like wings that support each other in flight they soar now in Heaven above.
On this Earth below they are linked each with each, like intertwining
   branches.
The physical souls of their essences keep, forever, in touch with each other,
Which far surpasses living on, if they'd had to live apart.[21]

A different situation is described by Liu Tianqian's "Reaping Wheat". The poet has walked out into the country and sees much of the farmwork being done by women, which puzzles him. Then an old man explains that the able-bodied men have all been taken away by the tax-collectors, to be tortured to make them pay up. There is some exaggeration and hyperbole in the old man's tale (since, presumably, torture does not have to take a long time), but it too would not have been anthologized if it had not contained a significant element of truth:

The drought-parched autumn, last year, had desiccated the sprouts.
Weeds grew in the corners of stony acres, which turned into scrubby wastes.
We filled our bellies with bark from trees, and chalk ground into flour,
Yet even before our tax fell due they were quick to abuse and chase us.

At dead of night they come to a village. They seize and tie up their victims.
We tremble like people with fever when we go past the magistrate's court.
Our guts, burning with hunger, turn round inside like a windlass,
As we drive our bodies on and on, in terror of being tortured.

                    . . . .

See that half-wit begging for rice? Beside that wall there. And yammering.
He's sold himself to a merchant, to drudge for him as his serf.
Once he'd his hands on the silver, the whole lot went to pay tax:
Two-thirds to clear the basic sum, 'official wastage' one third. . . . .

At the end the old man himself is seized by the tax-prompters, tied up in ropes, and carried away.[22]

There are many other examples of this sort of poem. One must not make too much of them as regards specific details, but they fairly certainly convey a widespread popular attitude towards taxes. Taxes hurt.

---

[21] See Elvin (1998), 172.
[22] Elvin (1998), 134.

The linkage of taxes with land ownership

We resume the main line of thought: The adherence of the Chinese state to the principle that the land tax and the levy of labour-services should be based on the quantity of land owned by a household is explicitly expressed in the comment of the Imperial Secretariat on a memorial from a censor in 1820 when its implementation was beginning to break down under the pressure of accumulated abuses, and needed to be re-established:

> When the soil is transformed into taxes, it follows automatically that it is right that the grain tax is adjusted in accordance with the [amount of] farmed land. According to the memorial from the said censor, there are poor people in Jiangsu province who do not own even enough land for a dwelling, and yet pay a land tax every year of from several ounces of silver to several tens of ounces; and that there are others who possess merely several mou[23] whose yearly tax is that for fields of from more than ten to several tens of mou. In all these cases [according to his memorial] this has been the result of the failure of the owners of lands, when they earlier sold them, to transfer [the obligation to pay] the entire amount of land tax from their household to that of the purchaser, so giving rise to conflicts and confusions.... Those households who at present possess a given area of farmland, must pay the [corresponding] given amount of land tax in accordance with the regulations. 'Empty taxes' that cannot be attached to [farmed land] are all to be shifted to be collected from others [who do have the appropriate amounts of land].[24]

The land was graded into three levels of quality, each grade being further subdivided into three, with a tax rating assigned to each. Differences in location and in crops harvested meant that there were innumerable variations in practice. In financial emergencies it occasionally happened that the grading system was ignored. This occurred in the late Ming, when an extra hundredth of an ounce of silver was levied on each mou of land, and this was repeated in 1661 under the Qing, but stopped the next year for fear of overburdening the population.[25] This additional burden raised a total of rather more than 577 million ounces of silver, which would have been of the (very) approximate order of magnitude of an extra 15 ounces of silver per household of five people in a total farming population of 200 million, and

---

[23] One mou was on average about 0.067 hectare under the Qing, and 0.07 under the Ming. There were wide local variations. See Excursus A for an example where the figure was 0.09.
[24] *QSLJJZLJY*, 695.
[25] *QSLJJZLJY*, 662.

almost immediately caused a political problem. From Excursus A it is easy to understand why.

Where the tax was collected in kind, which after the seventeenth century was mainly only the case when it was the grain that was shipped up the Grand Canal to supply the Capital, difficulties could be caused to this system by variations in the weather. In 1826 a memorial noted that:

> It is specified that for the tax-grain supplied to the Capital from Jiangsu province late-ripening rice is to be collected.... In Jurong county there are many hills, however, and but few polders. Last year, the summer rains were somewhat delayed, and the majority of the peasants in this county planted early-ripening rice. This caused the colour of the rice delivered to the government granaries to be irregular.[26]

This mattered more than it might seem, as early-ripening rice did not last so well in storage as the late-ripening kind. It was decided that henceforth the local officials would examine the weather each year to determine whether or not it was appropriate for the peasants to guard against a coming drought by planting early-ripening rice. This is unlikely always to have given decisions acceptable to locally experienced farmers who may be presumed to have preferred to trust their own judgements, and not to have to wait for a bureaucratic go-ahead.

Labour-services for water-system maintenance and construction and other tasks were, in principle, linked to the size of a landholding,[27] but in ways that varied from one region or even locality to another,[28] and were not always properly observed in practice. It is clear that poorer peasants sometimes greatly feared the loss of farm-labour that could be caused by these compulsory services. The linkage of labour-services and ownership of land emerges clearly from a report submitted by the governor of Gansu province in 1742:

> The lands of Gansu are located on the frontiers. In the past the arable was extensive and the population sparse. Since our present dynasty was established, vagrants have little by little gathered, but at the beginning of the opening up of new lands for farming, the least well-off commoners were frightened of the labour-service levies, and always borrowed the names of

---

[26] *QSLJJZLJY*, 696.

[27] For example *QSLJJZLJY*, 647, 661, etc.

[28] Thus in 1726, when the Board of Finance wanted to bring the labour-services required from residents of the reed-lands associated with salterns into the main land-tax system, they had to leave out those areas where the inhabitants owned too little land for this to make fiscal sense. *QSLJJZLJY*, 670.

the gentry [as their nominal landlords] for cover. When reporting the clearance of new land and the undertaking of cultivation, they placed themselves in the condition of tenants who paid rent every year.

They also feared that once they had developed their acreage to a mature condition they would have nothing to rely on [legally] in the future if they were suddenly dispossessed, so they established contracts to serve as guarantees entitling them to cultivate in perpetuity, and not to be deprived of their tenancies.[29]

Rents were quite often (though probably in rather less than the majority of cases) calculated on an acreage basis. This may be presumed to have had an impact on behaviour broadly similar to that of taxes on small landowners. An example of rents on a per-mou basis can be found in the report in 1746 of a censor who was in acting charge of the tax-grain transport:

The military colony fields of the Guard Area of Wenzhou have, according to a previous decision, paid a harvest rent [per mou] of 3 piculs for the land of the best quality, 2 piculs for land of middling quality, and 1.6 piculs for land of the lowest quality.[30] These quotas are far too heavy. Even in years when the harvest is abundant they cannot be met in full. It is now our considered decision that for every mou, in addition to paying in full [the grain for] the regular military rations, the best fields shall be charged an extra fee of four-tenths [of an ounce of silver], the middle-grade fields three-tenths, and the lowest-grade fields two tenths.[31]

Colony lands seem often to have been associated with heavy charges.

Since the foregoing quotations are drawn from the so-called '*Veritable Records*' of the Qing dynasty, they are excerpts or summaries put together, after the fall of a dynasty, from its much more voluminous official archives. To some extent they were recorded and preserved because the problems with which they dealt were exceptional. This point should not be forgotten, nor should the recent experience of several scholars that the full archival files on particular cases quite often reveal the events of an incident in a significantly different light from these extracts. They are used here, not for their often intriguing particularities, but to shed light on the general assumptions and practices underpinning the system.

---

[29] *QSLJJZLJY*, 680.

[30] The picul (*dan* or *shi*) was, properly speaking, a measure of capacity of 10 pecks (*dou*), each of 10 pints (*sheng*). The *sheng* in Qing times was close enough to 1 litre for rough calculations of small totals.

[31] *QSLJJZLJY*, 684–685.

These, then, were the very general features of the late-imperial land tax. When we look further, what emerges is that the experience of the burden of these taxes, which were in practice accompanied by varying arrays of additional surcharges and, more often than not, serious corruption and squeezing on the part of those who collected them, not infrequently both led peasants to abandon their homes, and the better-off to avoid investing in landholding. The imperial response made in 1703 to a report by an official from Hu'nan province sums up what was a bad but probably far from unique situation:

> Hu'nan is placed between regions that have distant borders, and We have heard that the accumulated [mal]practices of the officials high and low are continuing as before, with no limits to the many kinds of the illegal levying of taxes. If one calculates the levies of tax and the assignation [of labour-services] every year, there are some instances where they reach a multiple of several times the formally specified quotas. When the authorities raise taxes they further take an allowance for 'wastage' that is uniquely heavy when compared to those of other provinces. The common folk are so distressed by their exhaustion that they cannot endure it, and many of them leave [their homes] and migrate from one place to another.[32]

The emperor ordered that every effort be made to induce the vagrants to return home and take charge of their properties again.

Other areas also faced heavy surcharges. In 1672, in Jiaxing prefecture and Huzhou prefecture in the lower Yangzi valley, one of the most intensively farmed parts of the empire, the 'wastage' surcharges were, respectively, 80% (as mentioned earlier) and 55% of the nominal formal tax. They were both reduced in this year to 45%.[33]

On taxes as a disincentive to owning more land, a censor wrote in a memorial on the province of Yunnan in 1668 as follows:

> Of the various paths that the State has for developing financial resources, the most important is the opening up of uncultivated lands. This has been pursued for more than twenty years, but without results. This is the result of three disastrous circumstances. 1. Taxes and labour-services are too pressingly levied, and rich people consider that the possession of land is a burden. 2. There are no funds for persuading tenants to come [and sign contracts], and poor people regard having land bestowed on them as an affliction.[34]

---

[32] *QSLJJZLJY*, 667.

[33] *QSLJJZLJY*, 665.

[34] Perhaps also because of the lack of infrastructure and working capital.

3. The evaluation of the performance of the officials is too easy-going, and those in authority do not regard the opening of new lands for cultivation to be part of their duties....

Farmland comes in different grades. That which has only recently fallen into disuse should be charged tax after a grace-period of three years [following its reopening]; that which has been in disuse for a long period should be charged tax after a grace-period of five years; and that which is in an extreme condition of wilderness [before being developed for farming] should never be charged tax. If this is done, then the common people will be well-off for resources, and there will be numerous tenants cultivating newly opened fields....

People have different grades of wealth. Those who would [otherwise] migrate away should be endowed with [land run by] official 'manors'; those who are [merely] short of money should be loaned official cattle, and [provided with?] barrage-controlled reservoir-ponds and irrigation channels that are maintained by official funds. If this is done, then people will be well-off and those energetically opening up new land will be numerous.[35]

Thus the tax burden was important, but only part of a more complicated economic pattern. The memorialist's enthusiasm notwithstanding, it seems reasonable to imagine that once a peasant, whether tenant or owner, was over the difficult 'hump' at the beginning of making a new holding workable, he would generally be more concerned, other things being equal, with maximizing his returns from what he already had in functioning order than with tackling, or being a participant in tackling, another transition across a new 'hump.' Opening new land on any scale also tended to need the involvement of rich landowners, merchants, or the government because of the capital required.[36]

A simpler formulation appeared in the report of the governor of Guangdong province in 1671 about the military colony fields there that paid a land tax of 3 pecks for every mou, this being 'almost twice the rate on commoners' lands.' The people, he continued, 'fear the heavy burden of the land tax, and do not dare to admit it if they have opened new land for cultivation.'[37] By 1853, the estimated area of 'salty fields' along the Guangdong coastline not registered for taxation had reached 'some tens of millions of *qing*,' the qing being a unit of 100 mou. The authorities are said to have shrunk from the costs

[35] *QSLJJZLJY*, 64–65.
[36] *QSLJJZLJY*, 84, 87, 93, 97.
[37] *QSLJJZLJY*, 665.

and complexities of trying to survey such land, and everyone else concerned to be 'cringing in fear' at the possibility that this might happen.[38]

In the course of the eighteenth century, it seems to have become officially established that very small parcels of land did not have to be reported for taxation. In Guangxi province in 1741, for example, a holding of less than 1 mou of ordinary wet-field paddy or 3 mou used for dry-land crops did not need to be registered. For land of the poorest quality, the tax-free limit rose to 5 mou for paddy-fields and 10 for dry-land crops. A roughly similar rule applied in Shanxi and Shaanxi.[39] In Guizhou province at the same period the land so poor that it had to be left fallow in alternate years did not pay tax.[40] It seems likely that it would often have been economically rational to extract the maximum from an untaxed small unit than rather more from a larger but taxed one.

Overall, there is now a prima facie case that tax pressure was a significant variable in the planning of farm management in late-imperial China, though other factors would surely have also played a significant part. At the same time, it seems to have been becoming tacitly accepted by the state as time went by that it was not worth the effort to try to extract tax from land of the poorest quality. One possibility is that doing so would have simply discouraged most of the owners from cultivating it at all, but this can only be conjecture.

The long-term historical process of intensification in Chinese agriculture

Let us define a 'revolution' in the economic domain as a whole as 'a rapid shift in the basic interrelationships between newly modified technologies of production, transport and communication, on the one hand, and the dominant types of interpersonal and inter-institutional transactions, on the other.' For it to be of any significance, we need to add that it sets its stamp on a long subsequent period during which these changed interrelationships develop, mature, and strengthen their hold. If this definition is accepted, then there was only one economic 'revolution' in imperial Chinese history. This was that of the Song about a thousand years ago. It was marked by a massive extension of monetization, commercialization, credit-use, urbanization, printing, greatly expanded inland and marine water-transport, increased use of various forms of machinery (including water-driven spinning-machines) and a modicum of mass production (as of iron arrowheads). It was also characterized by the first major steps on the road towards agricultural intensification, however

---

[38] *QSLJJZLJY*, 700.
[39] *QSLJJZLJY*, 74, 680, 679.
[40] *QSLJJZLJY*, 680.

geographically patchy, and, in this particular case, towards the round-the-year use of on-farm labour.

If one wanted to argue for a second 'revolution,' it would have to be applied to the changes over the rather more than a hundred years from the late sixteenth to the early eighteenth century. The most distinctive features of these changes were social (i.e., transactional). They included the near-total disappearance of a previously substantial sector of unfree rural labour that had existed in some areas (especially the south), the related move of the larger landowners into towns and cities, a corresponding need to reshape the structure of local power in the countryside for such collective tasks as the maintenance of water-control systems, a shift towards an even greater reliance on small units of operation in agriculture, with a linked rise in the small-scale credit provided by pawnshops, plus an increasing density of the market network, partially underpinning some new institutions like the long-distance trading guilds linking merchants from a given locality. Characteristics of secondary, but still substantial, importance included the adoption of a number of crops from the New World, the cultivation of spring-ripening crops like broad beans and colza in rice-fields, and many small but useful technical improvements. The main argument against labelling this complex of developments a 'revolution' is that all of them, as I argued in 1975 when describing the complex of technical advances during this period, probably had the effect of stabilizing the existing system in a period of strains caused by population growth and diminishing amounts of technologically accessible natural resources. It did not create a qualitatively new pattern of economic activity.[41]

Rural social structure apart, most of the central features of the Song economy remained clearly visible in this second phase, even if it had been without question modified significantly over the centuries. In both periods economic society was vibrantly entrepreneurial and competitive, and with a high social mobility for premodern times, clearly greater in the second period. Motive power in both phases was based mainly on human labour, with waterpower and animal-power at times playing a useful ancillary role, while wind-power was a limited but interesting addition in the second period. In the management of medium-scale hydraulic systems, consortia of major landlords played a significant role in the earlier period, but a much-diminished role in the second.

As regards fuel, let us note, as a brief excursus, that coal was known, used, and valued from the beginning of the first period, but coal-mining technology remained unnecessarily primitive. The pits are sometimes said to have become too difficult and costly to work once they had gone down to a certain depth, and this presumably implied a need for better drainage, ventilation, and lifting

---

[41] Elvin (1996).

gear, and the protection of well-placed pit props. Some of the machinery needed for the first three operations already existed in China by the early eighteenth century, such as piston-and-cylinder pumps, and all of it before the middle of the nineteenth. What is interesting is that it seems never to have been used for the extraction of coal, even under the pressure of rising prices in eighteenth-century Beijing, where the fuel was regarded as a necessity of everyday life.[42] Late-imperial China's lag relative to early modern Europe in the cost-efficient extraction of coal and metal ores is interesting in that it is almost the only case of potential major economic importance where the low level of the technology used cannot be wholly ascribed to the absence of early modern science. The techniques were still at a point before the steam engine could have made a crucial difference. Nor was the cause environmental factors since there were substantial quantities of coal, seemingly accessible with only modest improvements in technique; and deposits of coal and iron were not in fact in economic terms always that far from each other, provided water-transport was available, which it often was.[43] The failure was probably one of uncertainty as regards government policy. The government sometimes asked merchants – people with capital and some knowledge of what was going on in the country as a whole – to take the lead in opening coal mines,[44] but often also tried to limit their numbers out of fear that, in spite of the acknowledged benefits of coal, the concentrations of miners might provoke local disturbances, especially if the mines were in the hands of 'bad gentry' or 'traitorous merchants.'[45] Yet using the somewhat better technology already available might actually have helped reduce this problem by creating longer-lived and more profitable businesses.

To return from our excursus: When China's rural economy over the millennium from the Song to the Qing is contrasted, in the broadest terms, and without reference to interregional differences, with that of late mediaeval and early modern Europe, three further mutually reinforcing differences need to be taken into account. (1) China had virtually no fallowing. (2) Other than near the frontiers, there were few herds of large animals, as opposed to handfuls of individual beasts such as cows and sheep, plus scavengers like pigs and poultry.

---

[42] *QSLJJZLJY*, 294 and 206. On the technology, see Elvin (1996), 67–68, 84, 88–92.

[43] Elvin (2001) gives more details.

[44] *QSLJJZLJY*, 203–204 gives a case for 1761 from Hami in the far northwest, noting that coal there was 'assuredly much more economical than firewood.' Page 296 gives another case in 1780 for Miyun not far from Beijing.

[45] *QSLJJZLJY*, 209 from Shanxi province in 1856 describes the breakdown of the state licensing system, extortionate behaviour by local government clerks, and the illegal reopening of mines that had been sealed up. The phrases in quotes come from the following item, from Guangdong province, which dates from 1863, and is a forceful expression of central government fears. Allowance has to be made for both of these being in the disturbed period of the mid-nineteenth-century rebellions.

(3) Compared at least to England and Wales, there seems to have been a lack of skill in managing woodlands as sources of a continuing supply of timber and firewood.[46]

That the Song was approximately the changeover point for the first two of these points emerges from some remarks of Chen Fu, the Song-dynasty agronomist:

> When the fields were divided up in ancient times, it was the system always to have some fallow land for pasture.... In this way stock-raising and grazing were properly provided for. The animals were large and fat, and did not suffer from skin infections.... In later times there have been no fallow lands for pasturage, and since then [cattle] have not been properly provided for.[47]

Tree plantations were known in mediaeval China, if we can include orchards under this heading. Plantations for same-age same-species timber-trees appear in the second of our two periods, most notably those of *Cunninghamia lanceolata* ('Chinese fir'). Timber-based local economies could be found in late-imperial times, sometimes run by large commercial companies with sophisticated premodern machinery. Most of these seem, however, to have been based on straightforward plundering,[48] resembling more the earlier days of the American lumberjacks than Gartenbau, though in a few places one can also find reforesting.[49]

Running through the last thousand years, at a variable speed, but only rarely stopping entirely, was the long-term development of the ever-more sophisticated agricultural intensification that is our central theme. The core of this process was the production, on the basis of a probably slightly falling area of lowland land per person, of a more valuable household output through (1) technological refinements like improved seeds or more efficient pumps, (2) subtle experience-based environmental adjustments to specific local conditions of terrain, vegetation, fauna, water-supply, and soils, and, annually, to the variations in the prevailing weather, (3) symbiosis and synergy in the choice of crops and their uses (including conservation of the fertility of the soil) notably by means of rotations, multi-cropping, and intercropping, (4) optimal use of available

---

[46] Both coppicing and pollarding were known in late-imperial China, but there is insufficient evidence to establish how widely they were used. It seems that the Chinese generally treated trees somewhat similarly to crops as regards planting and harvesting. See Nicholas Menzies (1985), especially pages 621–623 and 626–629. On the developing shortages of both timber and firewood, see Elvin (2004), chapter 3. For the British Isles, see Rackham (1986).

[47] Elvin (1996), 26.

[48] Elvin (1993), 36–37, and id. (2004), 37–39.

[49] Elvin (2004), chapters 3, 4, and parts of 9.

family labour-power and skills, including both women and children, the details depending on the family's current position in the family age-cycle, but also comprising such non-farm components as spinning and weaving as contributions to home-based artisanal industries, and (5) the exploitation of the market for the sale of farm and handcraft produce and of off-farm labour, plus the purchase of inputs like fertilizers and raw materials for home industries, such as raw cotton or cotton thread, and, finally, the renting in or out of assets like farmland and draught animals. Household planning had also to incorporate provisions for meeting the land tax, labour-service obligations on males, and the cloth tax on women. Finally, as the environmental buffer against natural and man-made disasters diminished with the clearing of ever more land, and the loss of woods, reducing access to wild plants, birds, fish, and game, an economic buffer to some extent replaced it – even if never fully adequate – as the meshes of the market economy grew finer.

Agriculture was thus embedded in an economic, social, fiscal matrix that extended far beyond the domain of farming in its restricted sense.

Chinese intensification in a comparative context

What did intensification mean in practice during the final period of the late-imperial economy, let us say roughly from 1700 on?

Since the Chinese normally thought of yields in terms of the volume of a given crop per unit area, it is not easy to find materials that let one estimate the crop in terms of seed sown to yield harvested. We have, for example, a reference from 1730 to a seed-to-yield ratio for wheat, *Panicum crus-galli*, and other dry-land crops in the far northwest of between 1:13 and 1:14,[50] and a more detailed account of about 1:36 for wet-field unhusked rice sown to husked rice put in the barn for Jiaxing some time in the eighteenth century.[51] There are, however, so few data of this sort that there is no reliable way at present of knowing how typical these ratios were. I would guess that the second represents good but not necessarily best practice for its area.

In the southern half of China a large proportion of fields were double-cropped, or even triple-cropped. Thus the average total annual output of a given area of farmed land was considerably larger than just its main cereal crop alone. In Table 11.1 (p. 274 above) I have rearranged in a simplified form the more important items in a southern farming calendar published by Bao Shichen in 1844 so that the possibilities for double-cropping are easy to see. The underlined entries show the time of planting in the final location (ignoring any prior

---

[50] QSLJJZLJY, 73.
[51] Translation of the data and numerical reconstruction in Elvin (2004), 208–209.

preparation in seedbeds) and the entries in bold type that of harvesting. The calendar is cyclical, and the vertical lines indicate the months when the crops were in the fields. Thus rice on the one hand and colza (*Brassica rapa*) or broad beans on the other could have been grown in the same fields at different times of the year. The same applied to late wheat and barley on the one hand and soybeans, millet, and maize on the other. It should be borne in mind when looking at this table that the sowing and cropping dates would have been somewhat variable, depending on the location of the farm and the year's weather. It is also worth noting that for several crops manure was applied in the course of the growing season, which seems to have been a new refinement during the second period. (It is marked with a '*' in Table 11.1.) Bao's comments on the triple use of land are worth quoting:

> Two harvests [of rice] are common in the southern regions. For the first harvest [the farmers] transplant the rice-shoots just as for early-ripening rice elsewhere. In the middle of the sixth lunar month, ten days before reaping, they broadcast seed beneath the stalks of grain ['Rice2' in Table 11.1]. When the first crop is cut, the sprouts of the second are four to five inches high. They are weeded with a hoe, as in dry farming. At the end of the eighth lunar month they are harvested, and barley planted as usual.... After the early-ripening rice has been harvested it is [also] possible to plant buckwheat for reaping in the eighth lunar month, and before this harvest to scatter-sow mud-beans beneath the grain-stalks in the same way as the second rice crop is sown. So long as there is plenty of fertilizing power, the fields are not damaged.[52]

Speaking for the moment only of settled lowland agriculture, since the farming and especially the transient cultivation of the uplands is still poorly documented, when Chinese nonstop multi-cropping and the continuing use of fallowing in northwestern European farming at this period are further taken into account, it is evident that, as of the eighteenth century, south China by a large margin, and northern China by a more modest margin, must have surpassed Europe both in cereal seed-to-yield ratios averaged over a run of years (rather than as measured by individual harvests) and in the mean worth of farm output per unit of farmed area per year, using any reasonable measure of value. It is far from clear, however, if China was ahead or behind northwestern Europe on yields per hour of human labour. My guess would be that it was at least slightly behind, possibly substantially behind, because of the need to devote many more hours to the collection and preparation

---

[52] Elvin (1996), 75–76. Rough Western dates are in Table 11.1.

of the large quantities of fertilizer needed to maintain the productivity of the soil, and also in the South to dredging the irrigation and drainage channels and repairing their fragile mud walls, as well as those of the barrages of water-control systems. Figure 11.1 suggests the relatively wasteful use of around a third of the farmers' energy, but this question is, in a serious sense, still undecided.

It is fair to say that in the domain of agricultural intensification, the foundation of many other social and economic feature of late-imperial China, and the source of many late-imperial environmental pressures, the divergence between China and Europe had been a long time in the making.

Complementary considerations

Chinese commentators in Song times from the economic heartland in the lower Yangzi valley often associated the low technical level of farming in relatively remote and backward parts of the empire with a low local density of population. Thus in the twelfth century, one scholar wrote that in what is today Hubei province, 'The land is thinly populated, and they are slipshod in their methods of cultivation. When they sow, they do not plant out, a practice which is popularly known as "diffuse scattering".' Another commented on the same region that, 'The land is so sparsely populated that they do not have to bestow any great effort on their farming. They sow without planting or weeding. If, perchance, they have weeded, they do not apply manure.... They cultivate vast areas, but have poor harvests.'[53] It is likely that what appeared as carelessness to observers from the more developed regions was in fact efficiency in the use of labour-time. In the context of what has been suggested earlier about the probable pressures from unit-area taxation, it would be interesting to see if it can be determined if, in this early phase of rice cultivation in such places, the state was still imposing quotas per unit-area that reflected an earlier, and less-developed, stage of agriculture, and, if so, how they in due course undertook to increase them.

Near the other end of the approximately millennial development of Chinese Gartenbau, we have the well-known contrasting picture painted by the Jesuit fathers in the middle of the eighteenth century:

Many farmers have a refined knowledge of weather and time, or, in other words, of the sequence of seasonal change, each one of them as it applies to his own small area.... The little extra efforts and knacks, inventions and discoveries, resources and combinations, which have caused people to exclaim

---

[53] Elvin (1973), 120–121.

at miracles in gardens, have been transported on a large scale out into the fields, and have done marvels....

They are not content to determine what sort of manure is suitable for each soil. They go on to desire that account be taken of what has been harvested, and what is to be sown, of the weather that has gone before, and that chosen [for a particular operation]....

That which has struck us most forcibly in our observations of the Chinese is that man who farms one acre instead of three is neither short of time, nor overwhelmed with work.... It is by these means that heavy toil is kept to a minimum, and the small cares of cultivation become the norm, wherein he is assisted by the women and the children.... [The small lands] of peasant proprietors are of an astonishing fertility, in contrast with the large estates.[54]

The good fathers were inclined at times to romance a little about the virtues of the Chinese system,[55] but they were acute and sinologically literate observers. Their remarks about the part played by women and children in farmwork can be supported for the southern half of China by contemporary local gazetteers, poems on everyday life, and, for the early twentieth century, by photographs of women in the fields.[56] This use of female labour in farming, though at variance with ancient ideological norms, was an economically rational response to the perceived need for intensification.

The small size of the holdings owned by peasants in late-traditional times has often been associated with Chinese partible inheritance of lands among male heirs, and Muramatsu Yûji long ago documented for the close of the empire and the early years of the Republic the downward pressure on holdings of inherited lands caused by dividing them up in this way.[57] There is truth in this, but, for it to have any strong effect on intensification, the partible inheritance system needs to be placed a historical context of land shortage. When land can be opened easily or rented cheaply, it is labour power that is the determining factor in farm-family economics. It is also arguable that the most significant size as regards intensification is that of the unit of operation, rather than ownership as such, and this will be significantly affected by the labour power available to a family at any given moment, and its related renting in and out of land.

---

[54] Commented on in Elvin (1996), 13–14.
[55] As one of them so much as admitted, with a self-critical smile. See Elvin (2007), 14.
[56] Elvin (2004), xxvii, 209–211, and 213; Elvin (1998), 160–165; *Zhongguo nongcun diaocha ziliao wu zhong* [Materials from five surveys of Chinese rural villages] I, plates on unnumbered pages of photographs before the table of contents, sheets 16 and 17.
[57] Muramatsu (1949), esp. 305–306.

Last of all, to what extent were population dynamics – the age-pattern of female marriages, the age-pattern of mothers when their children were born, and the age-specific death rates – an independent variable? Recent research has shown three distinctive features for the lower Yangzi valley in the mid-Qing: a very high infant as compared to adult mortality, a mean age at marriage for females of about 17 years of age with cases clustered very tightly around the mean, and the ability to limit the birth rate within marriage after a fairly fast start to childbearing.[58] These suggest a straightforward historical interpretation of the celebrated Qing surge in population, insofar as it was not an artefact of improved statistics:

> The social impact of the desire to quickly and securely replace the household labour force increased its hold on the region as rural servitude (more preva- lent in south China than the north) weakened and then virtually vanished by the early Qing. Women and children were also of more use at an early age in high-intensity Gartenbau than in the older more extensive styles of agriculture. Hence support from the older generation for the almost univer- sally early female first marriage. A fairly rapid start to childbearing aimed at replacement as soon as possible before anything went wrong, but would be eased off once two or three children had reach an age between five and ten years, when child mortality had dropped dramatically, and the future was relatively assured. The Chinese were aware that too rapid childbearing, without spacing, was good for the health neither of a mother nor her chil- dren,[59] and also, we can probably assume, that more than a fairly limited number of children could have damaging effects on the household budget. For the moment this pattern can only be demonstrated as applying to the lower Yangzi valley, but here it fits in plausibly with everything relevant we know or can reasonably conjecture.

The theoretical model

The patterns sketched in the foregoing sections can be condensed into a rudi- mentary formal model. Figure 11.1 shows a hypothesized pattern of rising yields in a given unit area ('field') for increasing inputs of hours of labour. The 'non-intensive' and 'intensive' farming form a continuum, but a moment

---

[58] See Elvin and Fox (2009); and Elvin and Fox (2008). For data and programs, see Mark Elvin, Josephine Fox, and Tzai-Hung Wen, 'Qing Demographic History. The Lower Yangzi Valley in the mid-Qing', website http://gis.sinica.edu.tw/Qing Demography.

[59] Angela Leung, 'Autour de la naissance: la mère et l'enfant en Chine aux XVIe et XVIIe siècles,' *Cahiers internationaux de sociologie* 76 (1984).

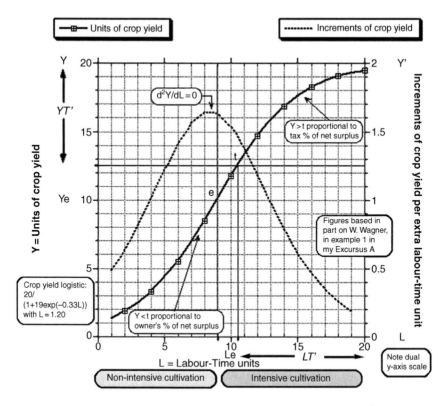

Figure 11.1 Illustrative relationship for North China in the early twentieth century of the input of owner-farmer labour-time to crop yield from a given area under premodern conditions

of transition from the first to the second can conveniently, if arbitrarily, be defined as the moment when the rate of growth of the continuous curve showing yields for the input of labour-time, having previously increased steadily, starts to decrease. Here the inflection is around e at point (9, 10.128) on the L and Y axes. This transition can also be seen more dramatically in the dashed curve which represents the size of the increase in yield for the last time-unit of labour input up to that point. Note that, in order to make it easier to read, it follows a different scale along the right-hand margin, though the units of measurement are the same.

The significance of this definition is this: if cultivable land were freely and readily available, and labour was the only input that was limited, and the farmer had a total of 18 units of labour-time available to him, then, since 9 units produce a yield of 10.128 units but 18 labour-units, applied to one-and-the same field, only 19.047 units of output, this transition point would roughly be where he would do better to shift half of his labour to an additional second field, so his total yield is 20.256 units. Of course in real life considerations of the actual or implied rental costs of the second field, and other factors, intervene.

The curve used here is a logistic, but it needs to be stressed that, while this is plausible, it is not based on empirical data. The parameters selected, and listed in the lower left-hand corner of Figure 11.1, are also plausible but formally arbitrary. The calculations for matters related to taxes are, however, based on data from the first case from Wagner cited in Excursus A. Another caution that needs to be made is that the curve shown is not symmetrical, first appearances notwithstanding. It is also likely to have been very significantly different in varying real-life circumstances, according to particular environmental, technological, and socio-economic factors.

Nonetheless, the basic economic logic, which it can now be used to illustrate, is simple:

The main curve (the solid line that goes from the lower left to the upper right) represents the output from the unit of land. It is represented in units of yield along the vertical or Y axis. It can be thought of either as physical output of a single crop or, more usefully, as the cash value of a complex mix of crops as shown in Table 11.1. As the input of labour increases (measured along the horizontal or L axis) the rate of yield for an extra input of a unit of labour falls towards zero. (A maximum possible output for given technology is set notionally at 20.) The ratio of extra marginal input to extra marginal output (dY/dL) rises to the point $L = 9$, where it becomes momentarily stationary ($d^2Y/dL = 0$), and then gradually decreases, as equal amounts of work bring in less and less return. We have defined this (that is, e in the graph) as the point of transition from non-labour-intensive cultivation to labour-intensive farming. Obviously the actual transition will have been less clear-cut in real life than this. The input of just a moderate amount of extra labour after point Le will bring in more in the field already opened up than just starting a new field whose cultivation is not done very thoroughly.

It was apparent, however, from the first of Wagner's cases that it is the net profit that is likely to have the strongest impact on peasant motivation, and not the total output. We have therefore to reconceptualize the meanings of the diagram. Net profit before tax was in this case about 15.88% of the total income. After tax it was about 12.5%. On the graph the amount of output needed for tax out of the pre-tax profit is shown by the arrowed line YT' and the amount of labour needed to produce it by the arrowed line LT'. Seen in this context the tax was equivalent to about 36% of the pre-tax profit, and a very serious burden. This farm was a relatively prosperous one, and the situation for most tenants was fairly certainly much less favourable.

A moment's examination of the scored line, showing the 'momentary'[60] ratio dY/dL, in the areas to the right of point t, shows – at least in one quite reasonable perspective – that the effect of the need to pay tax under these circumstances is 9.5 units of work, just below half of the total labour input, at a steadily falling efficiency as regards the production of output. Looked at another way, we could say that for an owner-farmer buying more land (valued in this area at this period at 2.8 times its proper economic cost in terms of annual output) in order to increase the productivity of labour would seem to have been to a significant degree checked by the loss that this would entail for ordinary farmers, without social immunity from overcharging, from tax levied on the basis of the farmed area.

To the extent that rent was paid on a per-unit-area basis the same would have held, *mutatis mutandis*, for tenants. Actual conditions, as has been noted, varied more widely as regards payment of rents than of taxes. Speaking generally though, rent would always have tended to have exceeded tax by a significant margin to allow the landlord, assuming he was effectively taxed, his own margin of profit.

Conclusion

The terms on which farmland was taxed thus seems likely to have had a more than trivial impact on the environment. If the hypothesis presented in this chapter stands the test of time, it seems hard to avoid the conclusion that penetrating to the heart of the origins of China's most pressing long-term environmental problem, namely her population, huge relative to her resources, and based on intensive agriculture, requires going far beyond environmental history as ordinarily conceived in order to do environmental history.

Much more work is of course needed on the earlier centuries of this process, during which the pattern described in these pages was originally crystallized. I have already begun to assemble materials for this, but it is a slow business.

Tarago, New South Wales 2007, and
Long Hanborough, Oxfordshire 2008

---

[60] The inverted commas are a reminder that in fact the calculation is based on steps of finite size and is not in fact continuous.

References

Amano Motonosuke (1962), *Chûgoku nôgyôshi kenkyû* [Studies on China's agricultural history], Tokyo: Ochanomizu shobo.

Elvin, M. (1973), *The Pattern of the Chinese Past*, Stanford: Stanford University Press.

Elvin, M. (1982), 'The Technology of Farming in Late-Traditional China,' In R. Barker and R. Sinha (eds), *The Chinese Agricultural Economy*, Westview: Boulder CO.

Elvin, M. (1993), 'Three Thousand Years of Unsustainable Growth: China's Environment from Archaic Times to the Present.' *East Asian History*, 6.

Elvin, M. (1996), 'Skills and Resources in Late Traditional China,' In D. Perkins (ed.), *China's Modern Economy in Historical Perspective*, Stanford CA: Stanford University Press, 1975, Reprinted in Mark Elvin, *Another History: Essays on China from a European Perspective*, Sydney: Wild Peony.

Elvin, M. (1998), 'Unseen Lives: The emotions of everyday existence mirrored in Chinese popular poetry of the mid-seventeenth to the mid-nineteenth century,' In R. Ames, T. Kasulis and W. Dissanayake (eds), *Self as Image in Asian Theory and Practice*, Albany NY: State University of New York Press.

Elvin, M. (2001), review of K. Pomeranz, *The Great Divergence. China, Europe, and the Making of the Modern World Economy*, Princeton: Princeton University Press, 2000, in *China Quarterly*, 167, Sept.

Elvin, M. (2004), *The Retreat of the Elephants: An Environmental History of China*, New Haven, Yale University Press.

Elvin, M. (2007), 'Economic Pressures on the Environment in China during the 18th Century Seen from a Contemporary European Perspective: Insights from the Jesuit *Mémoires*,' In Shiba Yoshinobu (ed.), *Tôyô Bunko hachijûnen-shi* [Eighty years of the history of the Tôyô Bunko], 2, Tokyo: Tôyô Bunko.

Elvin, M. and Fox, J. (2008), 'Local Demographic Variations in the Lower Yangzi Valley during Mid-Qing Times,' In T. Hirzel and N. Kim (eds), *Metals, Monies, and Markets in Early Modern Societies: East Asian and Global Perspectives*, volume 1, Bunka-Wenhua. Tübinger Ostasiatische Forschungen. Berlin: Lit Verlag.

Elvin, M. and Fox, J. (2009), 'Marriages, Births, and Deaths in the Lower Yangzi Valley during the Later Eighteenth Century,' In Clara Ho (ed.), *Windows on the Chinese World. Reflections by Five Historians*, Lanham MD: Lexington Books.

Elvin, M., Fox, J., and Wen, T.-H., 'Qing Demographic History. The Lower Yangzi Valley in the mid-Qing,' website http://gis.sinica.edu.tw/QingDemography.

Huang, R. (1974), *Taxation and Government Finance in Sixteenth-Century China*, London: Cambridge University Press.

King, F. (1911, revised 1949, reprinted 1972), *Farmers of Forty Centuries*, London: Cape.

Leung, A. (1984), 'Autour de la naissance: la mère et l'enfant en Chine aux XVIe et XVIIe siècles,' *Cahiers internationaux de sociologie*, 76.

Maddison, A. (2007), *Chinese Economic Performance in the Long Run*, Second edition: Paris: Development Centre of the Organisation for Economic Co-operation and Development.

Marks, R. (1998), *Tigers, Rice, Silk, and Silt: Environment and Economy in Late Imperial South China*, New York: Cambridge University Press.

Menzies, N. (1985), 'Forestry,' In J. Needham, *et al.* (ed.), *Science and Civilisation in China*, volume 6, Part 3, Cambridge: Cambridge University Press.

Muramatsu, Yûji (1949), *Chûgoku keizai no shakai taisei* [The social structure of the Chinese economy], Tokyo: Tôyô keizai shimpôsha.

Nankai University Department of History (ed.) (1959), *Qing shilu jingji ziliao jiyao* [Digest of materials on economics from the Qing-dynasty 'Veritable Records': *QSLJJZLJY*], Beijing: Zhonghua shuju.

Rackham, O. (1986), *The History of the Countryside*, London: Dent.

Shiba Yoshinobu (ed.) (2007), *Tôyô Bunko hachijûnen-shi* [Eighty years of the history of the Tôyô Bunko], volume 2, Tokyo: Tôyô Bunko.

Wagner, W. (1926), *Die chinesische Landwirtschaft*, Berlin: Paul Parey.

Wang, Y.-C. (1973), *Land Taxation in Imperial China*, 1950–1911, Cambridge MA: Harvard University Press.

Zhongguo nongcun diaocha ziliao wu zhong [Materials from five surveys of Chinese rural villages] (1933; repr. Taiwan: Xuehai chubanshe, 1971).

12
Reconsidering Climate and Causality: Case Studies from Colonial Mexico

Georgina Endfield

Introduction: reconsidering climate and human history

It is nearly 50 years since Professor of Geography at Bedford College, London, Gordon Manley, reflected upon 'The revival of climatic determinism.' His paper, which appeared in the *Geographical Review*, was spurred by Gustav Utterström's work on 'Climatic fluctuations and population problems in Early Modern Europe.' For an Economic Historian accustomed to taking into account the multifarious factors that could influence human well-being, Utterström's focus on the role of climate was, in Manley's eyes, important. It has been suggested that, with the exception of the French Annales School, Historians had hitherto been somewhat indifferent towards studying the impacts of climatic change on society.[1] Utterström's was a radical argument to make within the field of Economic History at the time. Yet it was also controversial for the Geographical community, whose science had arguably fallen 'on hard times' in the stagnating wake of some of the extreme views of environmental and climatic determinism in the late nineteenth- and early twentieth century.[2] Utterström's work, though later criticised,[3] would in fact figure among a trenche of more nuanced climate histories, which, eschewing generalisations about the significance of climate in world history, did nevertheless highlight the importance of climatic factors for societies. Theirs, however, was a more tempered and redefined climatic determinism.[4]

---

[1] Grove (1988).
[2] Blaut (1999).
[3] Le Roy Ladurie (1972).
[4] For example, Le Roy Ladurie (1972); Appleby (1981); Pfister (1978); Lamb (1969); Lamb (1977); Lamb (1982); Bowden *et al.* (1981); Wigley *et al.* (1981); Ingram *et al.* (1981); Bryson and Padoch (1980).

Building on this work, a new breed of scholar interested in climate–society interactions in history has emerged.[5] By employing rigorous and innovative methodologies, these studies have successfully moved forward debates about the role of climate in human well-being. Such pioneering work is not only providing detailed regional climate histories but is offering important insights into how societies have coped with, have adapted and responded to climatic variability and weather events in the past through examinations of 'the subtle patterns of social effects rather than chronicling climatic crisis'.[6]

Anxieties over the emphasis placed on the role of climate in shaping human histories, however, have resurfaced in response to what Geographers, Coombes and Barber (2005), have referred to as a 'new paradigm', in climatic determinism. 'The past decade' they argue 'has seen a revival of environmental determinism' (p. 303), manifest in a number of palaeoenvironmental studies in which abrupt climate change is invoked to explain catastrophe and civilisation collapse. But many of the cases Coombes and Barber cite as being exemplary of this 'new' determinism, including research by well-respected scholars de Menocal (2001) and Hodell *et al.* (1995), could equally be upheld as offering ground-breaking insight into human vulnerability in the past. Moreover, these studies acknowledge the importance of considering social, cultural, demographic and political as well as climatic parameters and contexts that might have contributed to cultural discontinuity, making a clear case that a major climatic event is not necessarily the absolute cause of discontinuity or 'collapse'. Nevertheless, the branding of such work as 'deterministic' both betrays and helps to promote a pervasive fear within an academic community oversensitised to the potential pitfalls of engaging in historical climate–society research.

Such histrionics might result in something of a reluctance to engage in climate history investigations.[7] Yet, in a context of ever-increasing scientific and public concern over predicted climate change, and at a time when the effects of a variety of extreme and unusual weather events are being felt across the globe, and often tragically so, the need to investigate the way in which society is, has been and will be affected by climate is more pressing than ever. While it is now generally agreed that direct, one-to-one cause and effect relationships between climate and human action have provided something of a 'false lead' (McCann, 1999), this should not preclude studies which explore the human implications of climate change or variability. Furthermore, there are many key

---

[5] See, for example, Pfister and Brazdil (1999); Brazdil *et al.* (2005); Barriendos (1997); Barriendos and Danneker (1999); Brazdil *et al.* (2005); Grove (1988); Jones *et al.* (2001); Pfister *et al.* (2001).

[6] McCann (1999), 261.

[7] McNeill (2005), 179.

issues in this respect which still need to be addressed. Although there is little doubt that climate change poses one of the greatest threats to humankind worldwide, the precise impacts of predicted changes for different social, economic and ecological systems are less-clearly understood,[8] and the mechanisms by which climate change can influence societies remain obscure.[9] It is also vitally important to assess the degree of meteorological adversity that could precipitate serious social and economic dislocation in any given context, and to identify what types or levels of climatic stress impose the greatest obstacles or indeed present opportunities for successful adaptation and response within different communities.[10]

Knowledge of successes and failures in adaptation to historical climatic variability could provide clues as to the range of coping and adaptive strategies available to a particular society and may ultimately increase the ability to respond appropriately to the threats of predicted climate changes.[11] This chapter explores how society in Mexico has been affected by, but has also adapted and responded to, climatic variability and extreme weather events in the past. Attention focuses specifically on the implications of drought which, by affecting food security and social and economic well-being, has played a pivotal and influential role in Mexican history and pre-history.[12] Various colonial archival sources[13] are used to identify and then to explore the impacts of and responses to seasonal expected drought, single-year and multiple-year drought events in three different case study regions of Mexico in the colonial period (1521–1821). The case studies cover a variety of environmental, social, economic and political contexts and histories and located at key points along a north–south rainfall gradient. The Conchos Valley of southern Chihuahua in the arid north (average annual rainfall 350–400 mm per annum), the Valley of

---

[8] Schneider (2001).

[9] Hassan (2000).

[10] Ingram *et al.* (1981), 39; IPCC (2007).

[11] De Menocal (2001), 667; McCarthy *et al.* (2001); Meyer *et al.* (1998).

[12] Hodell *et al.* (1995); Hodell *et al.* (2005a); Hodell *et al.* (2005b); Conde *et al.* (1997); Liverman (1999); Florescano (1980).

[13] Documents were consulted in the following archives: Archivo General de la Nación, Mexico City (AGN); Archivo General del Estado de Oaxaca (AGEO); Archivo Historico Municipal de la Ciudad de Oaxaca (AHMCO); Archivo Privado de Col Bustamante Vasconcelos, Papeles de San Bartolo; Archivo Historico Municipal de Leon, Guanajuato (AHML), Archivo Casa de Morelos (ACM), Morelia, Michoacán; Archivo Historico del Estado de Guanajuato (AHEG); Museo Nacional de Antropología, Archivo Historico (MNA. AH). In all cases, archives will be referenced in the following manner: the abbreviation of the archival repository, the document group (*ramo*) consulted, the volume or box (Caja) number, the expediente (if applicable), and / or the page (*foja*) number, (denoted by fa (single pages)/ fs (multiple pages). (Page numbers may also be accompanied by f (*frente*) facing page or v (*verso*) reverse page).

Oaxaca in the wetter south (1500 mm per annum) and west central Guanajuato located in the Bajío, area of the Central Highlands of Mexico, a region of climatic transition (650 mm–1000 mm per annum). Each region developed distinctive settlement and land use characteristics during the colonial period.

Guanajuato became the 'breadbasket' of the colonial political economy, specialising in the production of cereals, especially wheat and maize. Extensive use of water storage, diversion and water management systems in this region provided the irrigation water necessary for cereal production. Although irrigation proved a successful strategy to prevent harvest losses during the winter dry season and single-year drought events, it seems to have been less effective at guarding against losses associated with prolonged or successive droughts. Moreover, farming in this region grew progressively reliant on these irrigation systems throughout the colonial period, with water management reaching its zenith in the second half of the eighteenth century,[14] such that when droughts did occur they could be devastating and, as the main grain producing area of the country, harvest losses in this region had the potential to affect the country at large.

In Oaxaca, there was a good deal of indigenous land retention following conquest. As a result, agricultural production continued to focus on the production of maize, beans and chile though wheat and livestock were introduced soon after conquest. Oaxaca, however, became one of the main producers of cochineal dye for the whole of the Americas, the dye being produced solely by the indigenous populations and managed by Spanish or Creole financiers.[15] There are far fewer references to drought in the colonial records for this region. It is not clear, however, whether this reflects actual drought incidence as the degree of indigenous land retention and the cochineal economy may have provided the predominantly indigenous population with a opportunities for economic diversity, meaning that the impacts of droughts were simply less significant or noteworthy.

In Chihuahua, Spanish colonisation was delayed by the coupled problems of frequent drought and indigenous unrest, though by the seventeenth century, there had emerged a lucrative livestock and mining industry across the region. Droughts in this region were regular, prolonged and severe and caused many problems for both sectors of the economy. They may in part have also triggered some of the instances of unrest noted in the archival sources.

The archives for these three regions reveal that a number of strategies were devised to address inter-annual fluctuations in seasonal rainfall and single-year droughts in pre-Hispanic Mexico, many of which were maintained into

---

[14] Murphy (1986).
[15] Baskes (2000).

the colonial period. These adaptations proved successful in buffering society against normal, expected (seasonal) periods of drought and individual years of hardship, but proved ineffective during more prolonged droughts, or when the impacts of drought were exacerbated by contextual circumstances, or were combined with other coeval natural hazards. Such episodes of crisis tested social resilience and, at times, exceeded the capacity for routine adaptation, resulting in widespread social, economic and ecological crisis. Yet they also at once afforded opportunities for greater social cohesion and social instability. Thus while periods of harvest crisis stimulated immediate responses at the institutional, community and individual level, ranging from the marshalling of social networks and projects geared towards community engagement to humanitarian aid and individual acts of charity, in some situations they also resulted in the disturbance of social harmony, legal conflicts, riots and unrest. As shall be highlighted, experience of climate variability at a range of scales may have nonetheless contributed to fundamental developments in agrarian experimentation, urban infrastructure and environmental health.

Living with drought: long-term adaptation and 'the subsistence ethic'

There existed a wide variety of traditional-risk avoidance and adaptation strategies in pre-Hispanic Mexico designed to hedge against the impacts of drought.[16] Irrigation represents one of the most obvious examples in this respect and different cultures across the country developed quite sophisticated systems to manage, store and transport water for this purpose.[17] Farmers practiced mixed farming and may have also traditionally kept seed and grain[18] harvested from their own crop from one season to the next, a strategy that prevails in some parts of the country today.[19] The organised storage and trade of grains, food products and other saleable assets was another subsistence strategy representing an important buffer against expected seasonal shortages and a strategy that was practiced both in pre-Hispanic Mexico and also in Spain on the eve of conquest.[20] Periods of agrarian crisis also spurred a range of non-economic, sometimes reciprocal, relationships between patrons and clients, rich and poor. In the Valley of Oaxaca, for example, food deficits in one location were often met by transporting goods produced elsewhere in the region,[21] while the Aztec

---

[16] Wilken (1987).
[17] Doolittle (1990).
[18] Grain refers to kernels in general and can be used for consumption purposes, for animal feed or to trade. In contrast, seed is specifically for planting.
[19] Badstue *et al.* (2006).
[20] Blaikie *et al.* (1994), 66.
[21] Sanders and Nichols (1988).

Empire distributed maize from central granaries during periods of harvest failure and food scarcity. Such moral economic responses offered, in Scott's (1976)
terminology, a 'subsistence ethic' in that they were intended to provide a minimum level of subsistence and food security during times of want. These stores
and trading networks, however, proved to be insufficient during the most severe
times of prolonged dearth. The protracted drought of 1450–1454, for example,
was so severe that more desperate survival strategies were resorted to. People
'sold' themselves and their children in labour and slavery,[22] and were forced
to turn to so called 'famine foods' and liquids derived from cacti, agave and
mesquite fruit.

Many of these adaptive, coping and survival strategies continued to operate after conquest. There was an expansion in the use of irrigation, not least
because of the Mediterranean varieties of what that the Spanish introduced
required water during the dry winter months. Permanent and ephemeral
watercourses, rivers and arroyos all began to be exploited and it is clear that
groundwater was also tapped in many locations, though perhaps more so during the later colonial period.[23] Flood waters were stored for use during the dry
season, a strategy referred to in the documentation as medio riego.[24] Actual
drought and water scarcity may have necessitated co-operative efforts between
different sectors of the community. This seems to have been the case in areas
where demands for water for irrigation were high. There had already been
substantial water management for irrigated wheat in the Valle de Santiago in
southern Guanajuato by the beginning of the seventeenth century.[25] Water was
managed effectively between different stakeholders in the region.[26] By the second half of the seventeenth century, water was divided between landowners
according the 'tanda' system, which consisted of shifts of 'turns' in the use
of irrigation water, usually measured in days of water.[27] Water judges were
appointed to adjudicate fair water distribution according to the agreed proportions and water continued to be distributed relatively efficiently between
different users, though, as was the case with other regions of the country and
elsewhere in Central America,[28] there was bitter legal wrangling over water
rights, particularly in the eighteenth century.

---

[22] Hassig (1981); Therrell *et al.* (2004).

[23] See, for example, numerous references to 'waterlifts' or *norias* in AGN Historia, Vol. 72,
Exp. 9; AGN Civil, Vol. 73, Exp. 3; AGN Tierras, Vol. 514, Exp. 1, Cuaderno. 2, Fa. 47;
AGN Tierras, Vol. 618, Exp. 1, Cuad. 3, Fa. 61; AGN Tierras, Vol. 1353, Exp. 1, f. 69.

[24] AGN Tierras, Vol. 2705, Exp. 3, Fa. 1; AGN Mercedes, Vol. 10, Fa. 3.

[25] Gonzales (1904), 232–233; Murphy (1986), 65–66.

[26] AGN Tierras, Vol. 2959, Exp. 141, Fa. 20v.

[27] AGN Tierras, Vol. 2959, Exp. 141, Fs. 10v–16v.

[28] See, for example, Lipsett Rivera (1999); Endfield and O'Hara (1997).

The practice of maintaining a reserve of emergency food supplies, normally grains, to compensate for times of harvest loss and to hedge against price rises during periods of general scarcity together with the trading of foodstuffs from region-to-region remained important strategies for coping with localised harvest failures in the colonial period.[29] Government-sponsored grain markets or Reales Alhóndigas (royal grain stores) and granaries (pósitos) were also established in other principal cities across Mexico, but specifically in mining areas and close to the country's ports, those areas that represented the 'engines' of the colonial economy and where populations were dense. During years when harvests were poor or failed completely, the authorities, as Cope (1994:75) suggests, 'fearful of mob violence' stocked these city granaries as a matter of priority. In fact, actual experience of drought and harvest failure may have stimulated the establishment of a number of these facilities in the case study areas. The Alhóndiga in San Felipe el Real de Chihuahua (Chihuahua City), for example, was founded in the early 1730s after a period of worrying harvest failures in the mid-1720s.[30] Similarly, in Oaxaca in 1729, preceding a period of grain scarcity that would affect the whole of the region in the early 1730s,[31] requests were made to build an alhóndiga 'like those they have in all the cities',[32] in order to bring an end to the 'the hardships that the poor people suffer.'[33]

Where a Real Alhóndiga existed, all grain merchants were officially obliged to sell their produce first through the granary store in order to ensure that a fair price was then offered to the public. All the maize distributed by the Alhóndiga on a given day was sold at a set price agreed upon by individual sellers and reflective of prevailing market conditions.[34] Moreover, if merchants or farmers withheld grains for longer than this, in order to artificially create scarcity and so raise prices, they could be ordered to sell it at a specified price.[35] Speculation and profiteering, however, presented a real problem and became particularly problematic around the middle of the eighteenth century.[36] Price regulations and rationing of grain purchases were implemented in Leon in 1756,[37] and by 1772, similar measures had been introduced in Guanajuato, though this did little to resolve the problem. Deliberate forcing of grain prices, competition and insider dealing between merchants and the monopolisation and retention

---

[29] García-Acosta (1993).
[30] AHMCH Caja 2, Exp. 12.
[31] AHMCO Actas de Sesiones del Cabildo 1728–1733.
[32] Carga = load.
[33] Juan Bautista de Fortuño, AHMCO Actas de Sesiones del Cabildo 1728–1733.
[34] Cope (1994).
[35] AHMCH Gobierno, Actas de Cabildo, Caja 32, Exp. 5.
[36] AHML Bandos C5, Exp. 29, 1761.
[37] Leon, 1756, AGN Ayuntamientos Vol. 196, Exp. 2.

of goods remained problematic in the later eighteenth century.[38] Moreover, as had been the case in the pre-Hispanic period, the granaries and grain stores provided little succour during the most severe and prolonged years of agrarian crises. A document recorded in Chihuahua in 1758, for example, after several years of drought, hunger and disease had taken their toll on the population of the area, charts how 'because of the droughts experienced in all areas, the transporting and trading of seeds and flour to the Real Alhóndiga of the Villa [of Chihuahua City] has also dried up'.[39] Another note from December 1760 highlights the local council's concern over the 'scarcity of seeds which appears to have been presented in the Real Alhóndiga',[40] while in 1764, after the combined effects of drought and frost had severely reduced the harvests, the small quantity of maize that existed in the Real Alhóndiga was described as being of 'low quality' but was nonetheless commanding a 'very high price'.[41] Alhóndigas and pósitos thus appear to have been only partially successful in addressing food scarcities, price speculation and profiteering on the grain market during periods of prolonged hardship.

As the breadbasket of the economy, the Bajío was a region to which the government could and indeed did turn for assistance when harvests failed.[42] The Bajío remained an important source of grains even when more widespread harvest crises led to scarcity across the region itself. In 1692, after a calamitous year of failed harvests, produce was sought for the populations of Mexico City from Celaya, Salvatierra, Silao and the Villa de Leon.[43] Special purchases of 45,000 fanegas of maize were made, financed by a donation from the capital's wealthiest silver merchants.[44] On other occasions, there were severe problems of food provision in cities actually within the Bajío. Following drought and harvest failures in the vicinity of Guanajuato in 1782, however, individual landowners were identified who could potentially help alleviate public suffering within the region.[45]

By the middle of the eighteenth century, the organised trade in food stuffs from region-to-region and the purchase of grains by local government authorities from less-affected areas to supply more hard-pressed communities and populations had become an important coping strategy for drought-afflicted populations, particularly those in the urban centres of Chihuahua. Maize was

---

[38] AHMCH Gobierno, Caja 31, Exp. 9.
[39] AHMCH Guerra, Caja 2, Exp. 1.
[40] AHMCH Gobierno, Caja 32, Exp. 5.
[41] AHMCH Notarias, Abastos de Grano, Caja 45, Exp. 4.
[42] AHML Communicaciones 19.
[43] Cope (1994).
[44] AGI Patronato, Leg. 226, no. 1, r. 20, Fs. 3v–4v.
[45] AGN Alcade Mayores, Vol. 5, Exp. 8, Fa. 11.

regularly purchased and transported from the nearby valleys of Basuchil, San Buenaventure and Carmen when harvests failed in the vicinity of Chihuahua City.[46] In the case of more widespread prolonged droughts, however, trade in grains from location-to-location was often itself disrupted, effectively prolonging the crisis in some regions. It is to these episodes that I now turn.

'Want, hunger and calamity: multiple-year drought and coping with crisis in the eighteenth century

Although there are many direct and indirect references to weather and weather-related events recorded in the colonial records of Mexico,[47] not all caused dislocation or contributed to widespread crisis. Drought in any one year might in fact have had only negligible impacts and may have even gone unrecorded if, in the following year, a good harvest was secured. Successive extreme events, however, or individual events occurring in combination with other natural hazards or weather events, or against a backdrop of ongoing social, economic problems could have amplified and cumulative impacts.[48] The most devastating impacts resulted where prolonged or multiple periods of inhospitable weather or weather-related events compounded other difficulties, specifically population pressure and resource stress. While long-term adaptations may have failed during such periods, there was a range of short-term institutional, community and individual coping strategies and responses that were adopted or developed.

By the eighteenth century a whole suite of demographic, social, economic and political conditions coalesced to render Mexican society apparently more vulnerable to climatic variability. Comparing the archival references, it is possible to identify a series of particularly devastating, widespread agrarian crises in this period which may have been associated with climate or weather events. Subsistence crises reported in 1710, 1724–1725 and 1741–1742, for example, were preceded by drought and harvest shortfalls, while that of 1749–1751 was associated with a combination of drought and disastrous frosts which ruined the sown crops in western and northern parts of the country. Between 1770 and 1773 there was prolonged drought, which led to harvest shortfalls and high grain prices in the north of the country. Problems in 1781–1782, meanwhile, were preceded by three years of drought and epidemic disease,[49] while the well-documented crisis of 1785–1786, which because of its widespread and prolonged impacts and repercussions became known as the

---

[46] AGN Ayuntamientos 173, cuaderno 6.
[47] Florescano (1986); García-Acosta (1993); García-Acosta (1997); Endfield *et al.* (2004a); Endfield *et al.* (2004b).
[48] Wigley (1985).
[49] Cooper (1965).

'year of hunger' (see next section), was the result of a combination of delayed rains in the first few critical months of the newly sown crops, succeeded by two exceptionally severe frosts which destroyed the harvests in the centre, west and north of New Spain.[50] Swan (1981) highlights a rise in the price of grain around the turn of the nineteenth century which she links to abnormal weather patterns which affected crop production across the agrarian heartland of central Mexico, while the crisis of 1810–1811 was the result of a terrible drought which lasted for the whole of 1809 and 1810 and affected all cereal-producing regions of the country.[51] The climate events of the eighteenth and early nineteenth centuries were not unique or significantly different to previous events, but exposed the extreme social and economic inequality that characterised much of the country by this stage. As shall be illustrated by the following examples drawn from the 1750s and 1780s, such successive events challenged and even exceeded the society's capacity to implement routine adaptive or coping strategies.

A combination of drought and frost contributed to agrarian crisis across the Bajío between 1749 and 1751. There were harvest shortfalls in September 1749 and by early November that year, food shortages were being reported in some parts of the region. Prices soared to unprecedented levels and food riots were reported as having taken place as a result.[52] Dramatic loss of harvests and the death of livestock left many individuals, even wealthy land owners,[53] without any means of survival and with no alternative but to sell off their possessions, property and land.[54] Drought induced harvest failure might have also been responsible for reported food scarcities in Oaxaca between 1746 and 1748,[55] when public granaries were depleted, prices rose and the local administration in Antequera (Oaxaca City) was forced to provide food and financial aid for the poorer sectors of society who faced starvation.[56] Prolonged drought in the middle of the eighteenth century, however, appears to have particularly affected people in the more arid north of the country. Drought is recorded in the years between 1749 and 1752, in 1755 and again in the later 1750s. This had serious implications for pasture availability[57] and coincided with a period of livestock pestilence.[58] There were shortages of grains, especially maize and

---

[50] Florescano (1969), 56.
[51] Hamnett (2002).
[52] AGI Mexico 1506 1749.
[53] AHML Notarias 1750–1757, Fa. 24.
[54] AHML Notarias 1750–1751, Fs. 144–51.
[55] AHMCO Actas de Sessiones del Cabildo 1746–1748, Fa. 54.
[56] AHMCO Actas de Sessiones del Cabildo 1746–1748, Fa. 54.
[57] AGN Mercedes Vol. 76, Fs. 95–100.
[58] AHMCH Notarias Abastos de Carne, Caja 36, Exp. 12; AHMCH Guerra Caja 1, Exp. 7; AHMCH Notarias Abastos de Carne, Caja 36, Exp. 12.

populations faced starvation.[59] As the problems persisted in the north of the country throughout the 1750s and early 1760s,[60] epidemics and pestilence took hold of the weakened population,[61] which in turn led to labour shortages, an abandonment of some of the mining operations and contributed to the migration of hundreds of people. Epidemics were also recorded in other parts of the country at this time,[62] some of which were attributed to the desperate consumption of poor-quality famine foods by the most poverty stricken, starving members of society.[63]

Further crises followed only a few decades later. Observations reported in national newspapers in the early 1780s highlighted an awareness of changing climatic conditions at the time. The Gazeta de México of March 10, 1784 reported how, from 7 December the previous year, 'evil winds from the south and south east have blown with greater frequency.' Moreover, it seems that the Spanish crown might have anticipated the possibility of an imminent crisis only a few months later. On the 10 May, 1784, in what can now be considered something of a portentous act, a royal order had been issued in Spain requesting that 'all the heads of the Indies send each six months notice of the weather experienced in these dominions: noting whether the rains had been scarce or abundant and noting also the nature of the harvests of fruits and other produce'.[64] The order preceded one of the most widespread and devastating famines in Mexican colonial history. A combination of drought in 1780, 1782, 1784 and 1785 and frosts in 1784 and 1785 contributed to the so-called 'year of hunger' between 1785 and 1786.[65] This period of crop failure stimulated famine,[66] epidemic disease,[67] death[68] and economic retardation.[69]

There were efforts to reduce social vulnerability to climatic and agrarian disaster through the extension and consolidation of social networks and local, regional and national trade links. Food and grains were transported across much greater distances to help those most needy areas, particularly in the north

---

[59] AGN Mercedes, Vol. 76, Fa. 137.
[60] Archivo Arzobispado de Chihuahua, Ramo Gobierno y Administración; Cofradías, 1755, Caja 3, Serie 1.3.3.
[61] AHMCH Guerra, Caja 2, Exp. 4.
[62] Cavo (1949).
[63] AHMCH Notarias, Abastos de Carne, Caja 42, Exp. 12.
[64] AGN Bandos Vol. 13, Exp. 47, Fa. 160.
[65] Florescano (1969).
[66] Alhondígas 15, Exp. 1; Alhondígas 10; Tributos 20, Exp. 15.
[67] Tributos 20, Exp. 15, Exp. 1; Reales Cédulas 134, Exp. 179; Gazeta de Mexico TII, no. 13 and no. 17.
[68] Tributos 2, Exp. 5.
[69] Gazeta de Mexico, TII, no. 13.

of the country,[70] and there are reports of local benefactors making charitable donations to assist the starving poor.[71] Some councils in the Bajío, which as the grain-growing heartland of Mexico was among the regions most hardest hit by the crisis, provided relief,[72] and took the unprecedented step of circulating the names of wealthy individuals who could provide food or financial aid.[73] There is also evidence of community engagement in public works projects focused on the development of large-scale irrigation or water projects, intended to bene-fit entire communities.[74] Disease epidemics are reported in various parts of the country during and following the crisis,[75] and the repercussions were described as 'deplorable' for agriculture but also by extension for other sectors of the economy.[76] The drought affected the availability of pasture, disrupting the live-stock economy in the north and central regions of the country. Mules were critical to providing power to run refineries and yet were commonly fed on maize. Shortages of this most fundamental crop, therefore, also had implica-tions for the functioning of the mineral sector. Three years after the first signs of drought, hundreds of thousands of people had perished, or had been inca-pacitated and were thus in no fit state to cultivate lands when the rains did arrive.[77] Tribute demands became impossible to satisfy in the wake of crop shortages,[78] while the droughts had served to diminish the amount of seed stock available for sowing.[79] This period of crisis had particularly devastating consequences for society in the agrarian heartland of central Mexico. Five years later, some towns in the Bajío were still struggling to meet tribute demands,[80] and economic losses resulting from the crisis were still being recorded in 1792.[81]

The crisis of 1785–1786 does not appear to have affected society in Oaxaca as dramatically. In some parts of Oaxaca, the grain harvests in 1786 were particu-larly good,[82] but local Spanish district magistrates and administrators provided credit for the indigenous cochineal producers in the region, against a future

---

[70] AHMCH Justicia, Caja 126, Exp. 10; See, for example, Archivo de Casa Morelos, Leg. 841; AGN Ayuntamientos Vol. 169, fs. 49; AGN Ayuntamientos 173, cuaderno 6.
[71] AGN Alhóndigas, Vol. 10, Exp. 5, fs. 250–53; AHML Alhóndiga, Exp. 8.
[72] AGN Alhóndigas, Vol. 10, Exp. 5, fs. 250–53.
[73] AHML Alhóndigas, Exp. 8.
[74] AGN Indios, Vol. 91; Gazeta de Mexico, 6th December, 1785, Number 52, pp. 449–50.
[75] AGN Historia, Vol. 72.
[76] AHMCH Guerra, Caja 5, Exp. 5.
[77] AGN Historia, Vol. 72.
[78] AGN Tributos, Vol. 20, Exp. 15, 2, 5.
[79] Florescano (1981).
[80] AGN Correspondendia de Virreyes, Vol. 183, fs. 364–65.
[81] AGN Alhondígas, Vol. 10, exp. 5, fs. 250–53.
[82] Cited in Florescano, 1981, Vol. II: 570; 572.

delivery of the product in the form of 'repartimientos de bienes' – or goods advanced on credit and repayable by either coin or kind.[83] This may have given indigenous groups in Oaxaca more financial power to purchase what small amount of grain was available.

Widespread agrarian crisis would return around the turn of the century. Portillo (1910) records how 'from the year 1801, locusts began to invade the Spanish colonies in south and then central America, affecting the Mexican realm in 1802'. Various different forms of anomalous weather, floods and epidemic disasters are also recorded across the country between 1801 and 1804, including repeated flood events in Oaxaca[84] and elsewhere.[85] 'Scarcity of rains' between 1808 and 1810 triggered another devastating and widespread agrarian crisis. There was 'a lack of maize, beans and other seeds of basic necessity' resulting in 'want, hunger and calamity' across the country.[86]

By the last eighteenth and early nineteenth centuries, however, population expansion and a growing resource monopolisation by an emerging elite across many parts of the country had effectively created a more marginal – and as Tutino (1986) has pointed out for central Mexico – an increasingly dependent underclass. This particular sector of the population was far more vulnerable to the impacts of climate variability and faced subsistence crisis when successive and combined weather events served to drastically reduce harvests several years running. Population expansion, gross inequalities in ownership and privileges of access to natural resources, loss of customary rights and the general erosion and undermining of traditional ways of life were common problems.[87] Certainly, recognition of differential social vulnerability to weather events may have been one factor which helped fuel the agrarian dissent which started in the heartland of the Bajío but which would eventually sweep across the country in the drive for Independence.

Drought, deprivation and 'tipping points': climate and the context of social unrest in colonial Mexico

Although undoubtedly amongst the most turbulent times in Mexican history, the social, class and political struggles of the Independence period represented

---

[83] In this context, the term 'repartimiento' refers to 'repartimiento de bienes', the purchase and sale of goods by Spanish magistrates to their Indio charges. This institution, which is essentially a trading relationship between Indio and Spaniard, is completely unrelated to the Indio labour draft referred to as 'repartimiento de Indios'. Baskes (2000), 186–87.

[84] Archivo Privado de Don Luis de Canstaneda. Maps.

[85] AGN Rios y Acequias, Vol. 1, Exp. 9, 214ff.

[86] AGN Abasto y Panaderas, Vol. 2, Exp. 7, Fs. 385–87.

[87] Hamnett (2002).

the culmination of a much longer and more complex record of dissent.[88] Adopting Tilly's (1996) terminology, a broad 'repertoire of contention' can be identified which included local rebellions aimed at eliminating particular grievances with the colonial administration,[89] and what Scott has referred to as 'everyday' or 'Brechtian' forms of resistance and class struggle that included 'foot dragging, feigned ignorance, slander, arson, sabotage'.[90] Real or perceived social and biophysical vulnerability to climate variability and extreme weather events, however, may have been implicated in some of these different forms of dissent.

Perhaps the most common and persistent manifestation of social tension was the lawsuit (*pleito*). Given the propensity to drought across Mexico, access to and provision of water was (and is still today) a contentious issue. Problems of sharing water resources between competing users frequently led to legal infractions between all cross-sections of society. Law suits between *indio* communities and neighbouring estate owners were the most frequently filed of all *pleitos*. This is perhaps to be expected since irrigation for large-scale estate cultivation often entailed the diversion of water from location-to-location[91] and in some cases this meant the removal of water upon which local communities had been reliant, chiefly for maize cultivation, for generations.[92] Some native communities in the central and southern regions of the country, including those communities in Oaxaca, gained a reputation for being especially litigious, flooding local law courts with claims of water monopolisation, illegal abstraction, deprivation and usurpation.[93]

References to actual, expected or predicted drought feature in many of the cases, though not all references are necessarily legitimate. To some extent, an awareness of the propensity for drought and its impacts was inevitably

---

[88] Hamnett (2002).

[89] Cited in Hamnett (2002), 74.

[90] Scott (1976), 28–29.

[91] AGNM 8 Fa 50 vta; AGNM 70 fa. 73; AGNM 33 Fa. 32;

[92] See, for example, AGNM 73 fa. 133 vta; AGNM Vol. 35 fa. 81; AGNM 16 Fa. 287 vta; AGNM 73 fa. 133; AGN Indios 20, Exp 85, Fa 1.

[93] Serulnikov, 1996: 189; See examples of disputes over water in Guanajuato: (AGN Tierras, vol. 674, Exp. 1 30ff; Follow the dispute in vol. 675, Exp. 1; AGN Tierras vol. 192, Exp. 1; AGN Tierras, vol. 586, Exp. 8; AGN Tierras, vol. 1872, Exp. 15 10f; AGN Tierras, vol. 671, Exp. 3, 18 ff; AGN Tierras, vol. 2901, Exp. 36; AGN Tierras, vol. 2959, Exp 141; AGN Tierras, vol. 988, Exp. 1,2 y 3, 516 ff; AGN Tierras, vol. 1110, Exp. 18; AGN Tierras, vol. 1166 Exp. 1 fs. 450; AGN Tierras, vol. 2963, Exp.116, f 246–308; AGN Tierras, vol. 1352, Exp. 1; AGN Tierras, vol. 1368), Oaxaca: (AGN Mercedes 8 fa 50 vta; AGN Merceded 70 fa 73; AGN Mercedes 33 fa 32; AGN Tierras 149, Exp. 5; AGNT 149, Exp. 5, fs. 166; AGEO Real Intendencia, Leg 1, Exp. 39; AGEO Real Intendencia, Leg 1, Exp. 27; AGEO Alcadías Mayores, Leg 7, Exp. 20; AGNT 211m Exp. 2, fs. 48r; AGNT 939 Exp. 1; 48r; AGNI 11, Exp. 313).

employed by litigants as a legal 'tool' to reinforce or refute claims of water shortage, monopolisation, deprivation and/or restitution. Indeed, there may have been some opportunistic use of 'drought knowledge' in order to support particular testimonies over resource deprivation, particularly in the second half of the eighteenth century.[94] By denouncing another's lack of title to water, for example, a landowner could effectively secure a grant of water for himself.[95] Individual landowners would often exploit this facility in water disputes, especially where their properties shared a water source. Yet in such cases knowledge of drought may have been used opportunistically. One example is a *pleito* raised by mill owner and Councillor of Atatlauca, Oaxaca, Sebastian Gonzalez Romero, in 1787. According to the document, the Indios of the town had deprived Romero of the water with which he hoped to irrigate his lands – a decision they justified by claiming that Romero had contravened previous water agreements by sowing his lands during a time of drought. Romero argued, 'this is false...the lack of water that the republic of Atatlauca claim has affected their crops is not verified'.[96] It is indeed difficult to ascertain whether the community was referring more generally to the dry season, a specific period of drought in 1787 or the devastating successive droughts and scarcity that prevailed in the early and mid-1780s. There is little other evidence of a drought in Oaxaca in 1787 *per se*. Rather, in contrast, there are independent references to 'a strong wet season' with 'excessive rains' in some parts of Oaxaca during the summer of 1787.[97]

Drought most certainly aggravated the difficulties of managing scarce water supplies in at least some of the legal cases.[98] There may, however, be more direct links between drought and physical unrest. Drought, famine, disease in epidemic proportions and even catastrophic flooding, for example, combined to cause massive loss of life, especially among the indigenous populations across the Bajío in 1692.[99] The situation resulted 'in the emergence of popular uprisings in Guanajuato',[100] and there are also well-documented instances of social unrest and 'grain riots' around the same time in Mexico City,[101] where food shortages would prevail until 1696.[102]

---

[94] Endfield *et al.* (2004a).
[95] AGN Mercedes, Vol. 70, Fa 14v.
[96] AGEO Real Intendencia, Leg 1, Exp. 39.
[97] AGN Salinas, Vol. 15, Exp. 14.
[98] AGN Indios, Vol. 5, Exp. 154; AGN Indios, Vol. 5, Exp. 598; AGN Tierras, Vol. 1110, Exp. 18.
[99] Berthe (1970); Orozco y Berra (1938), 242–248.
[100] Marmolejo L. 181.
[101] Cope (1994).
[102] Gibson (1964).

In the north of Mexico, it has been argued that prolonged episodes of severe drought, such as the so-called 'megadrought' of the middle of the sixteenth century, may have been influential in exacerbating the difficulties of a subsistence way of life and in ultimately stimulating unrest among the indigenous populations at that time.[103] Other periods of unrest may similarly be associated with the vicissitudes of climate and its implications, specifically famine and disease, in this more marginal part of Mexico. Drought and pests, for example, had damaged the corn harvests in 1615, immediately preceding the revolts of this period,[104] while Martin (1989: 435), has highlighted how the Tarahumara Indians of the sierras staged their first revolt after a smallpox epidemic in 1645. Other revolts in the 1640s, 1650 and 1652 came after several years of drought and successive new epidemics,[105] and may have contributed to a high level of population mobility.[106] Increased raiding by nomadic groups who had been settled in various mission stations, particularly in the sierra, is also thought to have been a response to drought, famine and disease during the 1670s, 1680s and 1690s.[107]

Archival records reveal that eighteenth century incursions may similarly be associated with prolonged drought and actually indicate that links were drawn between the two at the time.[108] In 1727, for example, when drought and harvest failure resulted in rationing of bread and some of the poorest members of society were reduced to begging from door-to-door for food,[109] the local law courts were dealing with cases of livestock theft and attacks by 'Conchos' Indios from the lower Conchos Valley in southeast Chihuahua.[110] The severe droughts of the 1750s were blamed for inciting 'the contention of the enemies'[111] and hostilities continued into the second half of the 1750s,[112] when haciendas and ranches began to be abandoned because of repeated attacks.[113] Despite local defence campaigns,[114] violent incursions on settlements were still documented throughout the 1760s,[115] and by 1772, in the middle of a prolonged three year

---

[103] Cleaveland *et al.* (2003).

[104] Deeds (2003).

[105] Aboites (1994); Deeds (2003).

[106] AGN 19, fs. 247–261.

[107] Deeds (2003), 86.

[108] Endfield and Fernández-Tejedo (2006).

[109] Martin (1996).

[110] AHMCH Colonial, Justicia, Caja 17, exp. 17.

[111] AHMCH Guerra Caja 1, Exp. 8; AHMCH Guerra Caja 2, exp. 1.

[112] AHMCH Gobierno, Caja 27, Exp. 4.

[113] Martin (1996), 25.

[114] Archivo del Ayuntamiento de Chihuahua, University of Texas at El Paso, Microfilm no. 491, cited in Martin (1996), 25.

[115] AHMCH Gobierno, Guerra, Caja 2, Exp. 4; AGN Jesuitas, Leg II,–9, Exp. 30.

drought, 'flying squads' of Spanish militia were being enlisted to provide some form of defence,[116] although the 'terrible drought' the troops encountered upon arrival hampered their efforts.[117] The desperate years of drought and famine that affected the country in the mid-1780s rekindled the activities of 'barbarous indians' in northern Mexico. In this period, however, the colonial administration began to 'buy off' the raiding *indios* by granting them food and cattle subsidies. This encouraged, to some extent, their settlement and hence lessened the risk of raids,[118] though attacks on settlements, military garrisons and livestock continued into the first two decades of the nineteenth century and contributed to the abandonment of some of the larger haciendas in the region.[119]

A number of plausible explanations for this unrest in the north of Mexico have been forwarded. Most centre on oppression, forced labour, mistreatment, poor working and living conditions.[120] Political and cultural vitalism to reassert religious and cultural life in the face of increasing loss of control to the Spanish might have also played a key role.[121] The archival evidence presented above, however, suggests that drought, particularly when prolonged, might have been more influential than has traditionally been assumed in triggering social unrest, by exacerbating the general difficulties of survival on this marginal northern frontier.

Rethinking climate's agency: context, crisis and 'fundamental learning'

Drought was and is a regular feature of the Mexican climate and one to which society across the country has become accustomed. Pre-Hispanic and colonial societies in Mexico developed strategies to 'buffer' themselves against normal climatic variability, where 'normal' is defined by change on recent and short-term timescales which can be experienced without seriously affecting or disrupting society.[122] As illustrated, water and food storage, and trade in grain, for example, ensured that communities were afforded some protection against the impacts of seasonal and some inter-annual variations in rainfall. Long-term experience and awareness of seasonal and inter-annual changes in rainfall and periods of drought may have stimulated co-operative water-sharing strategies, but may have also been adopted as a legal tool in water disputes, where it

---

[116] AGN Presidios y Carceles, Vol. 4, ff. 154–160.
[117] AGN Presidios y Carceles, Vol. 4, f. 172; AGN Presidios y Carceles, Vol. 4, ff. 205–208; fs. 340–352.
[118] Katz (1988).
[119] Curtin *et al.* (2002).
[120] Cramaussel (1990); Deeds (1989); Martin (1996).
[121] Gradie (2000), 4.
[122] Erickson (1999).

could be used to either support or challenge charges of water theft, diversion, mismanagement or monopolisation. Some grain producers and traders may have also capitalised on drought knowledge and experience of harvest failure and grain scarcity to artificially inflate prices.

Low-frequency climatic variations with larger amplitudes, however, challenged many routine coping strategies and occasionally exceeded the capacity for routine human adaptability, contributing to socio-economic dislocation and life loss. Year-on-year droughts such as those documented in the 1750s and 1780s, or drought combined with other weather phenomena or volatile social-economic or political circumstances such as those which affected Mexico between 1808 and 1810, led directly to harvest failure, food scarcity and famine and indirectly to land abandonment and migration. It is open to conjecture whether disease epidemics can, or should, be directly attributed to weather conditions and food scarcity, or to the socio-economic organisation of the society. On at least some occasions, however, it is possible that climatic extremes may have interacted with ecological and socio-economic conditions to magnify the impact of infectious disease.[123] While social capital responses in the form of community projects may have become important in some regions during such periods, it is equally clear that some drought events may have provided something of a 'trigger' to already difficult social, economic, or political circumstances, rendering situations untenable for at least some sectors of society, notably the poorest, and predominantly indigenous communities.[124] The social upheaval associated with the 1692 riots in Guanajuato and Mexico City and the attacks recorded in Chihuahua, in the seventeenth and eighteenth centuries represent cases in point.

Experience of crisis and calamity, however, may have also improved the knowledge of risk among affected communities, increasing their awareness of their own vulnerability, and contributing to 'fundamental learning' in this respect.[125] This is particularly true of unusual and unexpected events which, under certain circumstances, might have provided a 'window for positive change and impetus for remedial action'.[126] Yet it also clear that inter-annual and even seasonal changes in water availability stimulated adaptive and strategic responses at a range of scales. In 1748, for example, at the start of a period of drought, harvest crisis and epidemic disease that would affect much of the north of the country into the 1750s and early 1760s, funding was sought to survey for water sources in the vicinity of San Felipe el Real de Chihuahua.[127]

---

[123] Ouweneel (1996), 91.

[124] Liverman (1990); Liverman (1999).

[125] Pfister, pers.comm.

[126] Streets and Glantz (2000), 100.

[127] AHMCH Gobierno, Caja 19, Exp. 3.

Potable water was required for the growing urban population, but also for the 'much needed' hospital in the town and in order 'to avoid the general illnesses associated with the original river because, being small, it carried the transport of metal wastes, it was used to wash the clothes of the infirm, [it was used] as a mud bath for the dirty animals, horses, mules and cattle so that it never arrives clear because it is so full of noxious filth'.[128] Built by Tarahumara Indians from the sierra in the 1750s, in return for payments of maize, meat and salt,[129] the aqueduct that brought water to the centre of the city went some way to resolving these problems.[130]

Climate events and subsistence crises might have also driven a range of more experimental agrarian strategies. The most common time for sowing wheat in the Bajío, for example, was in the spring in order to capitalise on the summer rains when the crop was maturing. Harvest losses due to drought, or more commonly the late arrival of the rainy season, as illustrated, was regularly recorded. Some agricultural communities in central Mexico responded by attempting to grow a second 'experimental' wheat crop in the late summer and autumn in the hope of late or even winter rains. The seedlings were known as trigo aventurero and some farmers appear to have resorted to this strategy during dry years or years when the rainy season appears to have arrived later than normal.[131] There was also some experimentation when it came to grazing pasture. Shortage of sufficient pastureland and a lack of water during particularly severe drought periods regularly destabilised the livestock industry, particularly in the north of the country, with consequences for other sectors of the economy. When drought contributed to a shortage of grazing pasture in Chihuahua in January 1758,[132] there were calls for 'common pasturage' by some people. Lands normally considered unsuitable for grazing began to be exploited and there are references to cattle ranchers making use of salt flats and swamps in the area. Notwithstanding such strategies, 'total ruin among the livestock' was recorded.[133]

Agrarian experimentation also took on a more long-term, strategic and scientific form. Wheat crops across the country were repeatedly if selectively destroyed by a crop blight referred to as Chahuistle,[134] which is thought to be linked to warm, wet and stormy conditions.[135] Chahuistle affected wheat

---

[128] AHMCH Gobierno, Caja 19, Exp. 3.
[129] AHMCH Hacienda, Caja 30, Exp. 10.
[130] Martin (1996), 42.
[131] The pleito can also be traced in AGN Tierras Vol. 675, Exp. 1; ACM , Leg. 834, Fa. 14.
[132] AHMCH Gobierno, Caja 30, Exp. 23.
[133] AHMCH Notarias, Abasto de Carnes, Caja 42, Exp. 2.
[134] *Chahuistle* is a folk term commonly used to describe wheat rust, and less commonly used to refer to insect or worm infestations.
[135] Humboldt (1811).

harvests in the Etla Valley of Oaxaca between 1696 and 1714,[136] leading to the abandonment of (wheat) flour mills in the region, a number of which, by 1714, were said to be in ruins. Wheat crops in the Bajío of the northern Central Highlands of Mexico were similarly affected on a number of occasions, most severely in 1700, 1706, 1711, 1718, 1735 and 1746, stimulating shortages of wheat flour and seed stock across the region.[137] There were some attempts to develop a remedy for this persistent scourge. A document from Leon in Guanajuato, dated 1782, details a 'secret recipe' which stated that it not only 'prevents chahuistle', but also 'increases the amount of harvest in proportion to the quality of the land'.[138] The recipe had been devised by Don Pedro Aspe, a French man who was concerned over the 'excessive prices for wheat' which had been ruined by 'the chahuistle contagion'.[139] Dissemination of this hitherto 'secret' information was a key development in social adaptation to climate variability in this central region and hints at evidence of longer term and possibly international research and development.

Deriving insights about the contemporary and future coping capacity and adaptability to climate changes from such historical examples is, of course, problematic.[140] The fact that past societies differed markedly from those in the modern world makes simple analogies unrealistic.[141] Moreover, historical documents such as those used in this investigation are rarely free from subjectivity and the interpretation of climate information, let alone details of responses or adaptation, using such sources can be problematic. Such examples do illustrate, however, as Manley suggested back in 1958, that climate has and does impose a 'constraint' on society, and it is also clear that climate also represents an important influence in the course of human adaptation and behaviour.

Fifty years on from Manley's statements, increasing economic losses coupled with an escalating number of fatalities due to extreme weather events are stark and tragic reminders of society's vulnerability to climate change and have underscored the need for integrated human climate histories.[142] It is becoming increasingly clear that explorations of the impacts of and social responses to these changes are equally if not more important as climate reconstructions *per se* (Hassan, 2000). So, it is perhaps time for us to be less fearful of the 'changing yet lingering role of deterministic ideas in climatology',[143] and

---

[136] AGN Tierras, Vol. 310, Exp. 1.

[137] ACM, Leg. 835 (1735), Leg. 838 (1700, 1706), Leg. 847 (1746), Leg. 860 (1711, 1718).

[138] AHML Bandos, Exp. 8, 1782.

[139] AHML Bandos, Exp. 8, 1782.

[140] McNeill (2005), 178.

[141] Ingram *et al.* (1981), 5; Meyer *et al.* (1998).

[142] Meyer *et al.* (1998); Easterling *et al.* (2001); Kundewicz and Kaczmarek (2000); Easterling *et al.* (1999), 285.

[143] McGregor (2004), 237.

to acknowledge that investigations of climate–society interactions in historical perspective, and specifically those that address instances of successful and unsuccessful adaptation at a range of scales, are highly relevant to current global climate changes studies. With this in mind there is still so much work yet to be done.

References

Aboites Aguilar, L. (1994), *Breve historia de Chihuahua*. El Colegio de Mexico, Mexico, Fondo de Cultura Economica.

Appleby, A.B. (1981), 'Epidemics and famine in the Little Ice Age'. In Rotberg R.I. and Rabb T.K. (eds), *Climate and History*, New Jersey, Princeton University Press, 63–83.

Badstue, L.B., Bellon, M.R., Berthaud, J., Juarez, X., Rosas, I.M., Solano, A.M. and Ramirez, A. (2006), 'Examining the role of collective action in an informal seed system a case study from the central valleys of Oaxaca, Mexico'. *Human Ecology* 34(2), 249–273.

Barriendos, M. (1997), 'Climatic variations in the Iberian peninsula during the Late Maunder Minimum (AD 1675–1715). An Analysis of data from rogation ceremonies'. *Holocene* 7, 105–111.

Barriendos, M. and Dannecker, A. (1999), 'La sequia de 1812–18124 en la costa central Catalana. Consideraciones climaticas e impacto social del evento'. In Nadal, R., Jose, M. and Martin Vide, J. (eds), *La climatologia espanolla en los albores del siglo XX1*. Vilassar de Mar, 53–62.

Baskes, J. (2000), *Indians, Merchants and Markets. A Reinterpretation of the Repartimiento and Spanish-Indian Economic Relations in Colonial Oaxaca, 1750–1821*. California, Stanford University Press.

Berthe, J.P. (1970), 'La peste de 1643 en Michoacán'. In *Historia y Sociedad en el Nuevo Mundo de habla española*, Homenaje a José Miranda, México, Mexico City, El Colegio de México, 247–261.

Blaut, J. (1999), 'Environmentalism and eurocentricism'. *The Geographical Review* 89 (3), 391–408.

Blaikie, P., Cannon, T., Davis, I. and Wisner, B. (1994), *At Risk. Natural Hazards, People's Vulnerability and Disaster*. London, Routledge.

Bowden, M.J., Kates, R.W., Kay, P.A., Riebsame, W.E., Warrick, R.A., Johnson, D.L., Gould, H.A. and Wiener, D. (1981), 'The effect of climate fluctuations on human populations: two hypotheses'. In Wigley, T.M.L., Ingram, M.J. and Farmer, G. (eds), *Climate and History. Studies in Past Climates and Their Impact on Man*. Cambridge, Cambridge University Press, 479–513.

Brazdil, R., Pfister, C., Wanner, H., von Storch, H. and Luterbacher, J. (2005), 'Historical climatology in Europe – the state of the art'. *Climatic Change* 70(3), 363–430.

Bryson, R.A. and Padoch, C. (1980), 'On the climates of history'. *Journal of Interdisciplinary History* 10 (4), 583–597.

Cavo, A.P. (1949), *Historia de México*, anotada por Ernesto Burrus, prologo Mariano Cuevas, México, S.A. Mexico. Editorial Patria.

Conde, C., Magaña, V., Sanchez, O. and Gay, C. (1997), 'Assessment of current and future regional climate scenarios for Mexico'. *Climate Research* 9 (1–2), 107–114.

Cleaveland, M.K., Stahle, D.W., Therrell, M.D., Villanueva-Diaz, J. and Burns, B.T. (2003), 'Tree-ring reconstructed winter precipitation and tropical teleconnections in Durango, Mexico'. *Climatic Change* 59, 369–388.

Conde, C., Liverman, D., Flores, M., Ferrer, R., Auaujo, R., Betancourt, E., Villreal, G. and Gay, C. (1997), 'Vulnerability of rainfed maize crops in Mexico to climate change'. *Climate Research* 9, 17–23.

Coombes, P. and Barber, K. (2005), 'Environmental determinism in Holocene research: causality or coincidence'. *Area* 37 (3), 303–311.

Cooper, D. (1965), *Epidemic Disease in Mexico City, 1761–1813*. Institute of Latin American Studies, Austin, University of Texas Press.

Cope, R.D. (1994), *Limits of Racial Domination: Plebeian Society in Colonial Mexico City, 1660–1720*. University of Wisconsin Press.

Cramaussel, C. (1990), *La provincia de Santa Barbara en Nueva Vizcaya* 1563–1631. Chihuahua, Universidad Autonoma de Cuidad Juarez.

Curtin C.G., Sayre, N.F. and Lane, B.D. (2002), 'Transformations of the Chihuahuan borderlands: grazing, fragmentation and biodiversity conservation in desert grasslands'. *Environmental Science and Policy*. 5, 55–68.

Deeds, S.M. (1989), 'Rural work in Nueva Vizcaya: forms of labor coercion on the periphery'. *Hispanic American Historical Review* 69 (3), 425–449.

Deeds, S.M. (2003), *Defiance and Deference in Mexico's Colonial North*, Austin, University of Texas Press.

de Menocal, P.B. (2001), 'Cultural responses to climate change during the Late Holocene'. *Science* 29, 667–673.

Doolittle, W.E. (1990), *Canal Irrigation in Prehistoric Mexico: The Sequence of Technological Change*, Austin, University of Texas Press.

Easterling, D.R., Evans, J.L., Groisman, P.Y., Karl, T.R., Kunkel, K.E. and Ambenje, P. (2001), 'Observed variability and trends in extreme climate events. A brief review'. *Bulletin of the American Meteorological Society* 81 (3), 417–425.

Endfield, G.H. and O'Hara, S.L. (1997), 'Conflicts over water in the "Little Drought Age" in central Mexico', *Environment and History* 3, 255–272.

Endfield, G.H., Fernández-Tejedo, I. and O'Hara, S.L. (2004a), 'Conflict and co-operation: water, floods and social response in colonial Guanajuato, Mexico'. *Environmental History* 9(2), 221–247.

Endfield G.H., Fernández-Tejedo, I. and O'Hara, S.L. (2004b), 'Drought, and disputes, deluge and dearth: climatic variability and human response in colonial Oaxaca, Mexico'. *Journal of Historical Geography* 30, 249–276.

Endfield, G.H. and Fernández-Tejedo, I. (2006), 'Decades of drought, years of hunger: archival investigations of multiple year droughts in late colonial Chihuahua'. *Climatic Change* 75 (4), 391–419.

Erickson, C.L. (1999), 'Neo-environmental determinism and agrarain collapse in Andean prehistory'. *Antiquity* 74 (281), 634–642.

Florescano, E.L. (1969), *Precios del maiz y crisis agricolas en Mexico, 1708–1710*, Mexico, El Colegio de Mexico.

Florescano, E. (1980), 'Una historia olvidada: la sequia en Mexico'. *Nexos* 32, 9–13.

Florescano, E. (1981), *Fuentes para la historia de la crisis agrícola de 1785–1786*. Mexico City, Archivo General de la Nación.

Florescano, E. (1986), *Precios del maíz y crisis agrícolas en México*, 1708–1810. Ediciones Era, Mexico.

García-Acosta, V. (1993), 'Las sequias historicas de Mexico'. *La Red*, No. 1, 2–18.

García-Acosta, V. (ed.) (1997), *Historia y desastres in America Latina*. Red de Estudios Sociales en prevención de desastres en América Latina. Mexico, CIESAS.

Gibson, C. (1964), *The Aztecs under Spanish Rule*, California, Stanford University Press.

Gonzales, P. (1904), *Geografía local del estado de Guanajuato*, Tip. De La Escuale Ind Militar.

Gradie, C.M. (2000), *The Tepehuan Revolt of 1616. Militarism, Evangelism and Colonialism in Seventeenth Century Nueva Vizcaya*, Salt Lake City, University of Utah Press.

Grove, J.M. (1988), *The Little Ice Age*, London, Routledge.

Hamnett, B.R. (2002), *Roots of Insurgency: Mexican Regions, 1750–1824*, Cambridge Latin American Studies, Cambridge, Cambridge University Press.

Hassan, F. (2000), 'Environmental perception and human responses in history and prehistory', In McIntosh, R.J., Tainter, J.A. and McIntosh, S.K. (eds), *The Way the Wind Blows: Climate, History and Human Action*, New York, Columbia University Press, 121–140.

Hassig, R. (1981), 'The famine of one rabbit: Ecological causes and social consequences of a pre-Columbian calamity'. *Journal of Anthropological Research* 37, 172–182.

Hodell, D.A., Brenner, M., Curtis, J.H., Medina-González, R., Idelfonso Chan Can, E., Abornaz-Pat, A. and Guilderson, T.P. (2005a), 'Climate change on the Yucatan Peninsula during the Little Ice Age'. *Quaternary Research* 63, 109–121.

Hodell, D.A., Brenner, M. and Curtis, J.H. (2005b), 'Terminal Classic drought in the northern Maya lowlands inferred from multiple sediment cores in Lake Chichancanab'. *Quaternary Science Reviews* 24, 1413–1427.

Hodell, D.A., Curtis, J.H. and Brenner, M. (1995), 'Possible role of climate in the collapse of the Classic Maya Civilisation'. *Nature* 375, 391–394.

Humboldt, A. von (1811 [1973 reprint]), *Ensayo politico sobre el reino de la Nueva España*, estudio preliminar y notas de Juan A. Ortega y Medina, Mexico, Editorial Porrua S.A.

Ingram, M.J., Farmer, G. and Wigley, T.M.L. (1981), 'Past climates and their impact on man: a review', In Wigley, T.M.L., Ingram, M.J. and Farmer, G. (eds), *Climate and History. Studies in Past Climates and Their Impact on Man*, Cambridge, Cambridge University Press, 3–49.

International Panel on climate change, (2007). *Climate Change 2007*. Fourth Assessment Report, IPCC, Geneva.

Jones, P.D., Ogilvie, A.E.J., Davies, T.D. and Briffa, K.R. (eds) (2001), *History and Climate: Memories of the Future?* London, Kluwer Academic.

Karl, T.R. and Easterling, D.R. (1999), 'Climate extremes: selected review and future research directions'. *Climatic Change* 42, 309–323.

Katz, F. (1988), 'Rural uprisings in pre-Conquest and colonial Mexico', In Katz, F. (ed.), *Riot, Rebellion and Revolution. Rural Social Conflict in Mexico*, New Jersey, Princeton University Press, 65–94.

Kundewicz, Z.W. and Kaczmarek, Z. (2000), 'Coping with hydrological extremes'. *Water International* 25 (1), 66–75.

Lamb, H.H. (1969), 'The new look of climatology'. *Nature* 223, 1209–1215.

Lamb, H.H. (1977), *Climate: Past, Present and Future*, 2, London, Methuen.

Lamb, H.H. (1982), *Climate, History and the Modern World*, London, Routledge.

Le Roy Ladurie, E. (1972), *Times of Feast. Times of Famine. A History of Climate since the Year 1000*, London, George Allen and Unwin.

Lipsett Rivera, S. (1999), *To Defend Our Water with the Blood of Our Veins: The Struggle for Resources in Colonial Puebla*, Albuquerque, University of New Mexico Press.

Liverman, D. (1990), 'Drought impacts in Mexico: climate, agriculture, technology, and land tenure in Sonora and Puebla'. *Annals of the Association of American Geographers* 80 (1), 49–72.

Liverman, D. (1999), 'Vulnerability and adaptation to drought in Mexico'. *Natural Resources Journal* 39, 99–115.

Manley, G. (1958), 'The revival of climatic determinism'. *The Geographical Review* 48 (1), 98–105.

Marmolejo, L. (1967), *Efemérides Guanajuatenses o datos para formar la historia de la ciudad de Guanajuato* vol. 1. Guanajuato, Universidad de Guanajuato Press.

Martin, C.E. (1996), *Governance and Society in Colonial Mexico. Chihuahua in the Eighteenth Century*, Stanford, California, Stanford University Press.

McCann, J.C. (1999), 'Climate and causation in African history'. *The International Journal of African Historical Studies* 32 (2–3), 261–279.

McCarthy, J.J., Canziani, O.F., Leary, N.A., Dokken, D.J. and Kasey, S. (2001), *Climatic Change 2001: Impacts, Adaptation, and Vulnerability*. Contribution of Working Group II to the Third Assessment Report of the Intergovernmental Panel on Climate Change. Intergovernmental Panel on Climate Change.

McGregor, K.M. (2004), 'Huntington and Lovelock: climatic determinism in the 20th century'. *Physical Geography* 25 (3), 237–250.

McNeill, J.R. (2005), 'Diamond in the rough: is there a genuine environmental threat to security. A review essay', *International Security* 30 (1), 178–195.

Meyer, W.B., Butzer, K.W., Downing, T.E., Turner, B.L. (II), Wenzel, G.W. and Westcoat, J.L. (1998), 'Reasoning by Analagy', In Raynor, S. and Malone, E.L. (eds), *Human Choice and Climate Change, no. 3, Tools for Policy Analysis*, Columbus, OH: Batelle Press, 218–289.

Murphy, M.E. (1986), *Irrigation in the Bajío Region of Colonial Mexico*, Dellplain Latin American Studies, No. 19. Boulder, Westview Press.

Orozco y Berra M. (1938), *Historia de la dominación española en México*, con una advertencia de Genaro estrada, 2 vols, Mexico, Biblioteca Historica Mexicana de obras ineditas Núm 10, 242–248.

Ouweneel, A. (1996), *Shadows over Anáhuac. An Ecological Interpretation of Crisis and Development in Central México 1730–1800*, Albuquerque, University of New Mexico Press.

Pfister, C. (1978), 'Climate and economy in eighteenth century Switzerland'. *Journal of Interdisciplinary History* 9, 223–243.

Pfister, C. (in press), Natural disasters – catalysts for fundamental learning. In Pfister, C. and Mauch, C. (eds), *Natural Hazards: Cultural Responses in Global Perspective*, to be published by Lexington.

Pfister, C. and Brazdil, R. (1999), 'Climatic variability in sixteenth century Europe and its social dimension. A Synthesis'. *Climatic Change* 43, 5–53.

Pfister, C., Brazdil, R., Obrebska-Starkel, B., Starkel, L., Heino, R. and von Storch, H. (2001), 'Strides made in reconstructing past weather and climate'. *Eos-Transactions of the American Geophysical Union* 82, 248.

Portillo, A. (1910), *Oaxaca en el centenario de la Independencia nacional*, Oaxaca.

Sanders, W.T. and Nichols, D.L. (1988), 'Ecological theory and cultural evolution in the Valley of Oaxaca'. *Current Anthropology* 29 (1), 33–80.

Schneider, S.H. (2001), 'What is dangerous climate change?'. *Nature* 411, 17–19.

Scott, J.C. (1976), *The Moral Economy of the Peasant: Rebellion and Subsistence in South East Asia*, London, Yale University Press.

Serulnikov, S. (1996), 'Disputed images of colonialism: Spanish rule and Indian subversion in northern Potosi', *Hispanic American Historical Review* 76 (2), 189–226.

Streets, D. J. and Glantz, M.H. (2000), 'Exploring the concept of climate surprise', *Global Environmental Change – Human and Policy Dimensions* 10 (2), 97–107.

Swan, S.C. (1981), 'Mexico in the Little Ice-Age', *Journal of Interdisciplinary History*, XIV (4), 633–648.

Therrell, M.D., Stahle, D.W. and Acuña-Soto, R. (2004), 'Aztec drought and the "curse of one rabbit" ', *Bulletin of the American Meteorological Society*, September.

Tilly, C. (1996), 'Conclusion: contention and the urban poor in eighteenth and nineteenth century Latin America', In Arrom, S.M. and Ortoll, S. (eds), *Riots in the Cities: Popular Politics and the Urban Poor in Latin America, 1765–1910*, SR Books, 225–242.

Tutino, J. (1986), *From Insurrection to Revolution in Mexico*, Princeton University Press, New Jersey.

Utterström, G. (1954), 'Some population problems in pre-industrial Sweden'. *Scandinavian Economic History Review* 2, 103–166.

Wigley, T.M.L. (1985), 'Impact of extreme events'. *Nature* 316, 106–7.

Wigley, T.M.L., Ingram, M.J. and Farmer, G. (1981), *Climate and History*, Cambridge, Cambridge University Press.

Wilken, G.C. (1987), *Good Farmers: Traditional Agricultural Resource Management in Mexico and Central America*, Berkeley, University of California Press.

Part IV
'Things Human'

13

Destinies and Decisions: Taking the Life-World Seriously in Environmental History

Kirsten Hastrup

As I think we agree, the earth as a physical space did not develop independently of the development of ever more complex life forms. Conversely, the emergence of particular life forms would have been unthinkable without specific physical environments. In explaining one or the other, their entanglement and mutual formation is a key issue. This relationship can be studied at various scales, spatial or temporal, but physics and biology are never far apart when it comes to map the history of development. In classical evolutionist thinking, the evolution of various species was conditioned by adaptation to an environment, and survival was a function of fitness – being neither more nor less than a measure of the very capacity for survival in a particular habitat. In that sense, evolutionism itself is an instance of environmental history, albeit of a timescale of many millennia.

In this chapter, I shall be concerned with just one species, humans, and my timescale will be approximately one millennium. Yet the principle of the mutual conditioning still holds. My material will be drawn from Icelandic history, which I have studied in depth in order to comprehend some of the major fluctuations in the long-term development of Icelandic society.[1] A brief overview of the history of Iceland in traditional chronological and (sweeping) narrative fashion will set the scene for the more analytical perspective to follow.

Iceland was uninhabited until the late ninth century, when Norsemen settled there in a final wave of Viking migration. The settlers soon succeeded in establishing a coherent society by way of a legal system of which learned men were in charge. The middle ages saw this develop further; the famous Icelandic sagas survive as a testimony to an original 'free' society, to astute statesmanship and legal sophistication – not to mention the literary achievement itself. In short, society seemed to flourish, and the settlers thrived on each of their farmsteads,

---

[1] Hastrup (1985, 1990a, 1990b, 1998).

by itself representing a 'whole' society, by comprising all social categories. The economic basis was farming and animal husbandry (supplemented by some fishing) and in the absence of money, homespun cloth provided the standard of value, as measured in *álnir* ('ellens'), for lesser items. For larger items such as farmsteads or land, *kúgildi* ('cow's worth') measured the value, pointing directly to the dominant economic basis.

In the thirteenth century, social conflicts began to erupt, partly due to an increased population and to overexploited natural resources; the virgin forest of low birch had been cut down for housing and firewood leading to serious soil erosion. One of the results of the combined factors was an emerging class of small-scale peasants, cottagers and migrant farmhands. The social conflicts were mainly conflicts over land, and these were one of the reasons why the Norwegian king had a relatively easy play in making the Icelanders his subjects in 1262–64. The land had become scarce and impoverished, and one of the clauses in the treaty between the king and the Icelanders entitled them to six shiploads of grain each summer.

The fourteenth century saw a change from a largely self-sufficient farming society, where the farmstead itself was society concretised and writ small, to a market-oriented society where fishing became the major source of external income. This led to the establishment of more or less permanent settlements on the seashore, where migrant paupers and farmhands that had become redundant found their own solution in emerging fishing villages. Full-time fishing appeared, thus splitting up what was always a dual economy *within* the farmstead in a new social division of labour. Significantly, *vættir* ('pounds') of fish now replaced *álnir* of homespun as the dominant standard of value. The market for fish expanded to most of northern Europe, and the returns were very promising. In 1380, Norway and Iceland were both integrated into the kingdom of Denmark.

From 1402 to 1404, the Black Death ravaged the island and killed off between one third and half the population, decimating it to about 45.000 (from an estimated 80.000), and disrupting the process I have just described. This is where a combined history of environmental and social decline began. The population of Iceland became increasingly vulnerable as the so-called Little Ice Age encroached, reaching the climatic pessimum in late seventeenth century. From the extensive Icelandic annals we know that at least one fourth of the 400 years between 1400 and 1800 was locally classified as 'lean' years, marked by extensive famine and death – according also to the late eighteenth century chronicler, Bishop Hannes Finnsson.[2] Some of the lean years were owed to volcanic eruptions, covering the fields in ash; in a marginal and relatively

---

[2] Finnsson (1970 [1796]).

isolated farming area, such natural events were fatal. Other setbacks were owed to changing winds on the North Atlantic that some years prevented ships from abroad from landing with the necessary provisions of grain and timber.

In short, the period 1400–1800 in Iceland was a time of recurrent crises – social, economic, demographic, environmental, climatic, and so forth. In all domains of society there was a conspicuous setback by comparison to the earlier period of the settlements and the high Middle Ages. By this standard, the devolution was remarkable. For each of the domains mentioned, we might name a cause by pointing to what was the most spectacular and most unique event in the course of events. I am thinking of events like the Black Death, volcanic eruptions, the introduction of a Danish trade monopoly and so forth. In naming such causes – all of which will be familiar to historians as traditional explanations of the decline – we may have identified significant parts of the empirical reality, but we have *explained* very little beyond immediate local experience. We have been trapped in a narrative logic of temporal causation that relates only to the succession of events but cannot account for the links between them. Truth does not equal explanation; truth is not a predicate and adds nothing to the subject.[3] Working within a historical frame, rather than within a pre-historical or palaeo-anthropological horizon enables us take a close look at the 'environmental history', seen as a link between historical and natural events that allows for a different understanding of the complexities of causation. Thanks to the rich Icelandic records of people's actions and thoughts on the one hand, and of historical and natural events on the other, we are able to show how environmental history is but history, because the environment is never outside of the lived world. The point is that while the social and demographic decline in Iceland from 1400 to 1800 was certainly in part a consequence of natural catastrophes, such as climatic change and volcanic eruptions, of epidemics, such as plagues and smallpox, and of external commercial pressures, such as the Danish trade monopoly, it would be a mistake to see the Icelanders as simply *victims* of an externally produced destiny; they were also *agents* in the historical development, to which they contributed their culturally informed decisions.

In what follows I shall address the relationship between people and their environment in terms of a perceived tension in history between decisions made and destinies suffered; I want to show how these cannot be empirically separated and that causation in history involves both human agency and facts of nature. I shall draw on my earlier work yet here I shall seek to re-frame it by employing certain terms from a phenomenologically inspired anthropology as an organising device. The terms have been consistently developed by

---

[3] Hacking (1992), 14; see also Hastrup (2004).

Tim Ingold,[4] but are not exclusive to his work. While not adopting Ingold's view of things human in its entirety, I find his concepts extremely useful in providing a fresh analytical focus on a fascinating and highly suggestive piece of long-term history. The chapter will be guided by notions that reflect a life-world perspective upon the environment, such as dwelling, way finding and mapping. This will allow me to show why destiny and decision do not belong to separate registers.

Dwelling: At home in topography

In a recent article, I have suggested that anthropology is now on the verge of a *topographic turn*.[5] By suggesting this I allowed myself to sum up some of the trends that we have experienced over the past few years, partly in reaction to the latent idealism of the literary turn where discourses and narratives seemed to be afloat all over the place and constitute their own significant space. It also takes the phenomenological interest in the spatiality of bodily perceptions and projections to a more comprehensive conclusion than the one offered so far by theories of practice and experience. Merleau-Ponty says:

> Space and perception generally represent, at the core of the subject, the fact of his birth, the perpetual contribution of his bodily being, *a communication with the world more ancient than thought* ... Space has its basis in our facticity. It is neither an object, nor an act of unification on the subject's part; it can neither be observed, since it is presupposed in every observation, nor seen to emerge from a constituting operation, since it is of its essence that it be already constituted, for thus it can, by its magic, confer its own spatial particularization upon the landscape without ever appearing itself.[6]

A communication with the world more ancient than thought; people always live on the surface of the earth and are subjects to its texture, shape and fecundity. These features, writes Wendy James, 'provide a base line to our human lives, not only our pragmatic activities, but to our conceptual understandings of the organized qualities of differentiated space, and our orientation within it'.[7] Yet, as Merleau-Ponty saw, the base line is constituted along with subjectivity itself and therefore tends to recede from view.

The implication of the phenomenological perspective is that social and geographical space are conflated in experience. It is not possible to 'think away' the actual geographical location of social life; lives are always grounded.

---

[4] Ingold (2000).
[5] Hastrup (2005).
[6] Merleau-Ponty (1962), 254; emphasis added.
[7] James (2003), 213.

Movements in space inscribe social life on the land, and with time particular paths are cleared and certain directions presented as more natural than others. We might even say, again with Wendy James, 'that the experience of physical places and the journeys between them is one of the commonest underpinnings of the human being's characteristic sense of living in a "formatted" social space from early childhood and earliest memory'.[8]

It is this kind of argument that led me to tentatively name the present turn in anthropology as topographic. Topography should not be conflated with cartography and map-making in general; the topographic turn is distinguished by its taking the life-world perspective seriously, including the movements of the social agents, and the paths they carve out, physically and socially, through their way-finding. The concreteness and materiality of topography thus defies the abstract map (the territory as represented), and is closely linked up with experience and practical mastery of the environment.[9] This also implies an acknowledgement of the temporality of social life that is absent from the idea of map-making; social life takes time, if you wish. It not simply unfolds in time but is profoundly temporal.[10] Timing and temporality are features of any social choreography, that is the actual shaping and possible routinisation of social relationships and movements. In my view, environmental history can be understood productively as a variation upon this topographic theme, fundamentally acknowledging the fusion of geographical and social horizons, when seen from the perspective of the dwellers in a particular habitat.

Following Heidegger, Tim Ingold suggests a renewed emphasis on the *dwelling* perspective as different from the *building* perspective, when it comes to study humans in their environment.[11] People still build, of course, but instead of seeing the buildings as *containers* of human activities that may then be studied independently, the activities – also of building – belong to our dwelling in the world, and to the way we are. From the dwelling perspective 'home' takes precedence over the house; the perception of the home is from a position *within* it, rather than from an external bird's eye perspective. As suggested by Gaston Bachelard, home is where we start from; it is from there we dream about other possible worlds that may (or may not) be conquered.[12] Significantly, we never go *to* home; we simply go home, or maybe back home. For him there is no demarcation between house and home, the former is simply a metaphor for the latter. In the phrasing suggested by Dawson and Rapport, home is where one knows oneself best.[13]

---

[8] James (2003), 67.
[9] Ingold (2000), 239.
[10] Bourdieu (1990).
[11] Ingold (2000).
[12] Bachelard (1950).
[13] Rapport and Dawson (1998).

For the Icelanders of ancient times, the centre of their world was the *bú*, referring both to their farmstead and to their home, their fixed place in the world. This was from where they started – wherever they went. The ancient Icelanders knew themselves best at their farmstead, metaphorically representing the controlled inside of the cosmological order, where kinsfolk lived and honour and integrity was to be expected. *Innangarðs*, meaning both within the fence (*garðr*) and within the farm, people were inviolable. *Útangarðs*, outside the fence, one was at the mercy of the wilderness and its inhabitants, comprising both supernatural beings such as trolls and elves, and the more human, if exiled, outlaws and outliers. This reflected the distinction in Viking cosmology between *Miðgarðr* ('The Middle Yard') and *Útgarðr* ('The Outer Yard'). The cosmological order was clearly concentric; the inner circle was where humans dwelled – but also where they cultivated the infields (the *tún*) and kept their cattle, including the sheep during winter. The point is that there is no demarcation line between people and landscape along the line of culture vs. nature. The landscape is always perceived from a position within it, whether it is seen as familiar or hostile.

The dualistic (concentric) model also affected the perception of the tasks that were undertaken in the various spaces. Ingold proposes the term 'taskscape' as a replacement for landscape in order to shift the focus from *form* to *activity*, and to circumvent the implicit connotations of externality attached to the notion of landscape.[14] A taskscape is essentially social, or rather it dissolves the distinction between space and sociality. While attending to their tasks, people also attend to one another. *Innangarðs*, the tasks were (by definition) domestic; at home were kinsmen and cattle, and domesticated nature. Milking and tilling the soil linked the inner taskscape directly to the physical space of (certain) animals and (parts of) landscape; the taskscape was constituted both by place and by people. *Útangarðs*, fishing and other forms of hunting and gathering were relegated to an outside wilderness. In this line of thought we can see how the Icelanders – in times of feast as well as famine – did not simply exploit the environment, but were participants in a history of dwelling within a perceived space that incorporated an age-old distinction between inside and outside, friends and strangers, cattle and fish. Tasks, social relationships and arenas are mutually defined. As movements within or between such spaces become routinised, people's relationship with the environment becomes increasingly 'ceremonial'.[15]

What I am suggesting here is that since the earliest times in Iceland, the dwelling perspective implied a sense of sociality as tied to a particular social

---

[14] Ingold (2000), 195.
[15] James (2003).

centre and a particular set of tasks. The concentric cosmology conflated the sense of self with the innermost space and with farming activities. The *bú* was where the Icelanders knew themselves best. Using the life-world perspective, instead of seeing Iceland from outside or above it, means that our explanatory models must take such lived metaphors seriously.

When plague hit and climate changed, the Icelandic reactions to the disasters make sense in relation to this perspective. The inside taskscape was increasingly threatened; hay failed to grow and to dry during winters. Cattle starved and poverty struck. The population declined both as a consequence of various disasters, and to a very high infant mortality rate and a staggeringly low marriage ratio. When the first full census of Iceland was made in 1703 (numbering just above 50.000 people), only 27.8 % of women in the fertile age between 15 and 49 years of age were married. Facing this demographic situation, the Icelanders sought to concentrate their efforts at home, so to speak. The modest pattern of transhumance that had lifted the pressure off some of the infields during summer was given up, and fishermen were called home from the seashore to align themselves with the needs of the farmstead. I shall return to that below, here I shall simply state that when threats from the perceived outside intensified, the Icelanders went home and concentrated on domestic tasks. Turning their backs to the wilderness did not help externalising the actual 'environmental history' from history in general; quite the contrary. Even within the domestic taskscape, nature encroached; the fences separating the social from the wild disintegrated, the plough disappeared, tufts spread in the infields, the hay decayed in stacks, and nobody could afford to marry. Famine, poverty and frequent death resulted – intensifying the sense of external assaults.

Wayfinding: Movement and knowledge

In this section I shall focus on the movements of the Icelanders, here seen as wayfinding to keep in line with the life-world perspective chosen in this exposition.

> Human beings are restless creatures. They are always moving about. One of the curiosities of twentieth century anthropology, however, is that it has proceeded as though the human body were like a statue, forever fixed in bolt upright position, while all the activity is going on in the mind.[16]

This implies that we are only now beginning to come to terms with the fact that 'cultural variation' relates to ways in which people move. Instead of dividing

---

[16] Ingold (2003), 40.

up the world in distinct entities like mind, culture and nature to be studied by each their discipline, Ingold suggests a combined view on the human being as 'a locus of creative growth within a continually unfolding field of relationships'.[17] I doubt that we can totally avoid slicing up this continuous field in analytically distinct objects, but I believe that for environmental history the suggestion is productive because it allows for a unified focus on living persons in their environment.

This unified view also immediately opens up for new qualifications of landscapes as such. The landscape is never simply a backdrop to social action, but part of it. For instance, in a recent article, Morten Pedersen has shown how the reindeer-herding people of Northern Mongolia, the Tsaatang, perceive their landscape as a heterogeneous network of powerful places to which they are attached in social practice. The '*nomadic landscape* produces neither boundaries nor finite spaces. Instead, the Tsaatang economy of places produces something very different, namely a number of spatial centres – or points of reference – from each of which an infinite spatial realm takes its beginning'.[18] The point again is to stress that environmental differentiation cannot be separated from social evaluation and movement. A nomadic landscape is distinct from a sedentary one. Different people may find different ways, even within the same physical space, and see different landscapes as a matter of course. The (earlier) coexistence of nomads and peasants in parts of the Middle East is an example of this.

Wayfinding is to be understood as 'a skilled performance in which the traveller, whose powers of perception and action have been fine-tuned through previous experience, "feels his way" towards his goal, continually adjusting his movements in response to an ongoing perceptual monitoring of his surroundings'.[19] Wayfinding, therefore, is not reducible to the reading of a map; it is an activity that springs out of dwelling. The first point we can make, therefore, is that while the Mongolian nomads moved within a heterogeneous network of places, the Icelanders moved between two well-defined spaces, inner and outer, each defined by distinct tasks and social values.

If we acknowledge that all knowledge is and must be inherently local, it implies that 'knowing, like the perception of the environment in general, proceeds along paths of observation', as Ingold has it, and which he summarises thus: 'we know as we go'.[20] People's knowledge of the environment undergoes continuous formation in the course of their moving about in it. (This, naturally, also applies to scholars in the field or going through archival material.)

---

[17] Ingold (2003), 41.
[18] Pedersen (2003), 247.
[19] Ingold (2000), 220.
[20] Ingold (2000), 229.

Some movements are more momentous in terms of knowledge than others. In Iceland, the primordial journey across the sea is of major importance in the knowledge of self and landscape even today.

The original *landnám* (land-taking) is constantly referred to, whenever one engages in conversations about the land. It signifies the original domestication of the wilderness, and the roots of Icelandicness as something distinct from the shared Nordic past. The settlers were primordial Icelanders; they colonised virgin land and made a lawful society where before only wilderness had been. In one of the extant manuscripts of *Landnámabók* it is claimed that it was written in order to establish the noble ancestry of the Icelanders, who allegedly had been accused of being all of them descendants of slaves and robbers. In this aim, it certainly succeeded, even if the motive for writing was probably also a wish to keep track of land-ownership at a time when all available (inhabitable) land had been claimed and the population continued to grow.[21] Today the *landnámsöld*, the age of settlement, or the First Times, is not totally unlike the way the indigenous Australian landscape is referred to as the Dreaming.[22] Antiquity and authenticity are conflated in the sense of Icelandicness that is irreversibly tied to the land; the 'song-line' created by the ancestral past contributes to a shared consciousness of the original journey to Iceland, by which Iceland was created and entered into history.

The original journey towards Iceland marked a discovery, not unlike Columbus' discovery of the Americas some five centuries later. For the Norsemen, too, the discovery of the New World (which included their discovery of Wineland) was based on a 'knowing the unknown'.[23] From their oceanic practices, the Norsemen knew that new territories existed and could be colonised without profoundly affecting the Nordic cosmography. In a smaller way, the process was repeated when emigrants from Iceland established New Iceland in Canada (Winnipeg) in the 1870s, which has been recast as yet another *landnámssaga*.[24] The New World is not ontologically new, just another place in the same Old World.

Related to the landscape's being redolent with memories is the paradoxical fact that by fixing the ancestry of Icelandic society in the land, it becomes a timeless reference point. Like in Australian Yolngu ontogeny, place takes precedence over time in Icelandic notions of the Beginning.[25] Without it actually being claimed that the Icelandic landscape is totemic, it does articulate a distinctive comment on society, by way of historical poetics.

---

[21] Rafnsson (1974).
[22] See, for instance, Layton (1995), 213.
[23] Paine (1995), 47ff.
[24] Jackson (1919).
[25] Morphy (1995), 188.

At a more general level, 'a journey along a path can be claimed to be a paradigmatic cultural act, since it is the following in the steps inscribed by others whose steps have worn a conduit for movement which becomes the correct or "best way to go"'.[26] Paths create relationships and the more people have walked there, the greater the significance attached to the relation. The paths created by generations of people structure the experience of subsequent walkers, and the historical marks left by predecessors form the conceptual space of present-day travellers. One significant example is provided by the two-dimensionality of the terms of orientation almost all over Iceland, and certainly also at my field sites in the 1980s. Chasing the stray sheep, or driving towards more distant goals from Suðursveit, was invariably conceived of as taking one of two possible directions: suður or austur, south or east. 'South' covered a varied field of directions, largely towards the south-west, but due to the rocky and crumbled nature actually covering the entire compass; 'east' likewise, if generally heading towards the north-east. The point is that, due to the topography, and the ancient political geography of Iceland, most locations are en route – towards somewhere else. Places, outside of home, are perceived in terms of going somewhere else. The route, invariably, consists in a line connecting two or more points on a (perceived) circular path, embracing all of Iceland, and linking the four quarters of Iceland, established c. AD 965 as a judicial division. Thus, the recurrent reference to what appears to be a two-directional space is owed to the nature of Icelandic society having of necessity been built (more or less) along the coastline, and to the fact that the line thus created was punctuated by a political decision to subdivide the country into four smaller judicial units. They lasted only briefly, but they deeply marked the representations of the landscape in all quarters, where orientation is generally two dimensional.[27]

If this goes to show how movement and perceptions of space are linked up with past realities, it also takes us further towards a realisation of human movement impinging itself upon knowledge. When external forces (climate, trade, plagues, etc.) began to undermine the once flourishing society, the movements that had taken the Icelanders out into the wilderness in search for prey and supplementary grazing, became socially restricted, as we saw. Social life was contracted around the farmstead; the saeters were abandoned, and fishing restricted.[28] 'Moving out' or 'about' itself was seen as threatening, and lausamen (migrant labourers) were banned along with outlaws and foreigners; they did not belong inside. The Icelanders had become inward bound, and could no longer know what was outside.

---

[26] Tilley (1994), 31.
[27] Haugen (1957).
[28] Hastrup (1989).

Mapping: Re-enactment of movement

Mapping is 'the re-enactment, in narrative gesture, of the experience of moving from place to place within a region'.[29] Narrative gestures are of many kinds; one is a simple gesture of naming and through this gesture to fix a particular knowledge that can then be shared. The clearest example of this is the fixing of place names; due to the relatively recent settlements (on the verge of history), place names are very transparent and their meaning is common knowledge, as it were.

In addition to the naming of places, a performative gesture that reminds the Icelanders of their communality with earlier times are the thousands of cairns indicating old paths in the wilderness across vast and barren territories. Otherwise, the traces left by a thousand years of Icelandic history hardly amount to more than some moss overgrown ruins of individual farms, originally built from stone and turf, some with wooden gables, but all of them very modest by continental European standards. There are no castles, churches or feudal mansions standing from the Middle Ages. Every building in the country, except one or two eighteenth-century stone buildings in Reykjavík – since 1800 the administrative centre of the country – was made from nature itself, as it were, and once abandoned was swallowed back into nature. Timber was extremely scarce, and recycled until left for the fire. Therefore, *words* (including names) remain the most significant remains of the past; they are the ones that carry the message of antiquity, as attached to little knolls, rocks, ruins and cairns that are often barely visible.

Also more abstract terms such as compass directions contain a narrative re-enactment of moving from one place to another; *the áttir* (compass directions) are not simply north, north–east, east, south–east, south, south–west, west and north–west, as one might think; instead, they are north, land-north, east, land-south, south, out-south, west, and out-north. The notions of 'land' (*land*) and 'out' (*út*) in the composite terms reflect the Norwegian point of departure. Here Norway (*Norvegr*, or literally, the 'way north') was perceived as lying on a north–south axis, and land-north (*land-norðr*) was north-bound towards the landside, while out-north (*út-norðr*) was north-bound out towards the sea. A set of terms may contain a whole history of origin and be crucial for the sense of a shared identity. 'Through an act of naming and through the development of human and mythological associations such places become invested with meaning and significance.'[30]

The act of mapping, whether in naming or some other narrative practice, recalls the past. It cannot be mapped in just any way, of course. For it to make

---

[29] Ingold (2000), 232.
[30] Tilley (1994), 18.

sense, it must be grounded.[31] In the Icelandic world, this grounding of the collective remembering must be taken very literally. Remembering Iceland, and thus to perceive it in the first place, is not to retrieve its history in accurate detail, but to move in the space of momentous pastness, along ancestral paths, and in a shared sensorial field of tactility, sound, smell, taste, and vision. Within this space the individual is constantly reminded of the collectivity; the past is not a foreign country.[32] Quite the contrary: the landscape is a well-known history.

What I am arguing here is that history is not visible simply as an imprint upon the landscape in the form of human-made material structures or other physical traces left by the ancestors. Far more important are the mental imageries and social practices linking modern Icelanders directly to their ancestors walking along the same paths.

Another performative gesture of mapping movement or change is found in the making of laws. In the words of Clifford Geertz, law is a 'skeletonization of fact' to fit a particular genre of imagining society.[33] In Iceland, once founded by an act of lawmaking and ever since subsiding to the decisions made at the Althing, laws were instrumental not only to imagining reality but also to map particular responses to shifting experiences. The laws were orally transmitted for the first couple of centuries, where the highest office held in Iceland was that of the Lawspeaker; words were the main vehicle of social cohesion. This continued also after reading and writing had become common; the legal deliberations and decisions were among the most important mappings of the experiential space. When disaster struck, the legal reaction was immediate and highly suggestive of the Icelandic self-perception.

Thus in a law from 1404 (right after the Black Death) – which incidentally is both the first and the last time any notion of fishermen (*fiskimenn*) surfaces in legal documents (until a much more recent date) – farm service was made compulsory, and the fishing villages became illegal settlements. If the fishermen did not take up residence – and work – at a farm, they were to be exiled.[34] At the very moment fishermen had appeared and been recognised linguistically, the category was subsumed under farming, and strict rulings as to when and for how long it was permitted to reside at the seaside were issued. Farm labour had become scarce due to the plague, but by all accounts fishing had already become the major source of income for the society as a whole in the fourteenth century, as mentioned earlier. It is therefore significant that in the self-understanding of the people, farming was still by far the dominant element

---

[31] See, for example, Shotter (1990), 133.
[32] Cf. Lowenthal (1985).
[33] Geertz (1983), 170.
[34] *Lovsamling for Island* (vol. 1: 34–35).

in Icelandicness (as defined by the legislating body at the people's assembly, the Althingi, dominated by the farmers). Fishing continued within the farming household, but fishermen disappeared into the general category of farmhands (*vinnuhjú*), defined by their position within the farming household (*bú*) headed by a free farmer (*bóndi*). Henceforth, fishing rights were contingent upon land rights and once again it was up to the landowner to organise fishing as part of the household economy. The *bú* still represented the core of Icelandicness.

The insistence upon the conceptual priority of farming can be inferred from laws stipulating that irrespective of catch, farmhands had to return to their farms by harvest at the latest. Not all were happy to comply, and the general assembly of farmers repeatedly had to remind people in general about the ground rules. During the fifteenth century, when the Icelanders probably still had a clear recollection of the surplus gained from the sea – and during which new plagues and other calamities hit – the court passed one law after another to make fishing less attractive. For example, fishing with more than one hook on the line was banned, explicitly because the farmers feared that if returns increased, fishing would become too attractive to their servants.[35] Sinker lines were likewise made illegal, and the use of worms as bait was prohibited. It was not until 1699 that some of these restrictions were lifted, sinker lines with several hooks again being allowed, but only during the 'season'; outside of this period they were still prohibited because of their allegedly damaging effect on farming.[36] By then, the Icelanders had lost their motivation, however – as well as the better part of their fishing fleet. A century later, in 1785, the enlightenment reformer Skúli Magnússon noted that lines with just one hook reigned supreme, and he took it upon himself to re-educate the Icelanders by making manuals about how to construct and use lines with several hooks.[37]

Judging from the legal documents (and corroborated in many other sources), there seems little doubt that farming was seen as a cultural 'fundamental' (Hastrup, 2006), and one that came increasingly into focus during long centuries of decline and poverty. It is the more remarkable, therefore, that farming skills deteriorated along with the fishing technologies. The plough went out of use, the fences separating the infields from the outfields disintegrated and with them the basic protection of the harvest from stray animals. The fences also had a symbolic function separating the home, where order reigned, from the uncontrolled wilderness beyond. In the eighteenth century this became a major issue in the reconstitution of Icelandic society. In 1776 the Danish King issued a decree demanding that the Icelanders rebuild their fences under threat of fines

---

[35] *Althingisbækur Íslands* (vol. I: 432–4).
[36] *Lovsamling for Island* (vol. I: 564–67).
[37] Magnússon (1944), 55–56.

and the promise of rewards. He also donated two spades to the country (covering 102.000 km^2!), one for the northern and one for the southern part of the island – that might serve as models for new ones to be used not only for fences but also for dealing with the tufts that had gradually destroyed the infields, due to the 'loss' of the plough. Barns that had protected the hay during the wet winters in the High Middle Ages likewise went out of use, and on the whole there is a sense of cultural amnesia, if you wish, which can not be explained by reference simply to natural disasters and other events, as indicated earlier.

To explain them means to establish the links between events, and this requires an attention to the ordinary, and the durable, including (in this case) the sense of Icelandicness that prevailed and which prompted particular responses to the events. One such response in Iceland was to concentrate labour in farming at the expense of fishing whenever crisis struck, making the economy ever more vulnerable and precipitating a negative spiral. We cannot ask contemporary Icelanders about their reasons for responding to circumstance as they did, gradually cutting themselves off from potential sources of supplementary income. But by paying attention also to other sources, such as literature, autobiographies, learned treatises on society and poetry, we find significant evidence of the Icelandic self-perception being a contributing factor in the decline. The self-perception both contextualised and intensified the legal and social process of marginalising the fishermen described above.

The most remarkable feature of this process was the reclassification of humans themselves; classification is another important act of mapping that we may now add to the narrative gestures of naming and lawmaking. In the Middle Ages *mennskir* ('humans') comprised all human beings, as distinguished from *ómennskir* ('non-humans'), which were such creatures as trolls and elves. The two categories belonged neatly to each their domain in the concentric model. Gradually, as we may gather from the sources, the category of humans became restricted to the farming population, while migrant labour, vagrants, paupers and others without a permanent foothold on a farm gradually became conceptually subsumed under the rubric of *ómennskir*, non-humans. So, in the seventeenth and eighteenth centuries, humanity itself was determined as a function of a sedentary farming life. To protect this fundamental of Icelandicness in view of the external threats (natural, commercial or epidemic) and the encroaching disorder, 'others' were marginalised – dehumanised, even. The social had shrunk considerably.

The details are many and the material is very rich, but space only allows me to say that alongside the sequential historical narrative of a particular succession of events, there was a strong sense of enduring cultural values, rooted in the myth of origin of Icelandic society, connected with the *landnám* (landtaking) by free farmers allegedly 'fleeing the tyranny of king Harald Fairhair', as the sagas have it. Amidst historical changes, the (structurally dominant) Icelanders strove for a return to founding principles. Their identity as well as their freedom was embattled, and to overcome equivocation, they recurrently recast themselves in terms of past values and ideas that became increasingly 'unrealistic' as (environmental) history moved on. I have suggested the notion of Uchronia for this implicit appeal to 'a history out of time'.[38] In this particular case, Uchronia was deeply rooted in the 'first' free society of settlers.

I would like to stress that in suggesting that the return to founding principles played a major role in the decline and recurrent disasters in Iceland 1400–1800, I am not reverting to an undue culturalism. Culture is no explanation. But neither is economy nor climate. One needs an approach that takes into account the combined features of the particular historical situation, including the prevailing self-perception, social differentiation, divergence of interests and the environmental challenges – also as these were perceived from within the lived world. In short, one needs an approach that may pay heed also to such intangible factors as the local sense of vulnerability in addition to the manifest climatic and economic changes.

By this token we may suggest that the social processes that gradually made Icelandic history were marked by *amplification*, adapting the notion recently suggested by Marshall Sahlins (2005) for my own purpose.[39] Sahlins shows how minor issues at a local level may turn into major political issues by being articulated with increasing force in an ever more abstract rhetoric. In the Icelandic case, the decisions made in response to epidemic or climatic disasters actually intensified the problems these had entailed in the first place. The experienced *events*, threatening parts of Icelandic society from outside, became globalised and structurally amplified by the countermeasures taken, whether in the form of legal process or in the form of conceptual redefinition of humanity itself. People who did not fit the fundamental criterion of Icelandicness became externalised to the non-human environment from where disasters emerged. The migrant fishermen became part of the problem, not the solution. The process of amplification added greater oppositions (such as humans vs. non-humans) onto lesser ones (farmers vs. fishermen), escalating the processes of change and decline as a matter of course.[40]

While the reaction was certainly spurred by disasters of various kinds, it also contributed to their long-term effects in its own right by over-communicating a monolithic world-view at the expense of a viable dual economy. Even farming and husbandry suffered as a result of this return to cultural 'fundamentals'.[41] On the whole, the net result was an increasing loss of flexibility, as defined

---

[38] Hastrup (1990a, 1992).
[39] Sahlins (2005).
[40] See Sahlins (2005), 24–25.
[41] Hastrup (2006).

by Gregory Bateson as 'an uncommitted potential for change'.[42] Like the acrobat on the high wire in Bateson's example who will inevitably fall down if the flexibility of the body is hampered, so the Icelanders hit the hard rock. It was seriously discussed in the eighteenth century whether Iceland was inhabitable at all, or whether one should resettle the islanders in Denmark. Local enlightenment reformers, such as the above-mentioned Skúli Magnússon, and their fellow islanders gradually proved otherwise, and around 1800 a new practical and political course was set, leading to an improved economy – facilitated also by slightly warmer weather conditions.

Map-making: The cartographic illusion

If mapping refers to a way of representing particular performative gestures, the actual making of maps transforms this to an abstract notation. In Tim Ingold's words: 'The map, like the written word, is not, in the first place, the transcription of anything, but rather an inscription. Thus mapping gives way to map-making at the point, not where mental imagery yields an external representation, but where the performative gesture becomes an inscriptive practice.'[43] There is a 'cartographic illusion' inherent in maps, according to Ingold. The processes of wayfinding and mapping, both of which are situated movements, are preconditions of map-making; yet they are bracketed and the map is read as a direct transcription of the world.

As scholars we also produce maps in the sense that through our inscriptive practices we fix and freeze what are at base social processes of knowing; in that sense any field of scholarship seeking to understand human life finds itself between experience and theory.[44] This is inevitable. Yet it should not make us forget that in history, the irreversible is not necessarily the inevitable. Environmental history is not an externally induced destiny to which people are victims; people are destined also to make decisions that propel environmental changes in certain directions. By their acts of decision making, people map their perceptions and knowledge onto the space in which they live. I am not suggesting that individual actions can prevent volcanic eruptions, but I am suggesting that local reactions to such events may account for the scope of the possible disaster.

I have suggested that a life-world perspective may actually add life to environmental history. It may point to important links between destinies and decisions, while evidently not answering all possible questions. The meta-message is a wish to keep the field open to all sorts of knowledge about things human.

---

[42] Bateson (1972), 498.
[43] Ingold (2000), 231.
[44] See Hastrup (1995) for the anthropological postion.

References

Althingisbækur Íslands (1912–1982), vol I–XV, *Acta comitiorum generalium islandiae*, Reykjavík: Sögufélagið.

Annálar Íslands 1400–1800. Annales Islandici. Posteriorum Saeculorum, I–IV. Reykjavík 1922.42: Hið íslenzka bókmenntafélag.

Bachelard, Gaston (1950), *The Poetics of Space*, ed. John Stilgoe, Boston: Beacon Press (1994).

Bateson, Gregory (1972), *Steps to an Ecology of Mind*, New York: Ballantine Books.

Benediktsson, Jakob, ed. (1968), *Íslendingabók, Landnámabók*. Reykjavík 1968: Hið íslenzka fornritafélag (Íslenzk fornrit, vol I).

Bourdieu, Pierre (1990), *The Logic of Practice*, Cambridge: Polity Press.

Fagan, Brian (2000), *The Little Ice Age. How Climate Made History*, New York: Basic Books.

Finnsson, Hannes (1970), *Mannfækkun af hallærum* [1796], Reykjavík: Almenna bófélagið.

Geertz, Clifford (1983), 'Fact and law in Comparative Perspective', In *Local Knowledge*, New York: Basic Books.

Hacking, Ian (1992), ' "Style" for Historians and Philosophers', *Studies in the History and Philosophy of Science*, vol. 23 (1): (1–20).

Hastrup, Kirsten (1985), *Culture and History in Medieval Iceland. An Anthropological Analysis of Structure and Change*, Oxford: Clarendon.

Hastrup, Kirsten (1989), 'Saeters in Iceland 900–1600. An anthropological analysis of economy and cosmology', *Acta Borealia*, vol. 6 (1): 72–85.

Hastrup, Kirsten (1990a), *Nature and Policy in Iceland 1400–1800. An Anthropological Analysis of History and Mentality*, Oxford: Clarendon.

Hastrup, Kirsten (1990b), *Island of Anthropology. Studies in Past and Present Iceland*, Odense: Odense University Press.

Hastrup, Kirsten (1992), 'Uchronia and the Two Histories of Iceland', In Kirsten Hastrup (ed.), *Other Histories*, London: Routledge.

Hastrup, Kirsten (1995), *A Passage to Anthropology, Between Experience and Theory*, London: Routledge.

Hastrup, Kirsten (1998), *A Place Apart. An Anthropological Study of the Icelandic World*, Oxford: Clarendon.

Hastrup, Kirsten (2004), 'Knowledge and Evidence in Anthropology', *Anthropological Theory*, vol. 4 (4): 455–472.

Hastrup, Kirsten (2005), 'Social Anthropology: Towards a Pragmatic Enlightenment', *Social Anthropology*, vol. 13 (2): 133–149.

Hastrup, Kirsten (2006), 'Closing ranks. Fundamentals in History, Politics and Anthropology', *The Australian Journal of Anthropology*, vol. 17 (2): 147–160.

Haugen, Einar (1957), 'The Semantics of Icelandic Orientation,' *Word*, vol. 13(3).

Ingold, Tim (2000), *The Perception of the Environment. Essays in Livelihood, Dwelling and Skill*, London: Routledge.

Ingold, Tim (2003), 'Three in one: How an ecological approach can obviate the distinctions between body, mind and culture', In Andreas Roepstorff, Nils Bubandt, and Kalevi Kull (eds), *Imagining Nature. Practices of Cosmology and Identity*, Aarhus: Aarhus University Press.

Jackson, Turner (1919), *Brotaf landnámssögu Nýja Íslands* (Winnipeg, 1919).

James, Wendy (2003), *The Ceremonial Animal. A New Portrait of Anthropology*, Oxford: Oxford University Press.

Layton, Robert (1995), 'Relating to the Country in the Western Desert', In Eric Hirsch and Michael O'Hanlon (eds), *The Anthropology of Landscape. Perspectives on Place and Space*, Oxford: Clarendon Press (pp. 210–231).

Lovsamling for Island (1853–1889), O. Stephensen and Jón Sigurðsson (eds), Copenhagen: Höst og Søn.

Lowenthal, David (1985), *The Past Is a Foreign Country*, Cambridge: Cambridge University Press.

Magnússon, Skúli (1944), *Beskrivelse af Gullbringu og Kjósar syslur* [1785], Jón Helgason (ed.), Copenhagen 1944: Munksgård (Bibliotheca Arnamagnæana, 4).

Merleau-Ponty, Maurice (1962), *Phenomenology of Perception*, London: Routedge.

Morphy, Howard (1995), 'Landscape and the Reproduction of the Ancestral Past', In Eric Hirsch and Michael O'Hanlon (eds), *The Anthropology of Landscape. Perspectives on Place and Space*, Oxford: Clarendon Press (pp. 184–209).

Paine, Robert (1995), 'Columbus and the Anthropology of the Unknown', *Journal of the Royal Anthropological Institute (incorporating MAN)*, vol. 1: 47–65.

Pedersen, Morten A. (2003), 'Networking the Nomadic Landscape: Place, Power and Decision Making in Norther Mongolia', In Andreas Roepstorff, Nils Bubandt, and Kalevi Kull (eds), *Imagining Nature. Practices of Cosmology and Identity*, Aarhus: Aarhus University Press (pp. 238–259).

Rafnsson, Sveinbjörn (1974), *Studier i Landnámabók. Kritiska bidrag till den isländska fristatstidens historia*, Lund 1974: Gleerup.

Rapport, Nigel and Andrew Dawson (1998), *Migrants of Identity. Perceptions of 'Home' in a World of Movement*, Oxford: Berg.

Sahlins, Marshall (2005), 'Structural Work. How Microhistories become Macrohistories and Vice Versa', *Anthropological Theory*, vol. 5 (1): 5–30.

Shotter, John (1990), 'The Social Construction of Remembering and Forgetting', In David Middleton and Derek Edwards (eds), *Collective Remembering*, London: Sage (pp. 120–138).

Tilley, Christopher (1994), *A Phenomenology of Landscape. Places, Paths and Monuments*, Oxford: Berg.

Afterword

Peter Burke

It has been a privilege and an education to read these essays before publication. In what follows I shall focus on a few of their common themes, viewing them from the standpoint of a cultural historian with a special interest in historiography, commenting on what has been written and occasionally adding examples of my own.

I

Just as the interest in price history followed the great inflation of the 1920s, while demographic history took off in the age of the baby boom, environmental history, otherwise known as *écologie historique, Umweltgeschichte* and so on, is the offspring of increasing awareness of the environment and of what humans are doing to defile or destroy it.

As Sörlin and Warde rightly say in their introduction, environmental consciousness was not the simple result of a sudden 'awakening' in the 1960s, following the well-known book *Silent Spring* by Rachel Carson. This 'ecological moment', or 'environmental turn', as different contributors to this book call it, was indeed a turning point, but it also needs to be replaced in a much longer intellectual sequence. As Warde suggests in his essay (Chapter 3) in this collection, farmers have long been aware of the environmental conditions in which they work, while the anthropologist Tim Ingold has made a similar point about nomads. Richard Grove has told another part of the story elsewhere in his *Green Imperialism*, emphasizing tropical islands as places where naval officers or colonial administrators became aware of environmental problems.[1]

Arranging the points made earlier in this book in chronological order, one might single out a series of moments. The first would be the 1860s, the time of

---

[1] Grove (1995).

the foundation of Yosemite State Park (later National Park) and George Perkins Marsh's discussion of the need to preserve forests in his *Man and Nature*.[2] A second moment might be the 1920s, when the *Journal of Ecology* was founded and geographers and sociologists such as Carl Sauer, Robert Park and Radhakemal Mukherjee made use of the concept. A third moment would be the 1960s, when the idea of global warming began to spread, the Friends of the Earth were founded, and a UNESCO conference launched the idea of the 'biosphere'. A fourth moment is the beginning of the twenty-first century, when the environment became a central concern for politicians and public alike rather than the concern of a minority (often dismissed by others as eccentric).

Historians have marched in step with these trends and environmental history might be divided into three phases: the age of pioneers, the age of consolidation and finally the age of synthesis.

II

Although environmental history may be viewed as the response of historians to the rise of environmental consciousness, the new approach emerged from a number of trends and movements, like a river from its tributaries. Two intellectual traditions were particularly important. The pioneers of environmental history tended to come from two established disciplines: geography and economic history. As Grove and Damodaran note, geographers such as Daryl Forde and Dudley Stamp, at work in the 1930s, were among the early practitioners of what they call 'de facto environmental history'. The study of the history of climate and of the earth's resources, especially trees and water, sheltered under the umbrella of 'historical geography'.

The second tradition out of which environmental history emerged was economic history, which was already academically established by the 1930s. The study of agriculture in particular, converging with historical geography, turned into a broader agrarian history or rural history in which the interaction between population and environment was central. One of the pioneers of this agrarian history was Marc Bloch, whose *Caractères originaux de l'histoire rurale française* was published in 1931. He was followed by scholars from Germany (Wilhelm Abel), The Netherlands (Bernard Slicher van Bath), Britain (Michael Postan) and elsewhere.[3]

At the cross-roads between these two traditions stand a few remarkable books. One of them is much less well known than it deserves to be: a study of the Northeast of Brazil by the sociologist–historian Gilberto Freyre. Drawing on the ideas of Sauer and Mukherjee, this book, *Nordeste* first published in 1937 (and translated into French, but not into English), focuses on the sugar plantations

---

[2] George Perkins Marsh (1864); On him, Lowenthal (1958: 2nd edn, 2006).
[3] Abel (1935: 2nd edn, 1966); Slicher van Bath (1960: English trans., 1963); Postan (1972).

of the humid coastal regions of Pernambuco and Bahia and their influence on the landscape and the life of the people and animals who lived there, emphasizing the destructive effects of monoculture.[4]

One of the first Europeans to appreciate Freyre's work was Fernand Braudel, whose famous book on the Mediterranean, defended as a thesis in 1947 and published in 1949, was another combination of economic history with a study in human geography or 'geo-history' in the manner of his master Lucien Febvre and of Paul Vidal de la Blache. A similar point might be made about Braudel's student Emmanuel Le Roy Ladurie, who showed his concern with the environment in his doctoral thesis on the peasants of Languedoc (published in 1966) as well as his *History of Climate since the Year 1000* (1967).[5]

The British equivalent of Braudel and Le Roy Ladurie was W.G. Hoskins, who described himself as a 'local historian' and wrote a major study of *The Making of the English Landscape* (1955). As the title of the study suggests, Hoskins viewed the landscape in a more dynamic way than his French colleagues. Where they looked for structures, he concentrated on change. What he did for Britain, the Italian Marxist Emilio Sereni did for Italy in a history of the landscape published in 1961.[6]

One might say that environmental history was the cuckoo in the nest of economic history, a baby that grew more quickly and became more independent than anyone expected. Disciplinary independence generally comes at the price of the annexation of territory from other disciplines, and in the last generation economic history has been squeezed on both sides, losing at least part of the field of consumption to social and cultural historians, while studies of production, involving the use of natural resources, have been annexed by historians of the environment. On the other hand, this challenge, like the challenge of the cultural approach, might be beneficial to economic historians, encouraging them to view their subject in a different way.

By the 1970s and 1980s, historical studies of the environment had become sufficiently frequent to allow us to speak of an age of consolidation. In the United States, Alfred Crosby, William Cronon and Donald Worster were among those who were placing the topic on the academic map, while Roderick Nash was one of the first to teach a course on the subject as well as publishing a survey article on American environmental history. Incidentally, Nash organized his course in reaction to an environmental disaster, the Santa Barbara oil spill of 1969. It would be interesting to know how many other historians turned to the study of the environment following similar events.[7]

---

[4] Freyre (1937).

[5] Braudel (1949: English trans., 1972–3); Le Roy Ladurie (1966: abbreviated English trans., 1971).

[6] Hoskins (1955); Sereni (1961).

[7] Crosby (1972); Cronon (1983); Worster (1980); Nash (1972); Nash (2007).

It is from the mid-1990s, if not before, that a take-off of environmental history becomes visible in retrospect. To speak only of studies in English, 1995, for instance, was the year in which Richard Grove published his *Green Imperialism*, and Warren Dean his study of the destruction of Brazil's Atlantic Forest. The journal *Environment and History* was founded in that year, while old hands such as Cronon and Crosby published new studies.[8]

We might call this period 'the age of synthesis' because monographs were joined by more general studies, usually on a national scale, like Madhav Gadgil and Ramachandra Guha on India, Jan Luiten van Zanden and Wybren Verstegen on The Netherlands, Christopher Smout on Scotland and Northern England and Mark Elvin on China.[9] Robert Delort and François Walter produced a history of the European environment.[10] Sverker Sörlin, Anders Öckerman, Joachim Radkau, J. R. McNeill, John Richards, Marco Armiero and Stefania Barca and Wolfgang Behringer all took the world as their province in works of synthesis.[11] Jared Diamond has done the same in two ambitious comparative studies.[12]

General histories have gradually come to devote increasing space for the environment. Where Arnold Toynbee's *Mankind and Mother Earth* (1976) is less environmental than the title suggests (following a chapter on the biosphere, the book offers a conventional narrative history of the world), *Civilizations* (2000) by Felipe Fernàndez-Armesto organized world history not by period but by types of environment – tundra, desert, prairie and so on.[13]

III

As a number of contributors to this volume remark, environmental history is both an international and an interdisciplinary enterprise. It has its own academic geography, linked to that of the environmentalist movement. North America is well represented, as we have seen, like other neo-Europes such as Australia and Canada – Keith Hancock's *Discovering Monaro* (1972) was an important early work of environmental history.[14] It is appropriate that both Australia and Canada are represented in this volume, by Libby Robin and by Matthew Evenden and Graeme Wynn. So is Germany, represented in this book by Holger Nehring, and especially Scandinavia (one thinks of the work

---

[8] Dean (1995); Cronon (ed.) (1995); Crosby (1995).
[9] Gadgil and Guha (1992); van Zanden and Verstegen (1993); Smout (2000); Elvin (2004).
[10] Delort and Walter (2001).
[11] Sörlin with Öckerman (1998: revised edn, 2002); Radkau (2000); McNeill (2001); Richards (2003); Armiero and Barca (2004); Behringer (2007).
[12] Diamond (1997); Diamond (2005).
[13] Toynbee (1976); Fernández-Armesto (2000).
[14] Hancock (1972).

of Gustaf Utterström, Thorkild Kjærgaard, Peder Anker, and among the participants in this volume, of Sverker Sörlin and Kirsten Hastrup).[15] The tradition of
contributions from India is also worth noting (Madhav Gadgil, Ramachandra
Guha, and in this volume, Vinita Damodaran).

As for interdisciplinarity, as Grove and Damodaran point out, the name 'environmental history' was appropriated from geology. The natural sciences that
have made a contribution to environmental history include biology, botany,
glaciology, meteorology and palynology as well as ecology itself.

In the humanities, the original nucleus of historical geographers and economic historians has been joined by individuals from a number of disciplines.
Archaeologists, more accustomed than many of their colleagues in history
departments to the study of the long term, have explained the collapse of
civilizations such as the Pagan in Burma or the Mayas in Yucatan by climate
changes such as drought. The involvement of anthropologists goes back at least
as far as Clifford Geertz's *Agricultural Involution* (1963), described in its sub-title
as a study of 'the process of ecological change in Indonesia'. Economics, especially the sub-discipline of environmental economics, Nick Hanley's field, is
obviously relevant too.

Interdisciplinary work may be viewed either a solution to problems or as a
problem in itself, more exactly as a situation that generates problems (addressed
most directly in Chapter 7 of this volume). A historian like myself may be
permitted to observe that a high proportion of histories of the environment
have been written by individuals without a formal historical training and that
this remains a weakness, however much it may be compensated by other skills
and strengths that historians lack. Practitioners of environmental history are
pulled in different directions: towards general history, with the risk of ignoring
relevant work in the natural sciences, or towards the sciences, with the risk of
ignoring other forms of history (the history of taxation, for instance, discussed
in Chapter 11 of this volume by Mark Elvin).

Interdisciplinary work on the frontiers between neighbouring disciplines
such as history and anthropology is relatively free from problems and has produced some notable achievements (among them Hastrup's studies of Iceland).
Crossing the divide between what C.P. Snow, half a century ago, called 'the two
cultures', is rather more difficult. For one thing, historians are rarely trained
in science or scientists in historical method. Another problem is that of language; are the languages of historians and scientists compatible, is translation
possible? Yet another problem is diversity of aims, with most scientists seeking generalizations, and most historians concentrating on particular cultures,
periods or even events.

There are also the practical problems of collaboration. Is team-work
inevitable? Will it involve so many meetings that there is little time left for

---

[15] Kjærgaard (1991, trans., 1994); Anker (2001); Sörlin (1991).

research? My own view on this matter is that small groups (from two to six or possibly eight members) can often work together creatively, but that larger groups are likely to become bogged down in bureaucracy. In other words, the model of a string quartet is more viable than that of an entire orchestra. In any case, even if polymaths are now virtually extinct, I continue to hope that there will be room for the individual as well as the team and that the individual scholar of wide interests has not yet become an endangered species. The example of Jared Diamond's move from zoology to history is a heartening one.

One more problem in interdisciplinary studies concerns its aims. We have much to gain by trading concepts as well as pooling information, but we should not try to eliminate differences between what Sverker Sörlin calls 'the cognitive patterns of disciplines'. Like languages, academic disciplines embody views of the world, and we would be impoverished if we lost any one of them. As Michel de Certeau, one of the last of the polymaths, put it, 'Interdisciplinarity does not consist in the elaboration of a totalizing bricolage, but on the contrary... in recognizing the need for different fields.'[16]

IV

One of the virtues of this volume is the joint concern of its contributors not so much with past achievements as with current problems (interdisciplinarity among them) and with future opportunities.

There is certainly no shortage of problems. Even the definition of the environment is problematic, as both Nehring and Warde point out. So is the choice of the appropriate units of study, noted by Evenden and Wynn, small scale or large scale, raising the further problem of the links between the two, since even islands such as Iceland are not insulated from broader changes. There are also problems of method, for example, the strengths and weaknesses of the comparative method, represented here by Robert Dodgshon on Scotland and Switzerland.

An even more acute question, raised by Elvin, is whether environmental history should be regarded as a separate field of study. Elvin himself prefers to describe it as 'everywhere and nowhere'. As the number of environmental historians expands, it is likely (thinking of what has happened to cultural historians in the recent past) that the field will fragment, with different groups pursuing different agenda, illustrating a recurrent pattern in the history of sub-disciplines that become fully fledged disciplines but then generate sub-disciplines of their own.

---

[16] L'interdisciplinarité ne consisterait pas à élaborer un bricolage totalisant, mais au contraire à... reconnoitre la nécessité de champs différents'. Quoted Dosse (2002), 287.

Three crucial problems recur in the essays collected here: the problem of determinism, the problem of events and the problem of culture.

Like the old historical geography, the new environmental history is a battle-ground between determinists and voluntarists, re-enacting the conflict between Friedrich Ratzel, for instance, and Lucien Febvre.[17] Today, the emphasis seems to fall on a kind of voluntarism, emphasizing in the past – as in the present – the importance of human action, especially the everyday management of resources. It seems more fruitful, though, not so much to take up a position on one side or the other as to attempt to distinguish the domains in which people can make a difference from the areas in which they cannot.

Another old problem is that of the relative importance of events and struc-tures (economic, social or political). Here too the debate may become sterile unless we focus on the links between the two. Fernand Braudel, for example, used to read events as clues to structural weaknesses, while his colleague Ernst Labrousse interpreted the French Revolution as the response to a subsistence crisis. Conversely, disastrous events such as droughts, plagues and earthquakes may lead to changes in society. Environmental historians are well placed to make a positive contribution to this ongoing debate, given their professional concern to question traditional definitions of the environment.

In these debates, the idea of culture – however difficult it may be to define – deserves to play a central role, since human responses to crises are mediated by cultures or mentalities – whether or not people believe, for instance, that a plague is sent by God or a famine is the fault of the ruler. As some contributors to this volume note, a history of the environment needs to include the story of perceptions of the environment or the place of nature in the collective imagi-nation. Retrospectively, Keith Thomas's *Man and the Natural World* (1983) may be viewed as a pioneering study in environmental history which will surely be joined before long by similar studies of other countries, contributions to what might be called the 'cultural history of the environment'.

As for the future opportunities, the one that I see most clearly, perhaps because I am among other things an urban historian, is the expansion of environmental history to include cities. Tim Cooper's study of waste and its disposal, together with books such as Ercole Sori's *La città e i rifiuti* (2001), on similar problems in Italian cities, show the way (incidentally, it is intriguing and perhaps worrying that so many studies of rubbish have appeared in recent years). On the margins of the social and the environmental, or of 'ecology' in the literal and the metaphorical sense, we might expect more studies of the 'ecological succession' of different groups of immigrants in the same quarter of the same city (Italians, Afro-Americans and Hispanics in Harlem, for instance),

---

[17] Ratzel (1897); Febvre (1922).

along the lines of the sociological studies of Chicago by Robert Park and Ernest Burgess.

In short, this volume makes a contribution not only to the history of the environment, but also to its historiography and to the history of thought about the environment. It helps readers move away from a simple dichotomy between traditional respect for the environment and modern vandalism to a nuanced picture of humans constantly interfering with the environment and of the negative consequences of some attempts at conservation, as in the case of the wilderness parks discussed by Bill Adams. The volume joins substantive contributions to more general surveys of the changing field. It contributes, as the editors hoped, to bridge-building between disciplines and also to a dialogue with other kinds of historian, whether they work on politics or culture. We shall surely hear more of this dialogue as environmental history moves from the margins towards the centre of the discipline.

Including the environment means 'expanding the realm of history', as the editors put it, in more than one way. Since it is impossible to write the history of the environment without drawing on a range of disciplines, the practice of environmental history encourages interdisciplinarity at a time when the often narrow professional training of historians propels them in the opposite direction. At a time when historical narrative is under debate, the voice of environmental historians needs to be heard, since they offer us a Grand Narrative of a new kind. Must this narrative be emplotted as tragedy, or is there an alternative ending?

References

Abel, W. (1935: 2nd edn, 1966), *Agrarkrisen und Agrarkonjunktur: Eine Geschichte der Land- und Ernahrungswirtschaft Mitteleuropas seit dem hohen Mittelalter*, Hamburg.

Anker, P. (2001), *Imperial Ecology: Environmental Order in the British Empire, 1895–1945*, Cambridge, Mass.

Armiero, M. and Barca, S. (2004), *Storia dell'Ambiente: Una Introduzione*, Rome.

Behringer, W. (2007), *Kulturgeschichte des Klimas von der Eiszeit bis zur globalen Erwärmung*, Munich.

Bloch, M. (1931), *Caractères originaux de l'histoire rurale française*, Paris.

Braudel, F. (1949: English trans., 2 vols., 1972–3), *The Mediterranean and the Mediterranean World in the Age of Philip II*, London.

Cronon, W. (1983), *Changes in the Land*, New York.

Cronon, W. (ed.) (1995), *Uncommon Ground: Toward Reinventing Nature*, New York.

Crosby, A.W. (1972), *The Columbian Exchange*, New York.

Crosby, A.W. (1995), *Germs, Seeds and Animals: Studies in Ecological History*, Armonk, NY.

Dean, W. (1995), *With Broadax and Firebrand: The Destruction of the Brazilian Atlantic Forest*, Berkeley.

Delort, R. and Walter, F. (2001), *Histoire de l'environnement européen*, Paris.

Diamond, J. (1997), *Guns, Germs and Steel: The Fates of Human Societies*, London.

Diamond, J. (2005), *Collapse: How Societies Choose to Fail or Survive*, London.

Dosse, F. (2002), *Michel de Certeau: le marcheur blessé*, Paris.

Elvin, M. (2004), *The Retreat of the Elephants*, New Haven.

Febvre, L. (1922), *La terre et l'évolution humaine*, Paris.

Fernández-Armesto, F.A. (2000), *Civilizations*, Basingstoke.

Freyre, G. (1937), *Nordeste*, Rio de Janeiro.

Gadgil, M. and Guha, R. (1992), *This Fissured Land: An Ecological History of India*, Delhi.

Geertz, C. (1963), *Agricultural Involution*, Berkeley.

Grove, R. (1995), *Green Imperialism: Colonial Expansion, Tropical Island Edens and the Origins of Environmentalism, 1600–1860*, Cambridge.

Hancock, W.K. (1972), *Discovering Monaro: A Study of Man's Impact on His Environment*, Cambridge.

Hoskins, W.G. (1955), *The Making of the English Landscape*, London.

Kjærgaard, T. (1991, English trans., 1994), *The Danish Revolution, 1500–1800: An Ecohistorical Interpretation*, Cambridge.

Le Roy Ladurie, E. (1966: abbreviated English trans., 1971), *The Peasants of Languedoc*, Urbana.

Le Roy Ladurie, E. (1967), *History of Climate since the Year 1000*, London.

Lowenthal, D. (1958: 2nd edn. 2006), *George Perkins Marsh, Prophet of Conservation*, Seattle.

Marsh, G.P. (1864), *Man and Nature: or, Physical Geography as Modified by Human Action*, London.

McNeill, J.R. (2001), *Something New Under the Sun: An Environmental History of the 20th-Century World*, New York.

Nash, R. (1972), 'American Environmental History: A New Teaching Frontier', *Pacific Historical Review*, 61, 363–72;

Nash, R. (2007), Interview in *Environmental History*, 12.

Postan, M.M. (1972), *The Medieval Economy and Society: An Economic History of Britain in the Middle Ages*, London, 1972.

Radkau, J. (2000), *Natur und Macht: Eine Weltgeschichte der Umwelt*, Munich.

Ratzel, F. (1897), *Politische Geographie*, Leipzig,

Richards, J.F. (2003), *The Unending Frontier: An Environmental History of the Early Modern World*, Berkeley.

Sereni E. (1961), *Storia del paesaggio agrario italiano*, Bari.

Slicher van Bath, B.H. (1963), *The Agrarian History of Western Europe, A.D. 500–1850*, London.

Smout, T.C. (2000), *Nature Contested: Environmental History in Scotland and Northern England since 1600*, Edinburgh.

Sörlin, S. (1991), *Naturkontraktet: Om naturumgängets idéhistoria*, Stockholm.

Sörlin, S. with Öckerman, A. (1998, revised edn. 2002), *Jorden en ö: En global miljöhistoria*, Stockholm.

Sori, E. (2001), *La città e i rifiuti*, Bologna.

Thomas, K. (1983), *Man and the Natural World*, London.

Toynbee, A. (1976), *Mankind and Mother Earth: A Narrative History of the World*, New York.

Worster, D. (1980), *Dust Bowl: The Southern Plains in the 1930s*, New York.

Zanden, J.L. and Verstegen, W. (1993), *Groene geschiedenis van Nederland*, Utrecht.

Index